BIOMEMBRANES

ACADEMIC PRESS RAPID MANUSCRIPT REPRODUCTION

Proceedings of the
International Symposium on Biomembranes
Madurai University, Madurai, Tamilnadu, India, December 11-15, 1973

BIOMEMBRANES

ARCHITECTURE, BIOGENESIS, BIOENERGETICS, AND DIFFERENTIATION

EDITED BY

Lester Packer

Department of Physiology-Anatomy
University of California
and Energy and Environment Division
Lawrence Berkeley Laboratory, Berkeley, California

Academic Press, Inc.

New York San Francisco London 1974

A Subsidiary of Harcourt Brace Jovanovich, Publishers

ACADEMIC PRESS, INC.
111 Fifth Avenue, New York, New York 10003

United Kingdom Edition published by
ACADEMIC PRESS, INC. (LONDON) LTD.
24/28 Oval Road, London NW1

LIBRARY OF CONGRESS CATALOG CARD NUMBER: 74-19997

ISBN 0-12-543440-5

PRINTED IN THE UNITED STATES OF AMERICA

CONTENTS

LIST OF CONTRIBUTORS

H. N. Aithal, *Department of Clinical Pathology, Victoria Hospital, and Department of Biochemistry, University of Western Ontario, London, Ontario, Canada.*

D. Balasubramanian, *Department of Chemistry, Indian Institute of Technology, Kanpur, India.*

A. J. S. Ball, *Department of Biological Sciences, Brock University, St. Catharines, Canada.*

R. Bastos, *Department of Chemistry, Indiana University, Bloomington, Indiana*

P. M. Bhargava, *Regional Laboratory, Hyderabad, India.*

B. M. Braganca, *Biochemistry Division, Cancer Research Institute, Parel, Bombay, India.*

M. Briquet, *Laboratoire d'Enzymologie, Université de Louvain, Heverlee, Belgium.*

M. Brody, *Department of Biological Sciences, Hunter College, City University of New York, New York City, New York*

P. Cantatore, *Istituto di Chimica Biologica, Università de Bari, Bari, Italy.*

M. Capano, *Department of Biochemistry, University of Bari, Bari, Italy.*

E. Carafoli, *Istituto di Patologia Generale, Università de Modena, Modena, Italy.*

P. Charlwood, *Department of Botany and Microbiology, University College, London, England.*

A. M. Colson, *Laboratoire d'Enzymologie, Universitè de Louvain, Heverlee, Belgium.*

M. Crompton, *Department of Biochemistry, University of Bari, Bari, Italy.*

D. W. Deamer, *Department of Zoology, University of California, Davis, California*

C. De Giorgi, *Istituto di Chimica Biologica, Università di Bari, Bari, Italy.*

T. S. Desai, *Biology and Agriculture Division, Bhabha Atomic Research Centre, Trombay, Bombay, India.*

K. Dharmalingam, *Department of Biological Sciences, Madurai University, Madurai, Tamilnadu, India.*

F. Feldman, *Department of Chemistry, Indiana University, Bloomington, Indiana*

F. Foury, *Laboratoire d'Enzymologie, Universitè de Louvain, Heverlee, Belgium.*

R. Gallerani, *Istituto di Chimica Biologica, Università di Bari, Bari, Italy.*

C. P. Giri, *Biochemistry and Food Technology Division, Bhabha Atomic Research Centre, Trombay, Bombay, India.*

A. Gnanam, *Department of Biological Sciences, Madurai University, Madurai, India.*

A. Goffeau, *Laboratoire d' Enzymologie, Universitè de Louvain, Heverlee, Belgium.*

A. G. Govindarajan, *Department of Biological Sciences, Madurai University, Madurai, India.*

Govindjee, *Departments of Botany, Physiology and Biophysics, University of Illinois, Urbana, Illinois*

P. Hadvary, *Biochemie im Fachbereich Chemie, Philipps-Universität, Marburg/Lahn, Germany.*

R. M. Janki, *Department of Clinical Pathology, Victoria Hospital, and Department of Biochemistry, University of Western Ontario, London, Ontario, Canada.*

J. Jayaraman, *Department of Biological Sciences, Madurai University, Madurai, Tamilnadu, India.*

K. Jayaraman, *Department of Biological Sciences, Madurai University, Madurai, Tamilnadu, India*

B. Kadenbach, *Biochemie im Fachbereich Chemie, Philipps-Universität, Marburg/Lahn, Germany.*

J. S. Kahn, *Department of Biochemistry, North Carolina State University, Raleigh, North Carolina*

C. K. R. Kurup, *Department of Biochemistry, Indian Institute of Science, Bangalore, India.*

Y. Landry, *Laboratoire d'Enzymologie, Université de Louvain, Heverlee, Belgium.*

H. R. Mahler, *Department of Chemistry, Indiana University, Bloomington, Indiana*

B. C. Misra, *Department of Chemistry, Indian Institute of Technology, Kanpur, India.*

N. Murugesh, *Department of Biological Sciences, Madurai University, Madurai, Tamilnadu, India.*

L. M. Narurkar, *Biochemistry and Food Technology Division, Bhabha Atomic Research Centre, Trombay, Bombay, India.*

M. V. Narurkar, *Biochemistry and Food Technology Division, Bhabha Atomic Research Centre, Trombay, Bombay, India.*

L. Packer, *Department of Physiology-Anatomy, University of California, Berkeley, California*

F. Palmieri, *Department of Biochemistry, University of Bari, Bari, Italy.*

S. R. Panini, *Department of Biochemistry, Indian Institute of Science, Bangalore, India.*

P. S. Perlman, *Department of Chemistry, Indiana University, Bloomington, Indiana.*

F. Platzek, *Institut für Molekularbiologie und Biochemie der Freien Universität Berlin, Berlin, Germany.*

E. Quagliariello, *Department of Biochemistry, University of Bari, Bari, Italy.*

T. Ramasarma, *Department of Biochemistry, Indian Institute of Science, Bangalore, India.*

M. Rath, *Institut für Molekularbiologie und Biochemie der Freien Universität Berlin, Berlin, Germany.*

H. J. Risse, *Institut für Molekularbiologie und Biochemie der Freien Universität Berlin, Berlin, Germany.*

H. Rogge, *Institut für Molekularbiologie und Biochemie der Freien Universität Berlin, Berlin, Germany.*

G. Rothe, *Institut für Molekularbiologie und Biochemie der Freien Universität Berlin, Berlin, Germany.*

C. Saccone, *Istituto de Chimica Biologica, Università di Bari, Bari, Italy.*

P. V. Sane, *Biology and Agriculture Division, Bhabha Atomic Research Centre, Trombay, Bombay, India.*

G. Schaefer, *Department of Biochemistry, Medizinische Hochschule Hannover, Germany.*

B. B. Singh, *Biology and Agriculture Division, Bhabha Atomic Research Centre, Trombay, Bombay, India.*

A. Stemler, *Departments of Botany, Physiology and Biophysics, University of Illinois, Urbana, Illinois.*

L. Susheela, *Department of Biochemistry, Indian Institute of Science, Bangalore, India.*

V. G. Tatake, *Biology and Agriculture Division, Bhabha Atomic Research Centre, Trombay, Bombay, India.*

E. R. Tustanoff, *Department of Clinical Pathology, Victoria Hospital, and Department of Biochemistry, University of Western Ontario, London, Ontario, Canada.*

N. Vasantha, *Department of Biological Sciences, Madurai University, Madurai, Tamilnadu, India.*

M. Vivekanandan, *Department of Biological Sciences, Madurai University, Madurai, Tamilnadu, India.*

D. Wilkie, *Department of Botany and Microbiology, University College, London, England.*

N. Yamanaka, *Department of Zoology, University of California, Davis, California.*

PREFACE

The interface at which chemistry and biology meet is at the level of organization of the membrane. Until relatively recently, good methods for the study of biomembranes were unavailable. However, in recent years this has all changed. First there was electron microscopy, then techniques of cell fractionation in combination with biochemical analysis which are still being refined. This has given way to the fine details of analysis opened up by genetic studies and by the application of an avalanche of sophisticated physical and chemical techniques, including the use of external and internal probes to sense the organization and environment of membranes. Combination of these techniques has afforded a particularly powerful approach for the resolution of the unique specificity of biomembranes which lend directionality to metabolism. Indeed, reconstitution of the functions of natural membranes, a logical outgrowth of the present rapid rate of development, is presently witnessing amazing progress.

The present volume grew out of a desire to bring together some of the diverse contemporary approaches to the study of biomembranes. It was also inspired by a desire to bring talents together through the forum of an "International Conference on Biomembranes" held at the Department of Biological Sciences, Madurai University, Madurai, Tamilnadu, South India. Points of view represented are of ongoing research by Indian scientists and also of authorities in the four areas of biomembranes with which this volume deals: Biogenesis, Architecture, Bioenergetics, and Differentiation.

This unique opportunity came about as a result of the superb organization and determination of the convenors at the Department of Biological Sciences, Madurai University: Drs. J. Jayaraman, Kunthala Jayaraman, and A. Gnanam, who were aided in their efforts by Professor S. Krishnaswamy, officials of the University and the State of Tamilnadu, and national support from the University Grants Commission and the Indian Atomic Organization.

Support by scientific organizations in the countries of the contributing scientists also helped make this collection of papers possible.

L. Packer

BIOMEMBRANES

PART I

BIOGENESIS OF MEMBRANES

Considering our ignorance (until very recently) about the molecular basis of the structure, architecture and function of biomembranes, it is perhaps surprising that the past few years should also have witnessed an intense and expanding effort at studying the even more complicated problem of their biogenesis. This may have been foolhardy, or perhaps just reflects the unbounded optimism that motivates all of us. But whatever the reasons, the results obtained have rewarded the efforts beyond any reasonable expectation. As the chapters in this section will demonstrate, not only are we beginning to understand the limits and the schedule of the collaborative intra- and extra-organellar events that culminate in membrane synthesis, but in consequence of this understanding we have even obtained some novel insights into membrane structure and function itself.

REGULATION OF THE MITOCHONDRIAL GENETIC SYSTEM AND ITS EXPRESSION

Henry R. Mahler, Philip S. Perlman, Fred Feldman and Roberto Bastos

INTRODUCTION - NATURE OF THE SYSTEM

For the past ten years or so, our group at the Chemistry Department at Indiana University has been interested in the biogenesis of mitochondria. In trying to find out how these essential and ubiquitous organelles of eucaryotic cells are specified and assembled we - as have an ever increasing number of research groups throughout the world - have had recourse to a simple, unicellular eucaryote, with a high respiratory capacity, which occupies a favored place in the biochemical literature: _Saccharomyces cerevisiae_ or baker's yeast. In addition to its many other advantages this proved a fortunate choice. For, as a facultative anaerobe, subject to strong catabolite repression and as a target of extensive and exhaustive genetic studies encompassing not only classical chromosomal, but also non-Mendelian, cytoplasmic (now identified as mitochondrial) systems of inheritance, this organism has provided us with possibly the most appropriate and certainly the most versatile experimental system imaginable (for recent review see 1-5).

Quite early on, Dr. Jayaraman, Dr. Tewari and we were able to obtain evidence for the presence in yeast mitochondria of a species of DNA separate and distinct from that found in the nucleus of the same organism (6-8). We also showed that the organelles appeared to contain what might constitute the basis of an autonomous system of gene expression, namely the capabilities for RNA and protein synthesis, the latter at least with properties quite different from those found in the cytosol of the same cell (6). Thanks to the work of a number of investigators in our own group and particularly in other laboratories, we now know that we were dealing with a general phenomenon. "Normal" eucaryotic cells contain within their mitochondria a semi-autonomous genetic system: it consists of i) a unique DNA, ii) RNAs transcribed from the former, together with iii) all the enzymes and other proteins required for the replication and repair of the DNA, its transcription into RNA, and the utilization of the RNA species for purposes of protein synthesis. It is only semi-autonomous since it requires the co-operative participation of the classical nucleocytosol system not only for the specification but also the biosynthesis of the bulk - and perhaps the totality - of these various proteins (1-5). Similarly, the number and amount of polypeptides furnished the mitochondria by their own translational system is severly restricted. They are found first of all only among the enzyme complexes of the inner mitochondrial membrane to which they contribute 20% or less of its total mass. Furthermore, even among inner membrane functions only cytochrome c oxidase, the membrane integrated, oligomycin sensitive form of ATPase and to a lesser extent the ubiquinone-cytochrome c segment of NADH:cytochrome c reductase appear to require participation of the intramitochondrial system for their elaboration. More recent studies, with isolated highly purified enzyme complexes, particularly in the laboratories of Schatz (5,9,10), Tzagoloff (11,12), and Bücher, Sebald and Weiss (13, 14) summarized in Table I suggest the following: i) The biosynthesis of the complexes for which a mitochondrial participation has been inferred represents a co-operative venture that requires contributions by both systems of gene expression on the molecular level; ii) Since the complexes shown are estimated to contribute approximately 50% to the total mitochondrial mass the intrinsic contribution can account for virtually all of the anticipated mitochondrial products. As a corollary very few if any major products of mitochondrial translation remain to be identified. This inference is confirmed by the overlap of those

TABLE I
MITOCHONDRIAL SPECIFICATION OF THE MITOCHONDRIAL RESPIRATORY CHAIN

Complex	Description	Molecular weight $\times 10^{-6}$	Mitochondrial % Synthesis	Component Mass $\times 10^{-6}$
I	NADH-ubiquinone reductase	0.7	0	
II	Succinate-ubiquinone reductase	0.2	0	
III	Ubiquinone-cytochrome c reductase	0.23	∿15 (1)	0.03
IV	Cytochrome c oxidase	0.20	∿50 (3)	0.10
V	Membrane bound, oligomycin sensitive ATPase complex	0.47	30 (3-4)	0.14
Totals		1.80		0.27

individual polypeptides of the inner membrane identified as synthesized by the particle with the corresponding entities isolated from highly purified complexes. In addition these and other studies (15-18) suggest that iii) at least for cytochrome oxidase and ATPase polypeptides of mitochondrial origin probably are not responsible for catalytic activity *per se* but instead fulfill an integrative or regulatory function. It is to problems of this sort, and in particular to possible intra- and extramitochondrial contributions in the regulation of the replication and differentiation of the organelle that we wish to address ourselves to in this chapter.

REGULATORY ASPECTS OF MITOCHONDRIOGENESIS

What controls the number of mitochondrial genomes per cell?

One of the questions that we hoped to settle was the relationship between nuclear ploidy (i.e., the number of nuclear chromosome sets) and the number of mitochondrial genomes per cell in a stable population. Although one might anticipate a coordinate response between the two numbers, and an early survey by Williamson (19) using several different haploid and diploid strains suggested the existence of such a relationship, there really is no sound *a priori* reason for such a supposition. To subject it to a critical test we obtained three sets of cell lines: two haploids, identical in their nuclear and mitochondrial genetic make-up except for mating type, and the isogenic diploid line derived from their conjugation. To eliminate any additional possible causes of variability we grew all cells under identical conditions, under complete absence of catabolite repression, and harvested them at identical cell densities. Cell numbers were determined by hematocytometer counts, total cellular DNA measured by standard methods, extracted, purified and subjected to analytical ultracentrifugation to equilibrium in a CsCl gradient in order to obtain the contribution of mitochondrial (mt)DNA to total cellular DNA (20). Results are summarized in Table II. It is evident that

TABLE II
MITOCHONDRIAL GENOMES IN ISOGENIC CELLS

Ploidy	DNA per cell	% MtDNA	Mitochondrial Genomes		
	$g \times 10^{14}$		per cell	per unit n-genome	per mitochondrion
2 (2)	5.40	12.6 (± 1.2)*	83	35	3.8
1 (4)	2.86	13.5 (± 1.3)*	47	38	4.7

One mt - genome = 5.0×10^{7} daltons (8.25×10^{-17} g)
One n - genome = 1.25×10^{10} daltons (2.7×10^{-14} g)
Number of experiments in parenthesis
*Standard deviation.

mitochondrial and nuclear ploidy are coordinate; a doubling in the latter produces a corresponding doubling in the former, which itself amounts to some 38 copies per unbudded cell. In other words the cell disposes of a device that rather precisely apportions the relative fractions of nuclear and mitochondrial DNA. The existence of such a device is also evident from the observations reported from several different laboratories, that stable cytoplasmic respiration deficient mutant (ρ^-) cell lines are of two types: those not containing mtDNA in measurable amounts and those containing mtDNA, lacking wild-type sequences to a greater or lesser extent, but present in normal amounts (1,4,21,22).

What regulates the number and mass of mitochondria per cell?

The availability of the three strains just described also permitted us to ask another set of related questions. What fraction of the total cellular, or better, of the total cytoplasmic mass (or knowing the density, what is equivalent, of the total volume), is contributed by mitochondria, how is this mass apportioned among individual organelles, and what is the effect of doubling the ploidy on these parameters? We obtained answers to these and corollary questions using serial sections (20). Some of the results, again with derepressed cells, are summarized in Table III. We find

TABLE III
MORPHOMETRIC DATA FOR SERIAL SECTIONS OF ISOGENIC CELLS

Ploidy	No. of cells	Total Volume ($\mu\ m^3$)			Mitochondria		
		Cell	Cytoplasm	Mito	Vol % (Cyto)	No. per cell	Vol per particle[b]
2	7[a]	26.0	21.7	3.10	14.2	22	0.14 0.13
1	5	13.8	11.6	1.64	14.2	10	0.16 0.15

[a] 3 with buds

[b] in $\mu\ m^3$; the first set of numbers is derived from the data in the table, the second from the actual cumulative distribution

that the mitochondrial volume (or mass fraction) is independent not only of nuclear (and mitochondrial) ploidy but of the stage of the cell cycle and the extent of budding of individual cells examined. The same holds true for the distribution of mitochondrial volume (and size) observed. These distributions are indentical for haploids and diploids - i.e., mitochondria in the larger diploid cell are of the same size as those in the haploid - and virtually identical in different cells regardless of the stage of (nuclear) division. Therefore mitochondrial division in S. cerevisiae is not tightly coupled to cellular or nuclear division. This confirms similar conclusions reported earlier from studies on the timing of replication of n- and mtDNA in synchronized cells (23,24). Combining the data in Table III with those in Table II we can conclude that an "average mitochondrion" in this strain contains some 4-5 copies of mtDNA, a concept that may or may not be devoid of any real significance, since we do not know the rules that control the distribution of mitochondrial genomes among individual organelles.

Finally, it should be mentioned that we have performed analogous studies with different strains, and under different physiological conditions. Mitochondria of an aneuploid (approx. triploid) commercial strain grown under similar derepressed conditions account for an approximately equal fraction of the total mass but their number is much greater (100-200 per cell). In both this, as well as the diploid strain described above, grown under conditions of strong catabolite repression mitochondrial number is greatly reduced ($\leq$ 4 individuals per cell) with a much less pronounced decrease in mass fraction (by a factor of 3-4 fold).

Are the two systems of protein synthesis independent?

Earlier studies, already described in various publications from our group (17,25-27), had shown that radioactive formate provided an excellent tracer for investigations of mitochondrial translation (Scheme 1). It has been used to demonstrate that i) labeling of nascent polypeptide chains is completely

SCHEME 1

FLOW OF FORMATE DURING PROTEIN SYNTHESIS IN MITOCHONDRIA

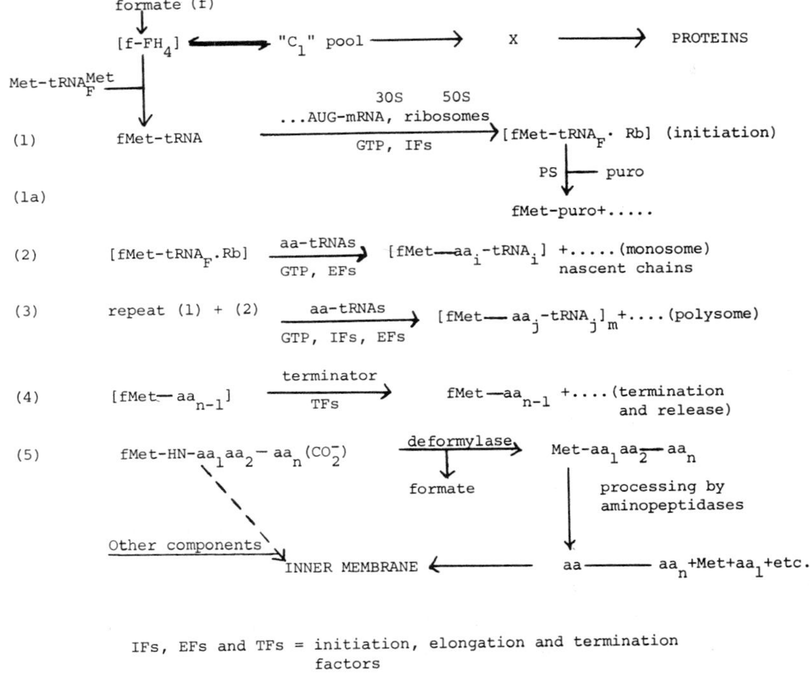

IFs, EFs and TFs = initiation, elongation and termination factors

PS = peptide synthetase (peptidyl transferase)

restricted to those being formed on mitochondrial polyribosomes; ii) a similar absolute specificity is exhibited in the formation of fMet-puromycin, which can then be used as a very convenient assay of mitochondrial initiation (or maximal translational capability), see Table IV; iii) the mitochondria of the strains usually employed by us appear to be devoid of fMet-polypeptide deformylase and therefore formate incorporation can be used to study the fate of mitochondrially synthesized polypeptides even after detachment from their site of synthesis, provided we can compensate or correct for any additional pathways of formate incorporation - "X" in the scheme. Under the

TABLE IV
MITOCHONDRIAL CHAIN INITIATION MEASURED BY FORMATION OF f*Met Puro

Experiment	Strain	C Source	Fraction	time (min)	cpm in fraction
I	D310-4D(ρ^+)	3% Lactate	Mt	0	<60
				5	2860
				10	5600
				30	24000
				60	49000
			Sup	10	650
				60	1890
II	IL8-8C($\rho^+C^RE^R$)	3% Lactate	Mt	10	5095
		2% Glucose			705
	IL8-8C/R5($\rho^-C^RE^R$)	2% Glucose			90
	IL8-8C/R2(ρ^0)	2% Glucose			96

conditions used by us the only such pathway can be shown to be provided by serine incorporated in the cytosol into polypeptides destined for the mitochondria. Since the formyl linkage in N-terminal fMet, unlike that of serine residues in polypeptides, is subject to mild acid hydrolysis, appropriate differential methods can be devised to distinguish the two alternatives. By this means we have demonstrated that at least two of the polypeptides present in our preparations of purified cytochrome oxidase, namely those with particle weights of ∿20 and ∿30 Kdaltons are of mitochondrial origin, thus confirming results obtained with the site-specific inhibitors of protein synthesis, cycloheximide (CH) for the cytosol and chloramphenicol (CAP) for mitochondria (9-13,16). One of the reasons for emphasizing studies utilizing our alternate approach was the supposition that the use of these inhibitors precluded investigations on the existence and nature of any interdependence between the two protein synthesizing systems during the construction of the mitochondrion.

An indication for the existence of such a coupling mechanism in the steady state was provided by studies on the inhibitor specificity in blocking formate incorporation into nascent chains on purified mitochondrial polyribosomes. Earlier, pulse-label (5 min) studies had indicated this process to be sensitive to CAP and ethidium bromide (Etd Br) but completely resistant to CH. When the length of incorporation was extended to 30 min. we observed that the process, although still inhibited by the mitochondrial inhibitors, had now become sensitive to CH as well. To check whether we were indeed dealing with a regulation of the mitochondrial by the cytosolic system we performed the following set of experiments: We used a temperature sensitive strain obtained from Prof. L. Hartwell that is incapable of performing polypeptide chain initiation in the cytosol - which involves a mechanism different from that of the mitochondria - at the nonpermissive temperature. We then looked for mitochondrial protein synthesis, using both histidine incorporation which measures total synthesis regardless of origin, and formate incorporation which measures the mitochondrial component. As is shown in Fig. 1a, shifting up to 36°, the non-permissive temperature leads to a prompt cessation of all protein synthesis not only in the cytosol, but - with a possible lag of <15 min - in the chains initiated in the mitochondria as well. The data of Fig. 1b indicate that mitochondrial initiation itself, as measured by the formation of fMet-puromycin, is unaffected

---

Fig. 1a. (following page)
Saccharomyces cerevisiae, Strain $ts^{-}187$, was grown at 23° on YM-1 3% lactate to midexponential phase, harvested, washed, and spheroplasts prepared but with all treatments at 23° rather than 30°. Spheroplasts were resuspended at 10 times the original concentration in YM-5 3% lactate supplemented with 1 M sorbitol, preincubated with adenine sulfate, L-methionine, and L-serine and labeled with 5 μc/ml ^{3}H-histidine and 0.5 μc/ml ^{14}C-formate at 23°. 10 ml aliquots were removed as shown, CAP (4 mg/ml), cyclo (100 μg/ml) and formate (1 mmole) added, the cells iced, and 10 ml unlabeled carrier cells treated in the same way added. Spheroplasts were transferred to the nonpermissive temperature after 60 minutes and further sampled as shown. Fractions were isolated, mitochondria, resuspended in 1 ml buffer A, 0.5 M sorbitol, and postmitochondrial supernatant (15 ml) were sampled as shown and analyzed.

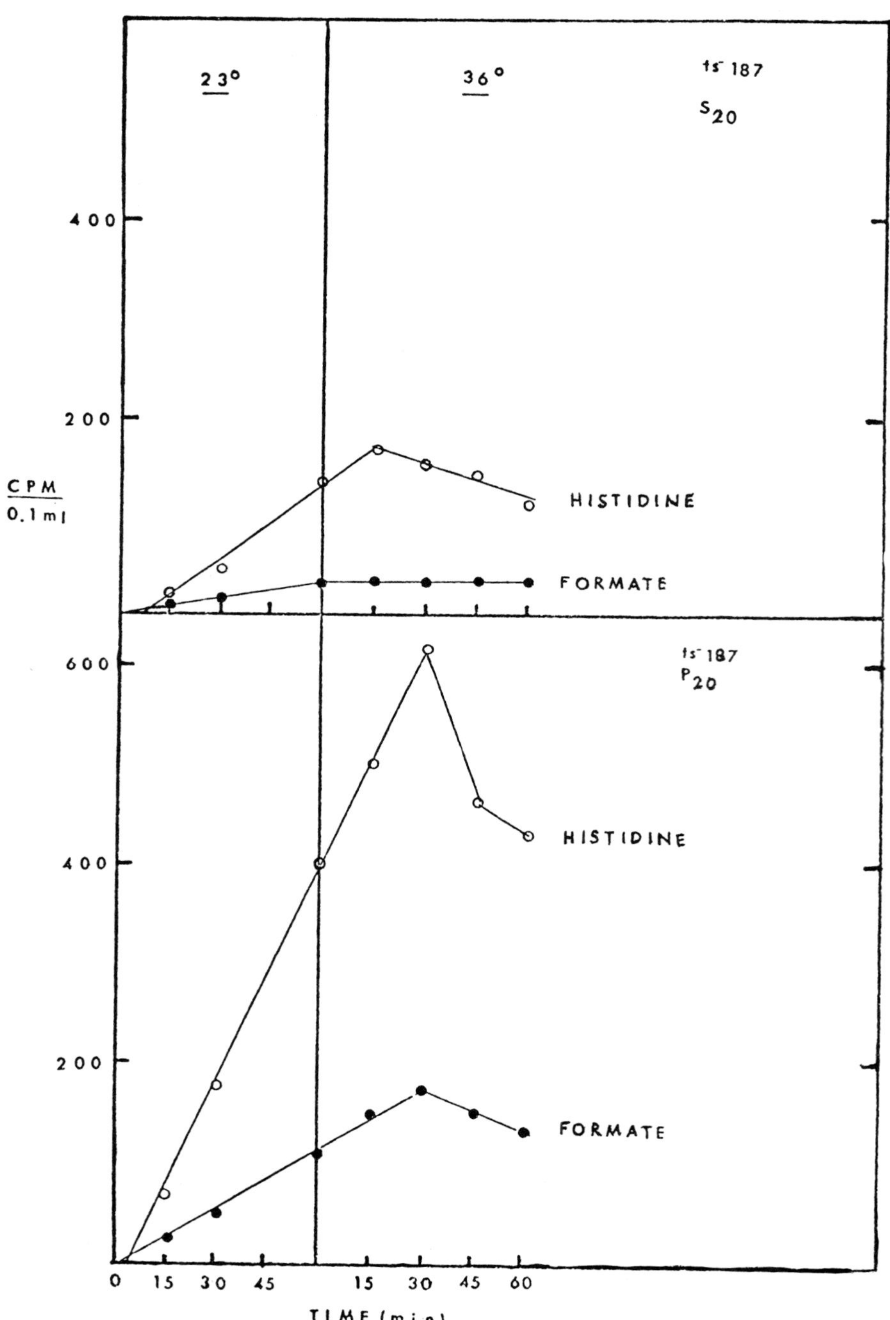

Fig. 1a. (Legend on preceding page)

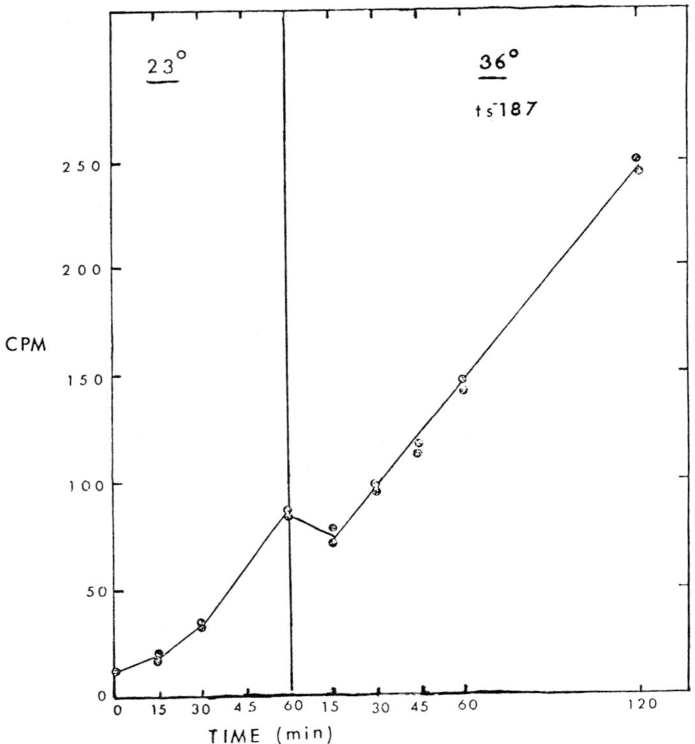

Fig. 1b. Saccharomyces cerevisiae, Strain ts⁻187, was grown and spheroplasts were prepared as in (27). Puromycin (0.5 mM) was added for 15 min followed by ^{14}C-formate (1 µCi per ml of cells). Ten-milliliter aliquots were removed as shown, chloramphenicol (4 mg per ml), cycloheximide (100 µg per ml), and formate (1 mmole) were added, the cells were chilled to 4°, and 10 ml of unlabeled carrier cells treated in the same way were added. Spheroplasts were transferred to the nonpermissive temperature after 60 min and further sampled as shown. Incorporation into fMet-puromycin was determined as before (26).

by the shift-up. This experiment indicates first that chain initiation in the two compartments does indeed involve at least one unique protein factor, and that the cytosolic system (or a product formed by it) exerts some form of translational control on the mitochondrial one at a stage subsequent to chain initiation. Finally we tested this last hypothesis by seeing whether prior synthesis of such a hypothetical product at the permissive temperature in the presence of CAP would alleviate the subsequent inhibition of the mitochondrial incorporation at 36°. This was found to be the case (Fig. 2).

Are mitochondrial products required for the release of catabolite repression?

A typical set of experiments indicating the parameters defining the derepression phenomenon is shown in Figs. 3, 4, and 5. We usually inoculate cells of our commercial strain at a concentration of 10^4-10^5 cell × ml^{-1} into a medium containing 1% glucose and allow eight doublings to take place before intiating the experiment, i.e., at t_o the cell concentration equals ∿3 × 10^6 × ml^{-1} or a dry weight of 0.21 mg × ml^{-1}. We then follow turbidity (A_{600}), whole cell protein (wcp), dry weight (dw), oxygen uptake (OU in μmoles ×mg^{-1} dry weight) and the various enzymes shown, measuring both their specific and total activity in the culture. Derepression as usually defined commences at the second arrow in Fig. 3 and at the double line in all subsequent figures, when the glucose concentration drops below 0.1% and cells enter stationary phase. However, it can be shown that, in fact, except for cytochrome oxidase all activities tested already increase non-isometrically (i.e., they become derepressed) at the first arrow, while cells are still growing exponentially for more than one generation. Analysis discloses that this first stage of derepression is derived exclusively by fermentation. This statement is based on the observation (Fig. 6) that all testable activities behave completely normally in the presence of Antimycin A. This compound is a highly effective respiratory inhibitor, the addition of which to cells growing on a respiratory carbon source leads to an instant cessation of all biosynthetic

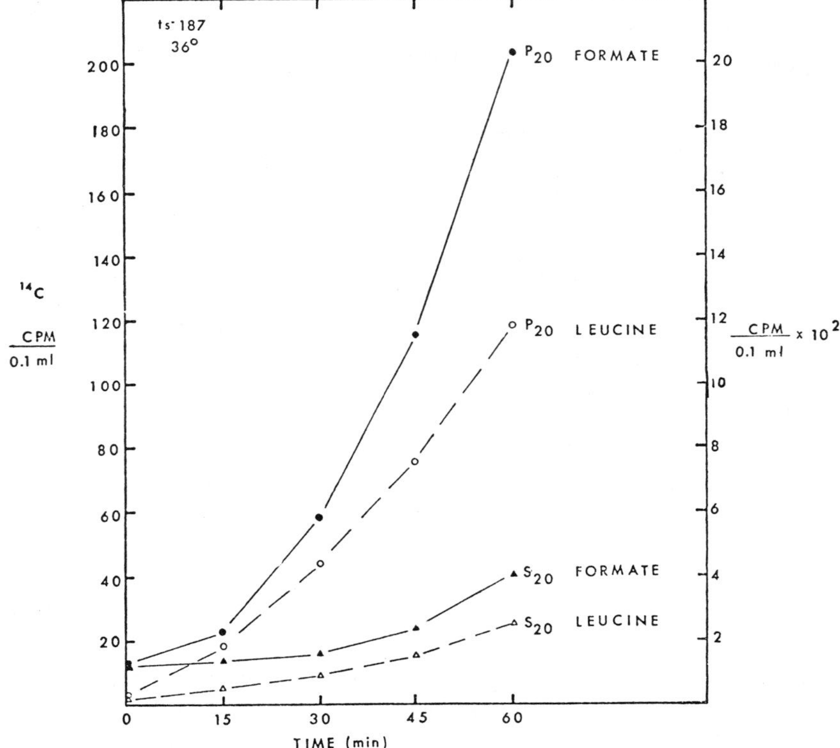

Fig. 2. Spheroplasts were prepared as in Fig. la. CAP (2 mg/ml) was then added and spheroplasts incubated for 18 hours with shaking at 23°. CAP was removed by centrifuging at 23°, the spheroplasts were washed and resuspended into fresh YM-5 lactate supplemented with 1 M sorbitol and preincubated 10 min with adenine sulfate, L-methionine, and L-serine. The spheroplasts were then transferred to 36°, preincubated 30 min at 36°, labeled and sampled as in Fig. 1A but with ^{3}H-leucine in place of ^{3}H-lysine.

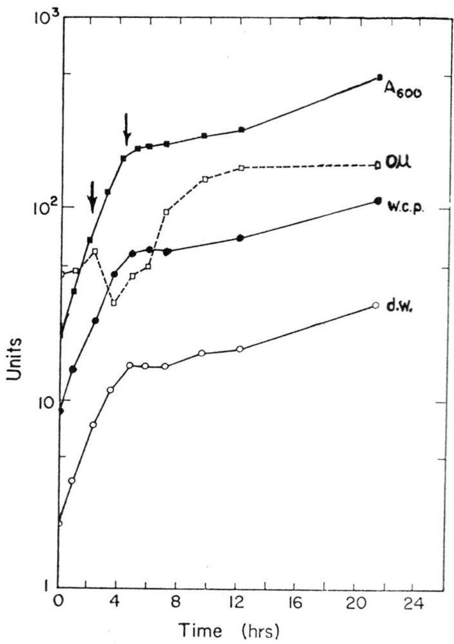

Fig. 3. Growth of Saccharomyces cerevisiae on 1% glucose (aerobic). Growth may be measured by any of the parameters shown: dry weight (dw) is the most easily reproducible one and was used routinely except as noted. Initial values were 0.212, 45, 0.086 and 0.22 for A_{600}, O.U., whole cell protein (wcp) and dw, respectively. The culture was initiated with derepressed cells (O.U. = 140) at 0.00028 mg protein/ml so that eight mass doublings had occurred by t_o on the plot. Levels of enzymes (cf. Fig. 4A and 5A) and O.U. cannot be explained by dilution of pre-existing enzyme during eight generations of growth. The bold arrow (>2h) denotes the beginning of the fermentative stage of derepression while the thin arrow (>4h) denotes the beginning of the oxidative stage of derepression.

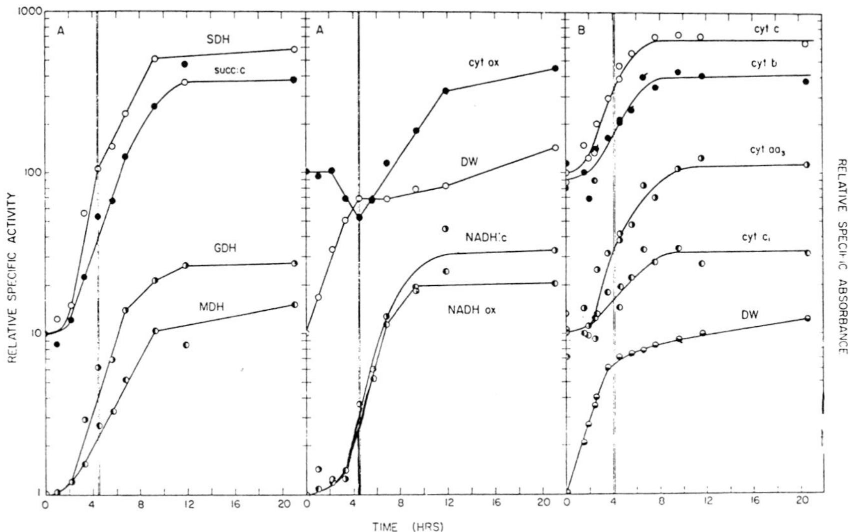

Fig. 4. Increases in specific activity of parameters in cells growing on 1% glucose. A, Enzyme levels: Cells whose growth data are presented in Fig. 3 were assayed for enzyme content in cell-free homogenates. The double line at 4-1/2 hrs indicates the beginning of stationary phase. The enzyme levels were relatively constant for the first three points; during the following two hours of exponential growth, all activities except cyt ox increased several-fold; this increase is termed the fermentative phase of derepression. During this period the cyt ox level decreased somewhat. In stationary phase, cyt ox level increased and all other enzymes continued to increase in specific activity. Initial values were 3.78, 203, 0.57, 1.28, 27.1, 1.46 and 5.8 nmoles$\cdot$min$^{-1}\cdot$mg^{-1} for GDH, MDH, SDH, succ:c, cyt ox, NADH:c and NADH ox, respectively. Initial dw was 0.223 mg/ml.

B, Cytochrome levels: In a separate experiment samples were harvested periodically and analyzed for cytochrome content at the temperature of liquid nitrogen. For cyt c and aa$_3$ a small increase occurs prior to stationary phase; all four cytochromes increase during stationary phase. Changes in cyt aa$_3$ and cyt ox levels do not correlate in time; the cells pass through a period in (continued)

(Fig. 4 continued)
which the ratio of cyt aa_3/cyt ox activity is high. Apparently the stoichiometry of respiratory enzymes is not fixed. Spectra were obtained with cells suspended in 30% glycerol-0.1 M phosphate buffer pH 7.4 at 13.7-18.5 mg cell protein/ml using a 2 mm path length. Initial values (absorbance/mg cell protein × 10^4) were 3.32, 20.6, 7.11 and 11.1 for cyt aa_3, b, c_1 and c, respectively. The first three values for cyt c and c_1 are only approximate due to an anomalous peak at 535-545 nm.

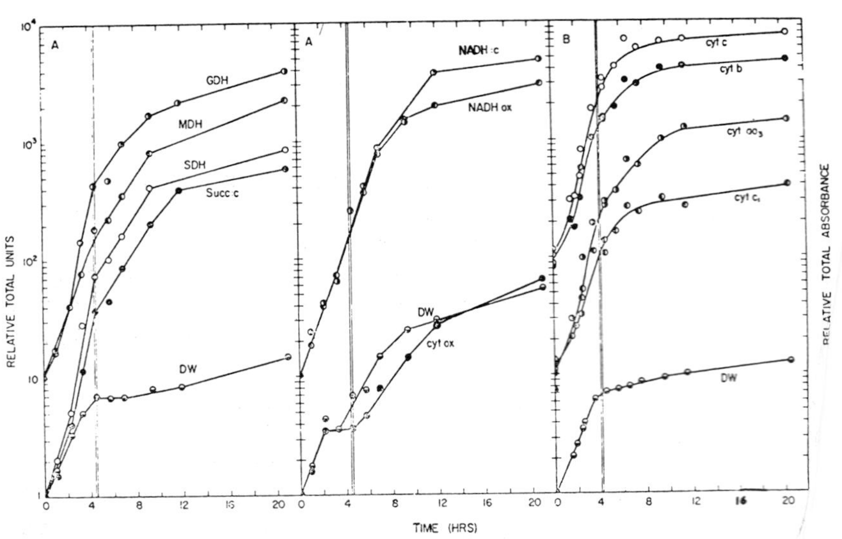

Fig. 5. Increases in total activity (per ml of culture) during growth on 1% glucose. Data from Fig. 4 were calculated as total activity and replotted. A, Enzymes: The change in biosynthetic rate during the fermentative phase of derepression is difficult to determine in this kind of plot. It is clear that the most active rate of synthesis (per hour) is attained just prior to stationary phase for all enzymes assayed except for cyt ox.
B, Cytochromes: The most active rate of synthesis (per hour) is attained prior to stationary phase.

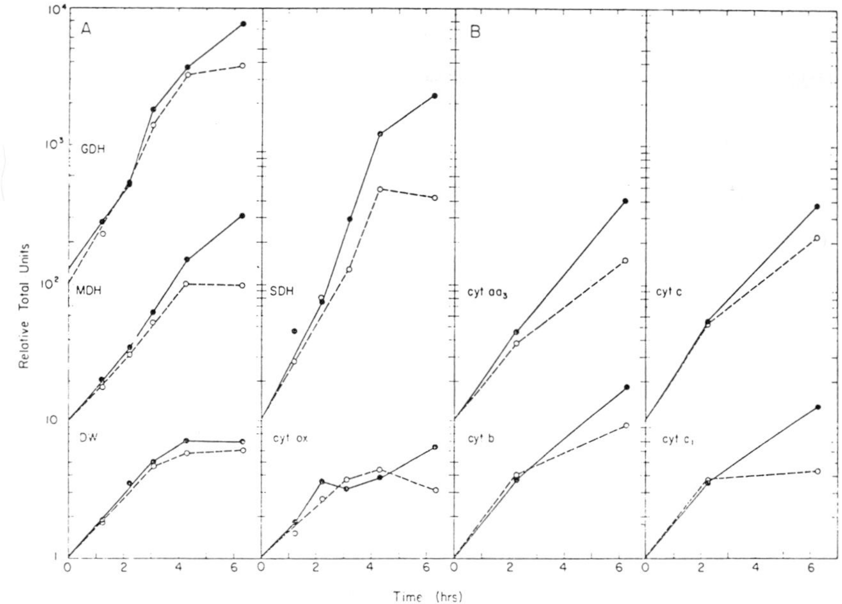

Fig. 6. Effects of antimycin A on derepression when added to repressed cells. Cells were grown on 1% glucose to A_{600} = 0.1 and transferred to sterile 1 liter flasks (500 ml each). When DW = 0.235 mg/ml Anti A was added to 1 µg/ml to half of the flasks. A, Enzyme levels: Initial values were 4.36, 306, 0.8 and 45 nmoles·min^{-1}·mg^{-1} for GDH, MDH, SDH and cyt ox, respectively. Substantial deviation from the control was found only during the last time interval (stationary phase). B, Cytochrome levels: Initial values were 5.6, 39.3, 22.4 and 17.9 absorbance units/ mg cell protein (× 10^4) for cyt aa_3, b, c and c_1, respectively. Cell samples were suspended to 22-24 mg protein/ml and a 2 mm path length was used.

processes including the elaboration of all enzymes under discussion (28). We have therefore called this first stage in derepression the fermentative phase. Its occurrence and the lack of effect of Antimycin A thereon permits us to conclude that neither functional (derepressed) mitochondria, nor the presence of an

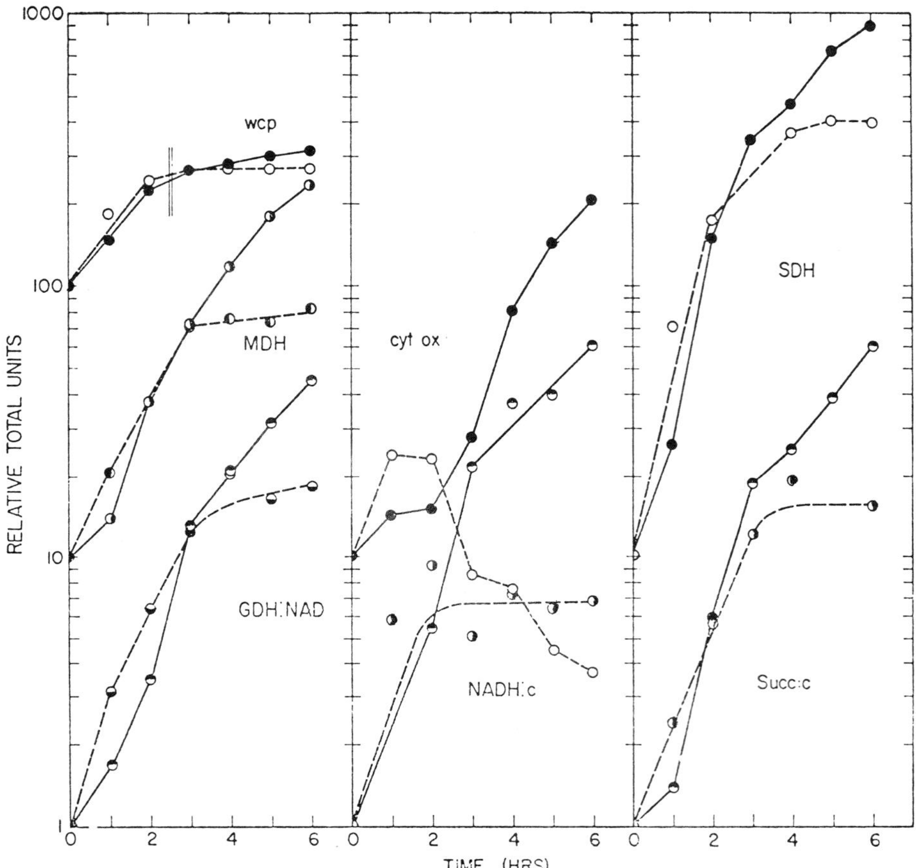

Fig. 7. Effects of CAP on the biosynthesis of "soluble" and respiratory enzymes when added to repressed cells. CAP was added to a final concentration of 4 mg/ml to cultures growing on 1% glucose when A_{600} = 0.65 (repressed). At one hour intervals for six hours aliquots of CAP-treated and control cells were chilled and harvested. Enzyme content was analyzed and converted to total units using whole cell protein/ml of culture as a measure of growth. Initial specific activities were 0.175 mg/ml, 10.3, 1.5, 1.27, 1.59, 294 and 6.15 nmoles$\cdot$min$^{-1}\cdot$mg^{-1} for WCP, cyt ox, NADH:c, SCH, succ:c, MDH and GDH, respectively. In this experiment the fermentative phase of derepression began at t = 1 hr, while the oxidative phase began at t = 3 hrs.

active respiration and energy transduction system is required for the initiation and maintenance of de-repression.

We have extended these studies to inquire whether products of the mitochondrial translational and transcriptional systems are required for the release from catabolite repression. We have used CAP as an inhibitor of mitochondrial translation, euflavine as an inhibitor of mitochondrial transcription, and Etd Br as an inhibitor with high selectivity against mitochondrial processes [whether exclusively transcriptional, or translational as well, is now open to some doubt (5,29)]. The results are shown in Figs. 7 and 8 and suggest that no mitochondrial products are required to bring about and initiate the fermentative phase of derepression.

Is mitochondrial energy transduction implicated in genetic activity?

A number of recent experiments (reviewed in 4 and 17) have suggested that mitochondrial mutagenesis by Etd Br, a highly efficient and specific process, may be modulated by mitochondrial respiratory and energy transducing functions. Possible direct tests of this hypothesis have become feasible by our discovery that Etd Br becomes covalently bound to mitochondrial DNA of susceptible cells as a prerequisite for, or at least coincident with, an early step in mutagenesis (30). More recent experiments have demonstrated that purified mitochondria incubated in a buffer mixture are capable of bringing about the same reaction. As shown in Table V, the reaction, using mitochondria isolated from actively respiring cells, exhibits an absolute requirement for divalent cations and its inhibition pattern is such as to suggest a direct link to the energy transducing device rather than pre-formed ATP.

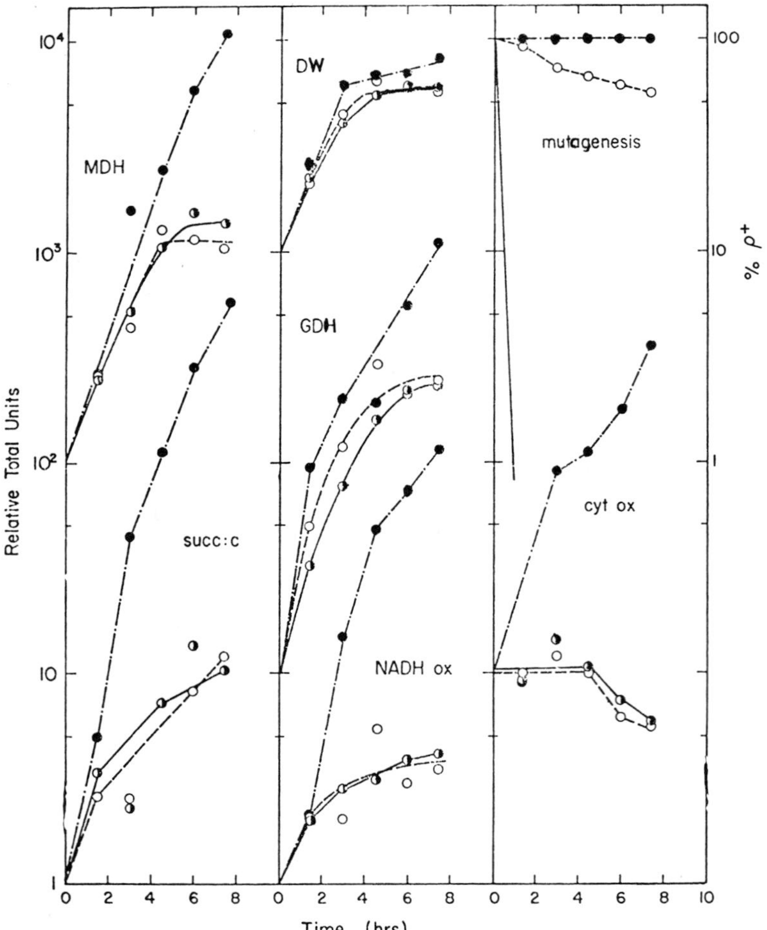

Fig. 8. Effects of transcriptional inhibitors on enzyme synthesis when added to repressed cells. The experimental design was the same as in the previous figures (5,6). Ethidium bromide was added to 50 μM and acriflavin was added to 5 μM. Initial DW was 0.25 mg/ml. Initial values were 352, 0.63, 8.7, 10 and 45 nmoles·min^{-1}·mg^{-1} for MDH, succ:c, GDH, NADH ox and cyt ox. Growth and mutagenesis data are presented in the upper portion of the figure.

TABLE V

KINETICS OF FORMATION OF MODIFICATION PRODUCT OF MITOCHONDRIA

Time (min)	Control cpm	Activity (% of Control)			
		CN^- (3mM)	CCCP* (10^{-5}M)	CCCP* (10^{-6}M)	EDTA for Mg^{2+}
20	1060	88	27		
40	2890	76	13		
60	3780	57	10		
90	3858	57	10		
120	3842	57	10	43	<5

Reaction by mitochondria from 100 ml cells isolated at early stationary from cells grown on lactate.
Incubated in 0.1 M sorbitol, 100 mM KH_2PO_4, 10 mM NH_4Cl, 10 mM $MgCl_2$, 0.1% BSA, 1.0 mM TES (pH = 6.5).
*Carbonylcyanide *m*-chlorophenylhydrazone.

REFERENCES

1. BORST, P. (1972) Ann. Rev. Biochem. 41, 334.
2. LINNANE, A. W., HASLAM, J. M., LUKINS, H. B., AND NAGLEY, P. (1972) Ann. Rev. Microbiol. 26, 163.
3. PREER, J., Jr. (1971) Ann. Rev. Genet. 5, 361.
4. MAHLER, H. R. (1973) CRC Crit. Rev. Biochem. 1, 381.
5. SCHATZ, G., AND MASON, T. (1974) Ann. Rev. Biochem. 43, in press.
6. MAHLER, H. R., TEWARI, K. K., AND JAYARAMAN, J. (1967) in Aspects of Yeast Metabolism, (Mills, A. K., ed.) p. 247, Blackwell, Oxford.
7. TEWARI, K. K., JAYARAMAN, J., AND MAHLER, H. R. (1965) Biochem. Biophys. Res. Comm. 21, 141.
8. TEWARI, K., VOTSCH, W., MAHLER, H. R., AND MACKLER, B. (1966) J. Mol. Biol. 20, 453.
9. SCHATZ, G. (1974) in Biogenesis of Mitochondria (Saccone, C., and Kroon, A. M., eds.) p. 477, Academic Press, New York.
10. MASON, T. (1974) in Biogenesis of Mitochondria (Saccone, C., and Kroon, A. M., eds.) p. 477, Academic Press, New York.
11. TZAGOLOFF, A., RUBIN, M. S., AND SIERRA, M. F. (1973) Biochim. Biophys. Acta 301, 71.

12. TZAGOLOFF, A. (1974) in Biogenesis of Mitochondria (Saccone, C., and Kroon, A. M., eds.) p. 405, Academic Press, New York.
13. SEBALD, W., MACHLEIDT, W., AND OTTO, J. (1973) Eur. J. Biochem. 38, 311.
14. WEISS, H. (1972) Eur. J. Biochem. 30, 469; WEISS, H. (1974) in Biogenesis of Mitochondria (Saccone, C., and Kroon, A. M., eds.) p. 491, Academic Press, New York.
15. MAHLER, H. R., PERLMAN, P. S., AND MEHROTRA, B. D. (1971) in Autonomy and Biogenesis of Mitochondria and Chloroplast (Boardman, N. K., Linnane, A. W., and Smillie, eds.) p. 492-411, North-Holland Publishing Co., Amsterdam.
16. SCHATZ, G., GROOT, G. S. P., MASON, T., ROUSLIN, W., WHARTON, D. C., AND SALTZGABER, J. (1972) Fed. Proc. 31, 21.
17. MAHLER, H. R., FELDMAN, F., PHAN, S. H., HAMILL, P., AND DAWIDOWICZ, K. (1974) in Biogenesis of Mitochondria (Saccone, C., and Kroon, A. M., eds.) p. 423, Academic Press, New York.
18. KOMAN, H., AND CAPALDI, R. A. (1973) FEBS Letters 30, 277.
19. WILLIAMSON, D. H. (1970) Symp. Exp. Biol. 24, 247.
20. GRIMES, G. W., MAHLER, H. R., AND PERLMAN, P. S. (1974) J. Cell. Biol. 61, 565.
21. NAGLEY, P., AND LINNANE, A. W. (1972) J. Mol. Biol. 66, 181.
22. FAYE, G., FUKUHARA, H., GRANDCHAMP, C., LAZOWSKA, J., MICHEL, F., CASEY, J., GETZ, D. G., LOCKER, J., RABINOWITZ, M., BOLOTIN-FUKUHARA, M., COEN, D., DEUTSCH, J., DUJON, B., NETTER, P., AND SLONIMSKI, P. P. (1973) Biochimie 55, 779.
23. SENA, E. P. (1971) Ph.D. Thesis, University of Wisconsin, Madison.
24. WILLIAMSON, D. H., AND MOUSTACCHI, E. (1971) Biochem. Biophys. Res. Comm. 42, 195.
25. DAWIDOWICZ, K., AND MAHLER, H. R. (1973) in Gene Expression and Its Regulation (Kenney, F. T., Hamkalo, B. A., Favelukes, G., and August, J. T., eds.) p. 503-522, Plenum Press, New York.

26. MAHLER, H. R., DAWIDOWICZ, K., AND FELDMAN, F. (1972) J. Biol. Chem. 247, 7439.
27. FELDMAN, F., AND MAHLER, H. R. (1974) J. Biol. Chem. 249, 3702.
28. MAHLER, H. R., AND PERLMAN, P. S. (1971) Biochemistry 10, 2979.
29. LEDERMAN, M., AND ATTARDI, G. (1973) J. Mol. Biol. 78, 275.
30. MAHLER, H. R., AND BASTOS, R. N. (1973) FEBS Letters 39, 27.

EFFECTS OF MITOCHONDRIAL INHIBITION IN *SACCHAROMYCES CEREVISIAE*

D. Wilkie and B. Charlwood

INTRODUCTION

For one reason or another, a wide variety of drugs primarily inhibit mitochondria when they enter living cells. In the case of antibacterial antibiotics, the explanation rests with the specific affinity of these compounds for reaction sites on the mitochondrial ribosome in presumed analogy with the bacterial ribosome (1). In the case of lipophilic drugs such as chlorimipramine and lampren, preferential inhibition of mitochondria is not directly associated with pathways of biogenesis, but with reactivity of these drugs with membranes (2,3). Whether there is a greater affinity of the lipophilic drugs for mitochondrial membranes or that the consequences of the reactivity are more disruptive for membrane-associated systems of the organelle compared with other membrane systems in cells, is a point worth establishing. Furthermore, it is also worth investigation whether an inhibitory effect on mitochondria affects characteristics of cells other than depression of energy metabolism.

Perhaps the most useful organism for studies of this kind is yeast, in which the standard technique in identifying preferential antimitochondrial activity in the first instance is failure of cells to grow in non-fermentable medium at drug concentrations that permit growth in fermentable medium. This distinction is possible because of the fact that *Saccharomyces cerevisiae*, the yeast commonly used, is a facultative anaerobe and can grow on the energy provided by glycolysis alone. This feature also allows the isolation and culture of respiratory mutants of the class known as petite (ρ-) having no respiratory chain and with no capacity for oxygen uptake in metabolizing substrates such as ethanol or glycerol. The lack of mitochondrial function in petite mutants, in other words, seems to confer no obvious disadvantage on cellular processes that are involved in general cell growth and division, provided glucose is available in the growth medium.

PHENYLBUTAZONE

The usefulness of the yeast system in detecting antimitochondrial activity of drugs is exemplified in our recent experiments with phenylbutazone (Butazolidine, Geigy). The drug is clinically used as an anti-inflammatory agent although its mode of action is not understood. As in the case of chlorimipramine and lampren, phenylbutazone is a lipophilic compound and as such will react with membranes.

In 4 haploid yeast strains tested (see Table I), the drug was found to be markedly more inhibitory to growth on non-fermentable agar medium compared with

TABLE I
INHIBITION OF OXYGEN UPTAKE BY PHENYLBUTAZOLIDINE IN YEAST CELLS*

Butazolidine (M)	Percent inhibition of oxygen uptake**			
	Strain			
	917	45B	D6	41
2.1×10^{-3}	9.3±0.8	3.5±2.1	31.1±0.1	44.2±4.0
4.3×10^{-3}	31.0±1.2	9.2±0.2	39.7±3.2	66.9±1.1
6.4×10^{-3}	51.6±2.8	16.7±1.1	54.1±3.1	76.6±0.8
8.5×10^{-3}	58.3±0.9	27.7±1.7	66.5±2.8	83.4±1.8
1.1×10^{-2}	62.8±1.2	35.0±0.1	76.9±3.9	85.5±1.3

*Approximately 10^6 cells/ml in 0.1 M phosphate buffer, pH 6.8, in a Clark-type oxygen electrode.

**Average of 3 experiments.

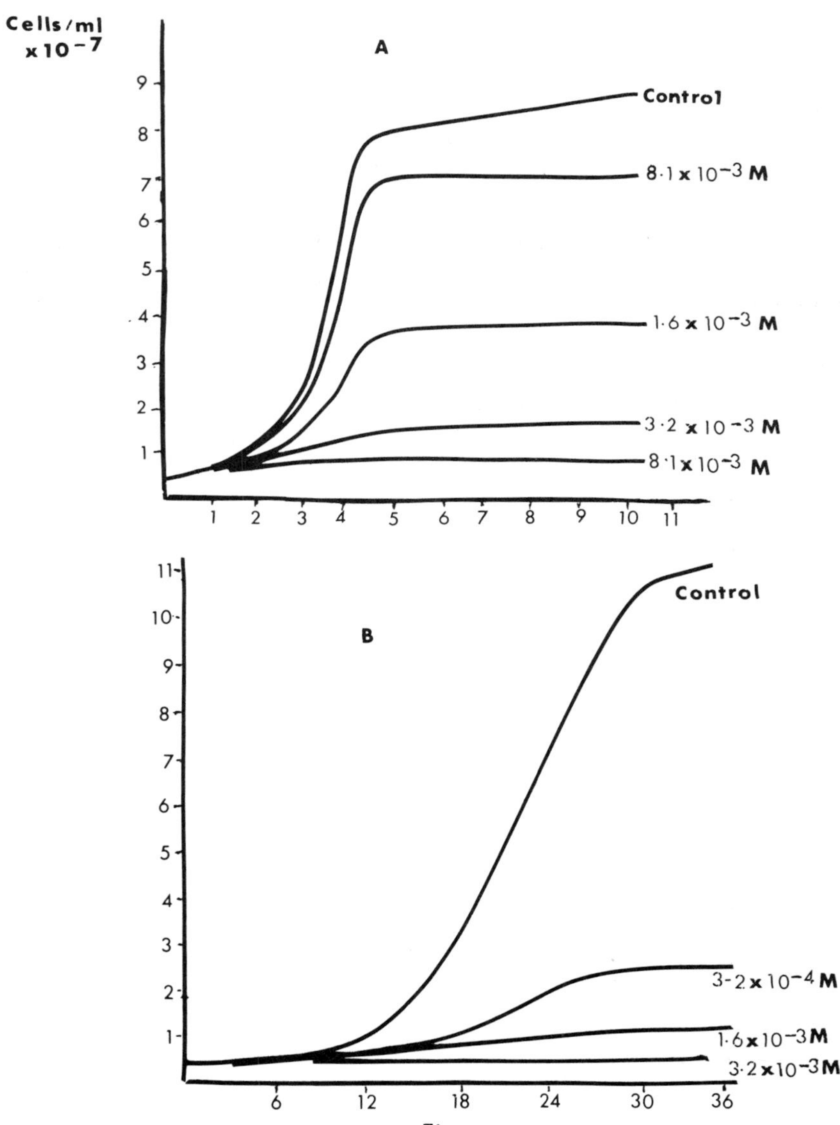

Fig. 1. Growth curves of strain 917 in the presence of phenylbutazone. A, glucose medium (YED). B, glycerol medium (YEG).

glucose medium under standard conditions. A more accurate assessment of the effect of the drug was obtained by measuring growth inhibition in liquid culture. Using strain 917, it can be seen (Fig. 1) that a drug concentration of 1.6×10^{-3} M in a yeast-extract, glucose medium (YED) inhibits growth about 60%; one-fifth of this concentration inhibits about 75% in glycerol medium (YEG). In all experiments, growth medium was buffered with 0.1 M phosphate buffer at pH 6.8. A 4- to 10-fold difference was seen in this respect in the other three strains. These results were more sensitive to the inhibitory effects of the drug than cellular processes.

Inhibition of oxygen uptake by the drug was found to correlate with the degree of sensitivity of the

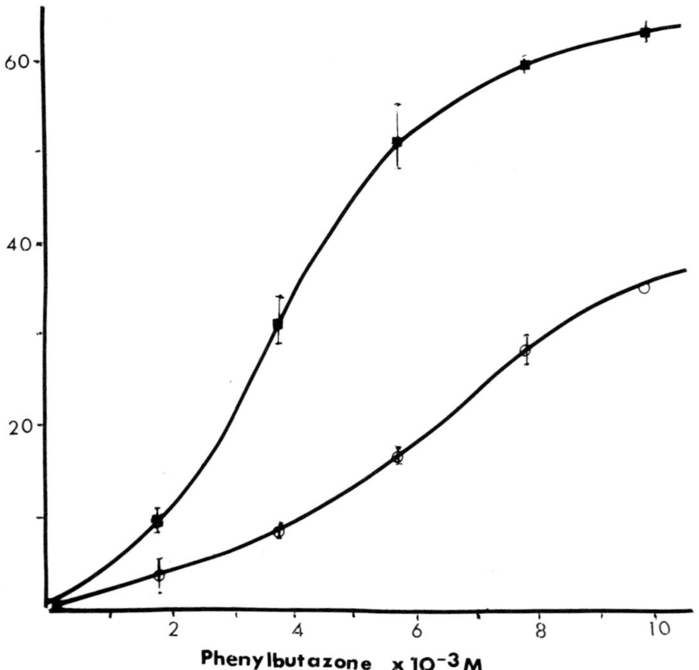

Fig. 2. Inhibition of oxygen uptake in strains 917 and 45B by phenylbutazone. ■——■ 917; o——o 45B.

yeast strains to growth inhibition in glycerol medium (Table I). These results indicate that the drug directly inhibits mitochondrial activity. In conditions of obligate mitochondrial function, i.e., in glycerol medium, this would lead to the arrest of cell growth. The inhibition of oxygen uptake is shown graphically in Fig. 2 in the case of strains 917 and 45B.

The effect of the drug on protein synthesis was studied in the case of strain 917. Cells were pulse labelled with ^{14}C-leucine, proteins precipitated and radioactivity measured as described in (4). There was a marked depression of cellular protein synthesis at lower drug concentrations in glycerol medium compared with cells metabolizing glucose (Fig. 3). Whereas the

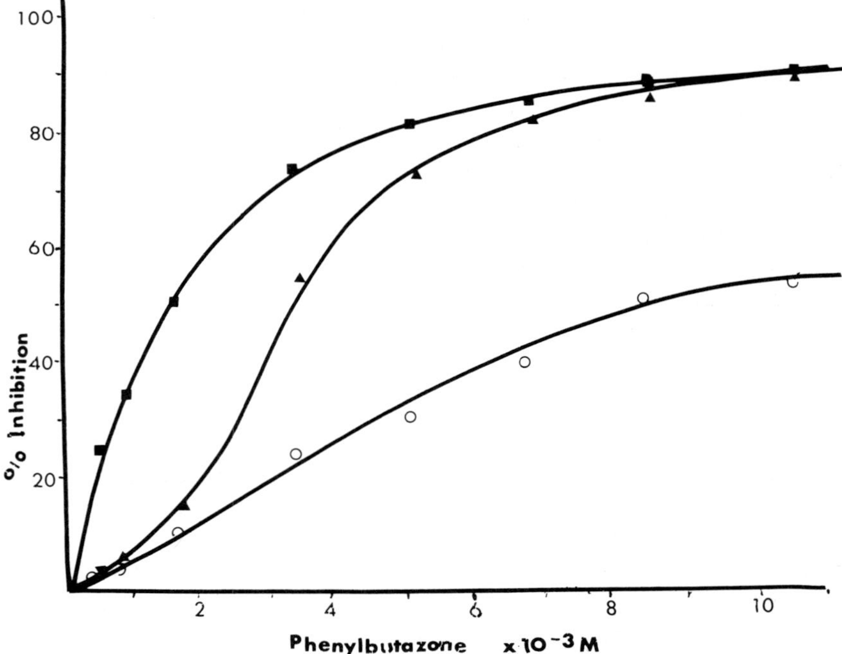

Fig. 3. Inhibition by phenylbutazone of uptake of ^{14}C-leucine in strain 917 and 917 ρ- (petite). ■——■ 917, glycerol substrate; ▲——▲ 917, glucose substrate; o——o 917 ρ-, glucose substrate.

inhibition in YEG medium may be ascribed to a primary restriction of energy metabolism, in glucose medium inhibition of protein synthesis may be considered as a manifestation of a more general effect on cellular processes presumably associated with membrane function.

In petri-dish cultures of strain 917 on glucose medium, it was found that isolated colonies that came up on inhibitory concentrations of the drug (5×10^{-3} M) were generally petite and with a frequency of about 1×10^{-2}. Since the spontaneous frequency of petites in this strain is about 3×10^{-2}, it appeared that a good proportion of petite mutants had a selective advantage on glucose medium in the presence of the drug. This indicates that in these petite cells, cellular processes were less inhibited by the drug than in respiratory competent cells. This was substantiated in the studies on cellular protein synthesis in which cells of a petite mutant were much less inhibited in this respect by the presence of the drug than wild-type parental cells (Fig. 3).

In view of the lipophilic nature of butazolidine, the results may be summarized in the following way: 1) reactivity of the drug with membranes is more crucial in the case of mitochondrial membrane systems than membrane associated processes in the cell; 2) a genetic change in the mitochondrion can lead to an alteration in the sensitivity of cell membranes.

IMIPRAMINE

We have presented evidence recently that the change to cytoplasmic respiratory deficiency (the petite condition), can result in an increase in the tolerance of yeast cells (which are sensitive when respiratory competent) to the membrane-reactive drug imipramine (2). Since not all petites show this effect, then respiratory deficiency *per se* is not responsible for the cellular alteration. It also implies that a translation product of mitochondrial DNA is not involved since it is generally accepted that petite mitochondria do not synthesize proteins.

In those petites that result in cellular resistance to imipramine, genetic analysis indicates cytoplasmic control of the resistance. Since various degrees of loss of genetic information occur in petites mitochondria, the conclusion is drawn from these observations that mitochondrial DNA carries factors involved in the expression of membrane characteristics of the cell. Put another way, a heritable change in cellular tolerance to imipramine is brought about by a genetic change in the mitochondrion. In all cases studied so far, pulse labelling of cells with radioactive imipramine has shown a correlation between the degree of binding of the drug to membrane fractions and the level of drug sensitivity of cells.

DISCUSSION

The results with butazolidine parallel those with imipramine, particularly with regard to alteration in cellular tolerance brought about by the petite mutation. The working hypothesis we have adopted is that a transcription product of mitochondrial DNA is involved in the determination of membrane characteristics of the cell, particularly the plasma membrane. This product may act in the regulation of nuclear genetic information, an idea recently put forward by Barath and Kuntzel (5) with regard to the synthesis of mitochondrial RNA polymerase.

Membrane-reacting compounds are clearly useful probes in studies of organelle-cellular interactions.

REFERENCES

1. GRIVELL, L. A., REIJNDERS, L., AND BORST, P. (1971) Biochim. Biophys. Acta 247, 91-103.
2. RHODES, P. M., AND WILKIE, D. (1973) Biochem. Pharmacol. 22, 1047-1056.
3. LINSTEAD, D., EVANS, I., AND WILKIE, D. (1974) in The Biogenesis of Mitochondria (Kroon, A. M., and Saccone, C., eds.) pp. 179-193, Academic Press, New York.
4. EVANS, I., LINSTEAD, D., RHODES, P. M., AND

WILKIE, D. (1973) Biochim. Biophys. Acta 312, 323-336.

5. BARATH, Z., AND KUNTZEL, H. (1972) Nature new Biol. 240, 195-197.

STABLE PLEIOTROPIC CHROMOSOMAL MUTATIONS WITH MODIFIED MITOCHONDRIAL ATPase AND CYTOCHROMES IN *SCHIZOSACCHAROMYCES POMBE*

A. Goffeau, A. M. Colson, Y. Landry, F. Foury, and M. Briquet

In yeast, single gene nuclear mutations causing disappearance of the cytochrome aa_3 absorption peak are often accompanied by other mitochondrial deficiencies. Ten years ago, Sherman and Slonimski reported the existence of chromosomal respiratory-deficient strains of *Saccharomyces cerevisiae* with multiple cytochrome deficiencies. These single-gene mutants were classified as p or pet strains (1). Subik and collaborators described single-gene nuclear mutants deficient in both cytochromes aa_3 and b which in addition had lost the oligomycin-sensitivity of their mitochondrial ATPase (2). They proposed that these pleiotropic effects were the result of a deficient mitochondrial protein synthesis. Similar mutants were reported in *Schizosaccharomyces pombe* by Goffeau *et al*. (3,4), who also described another class of nuclear mutants where multiple cytochrome deficiencies were accompanied by an almost total loss of mitochondrial ATPase activity. A mutant of *S. cerevisiae* with similar deficiencies was recently described by Ebner and Schatz (5), who demonstrated a functional but abnormal mitochondrial protein synthesis in this respiratory-deficient strain.

All the pleiotropic nuclear respiratory-deficient mutants which have been genetically analyzed so far are produced by single-gene mutations. They must thus be the result of a single primordial modification in a central process controling the expression of several distinct proteins of the inner mitochondrial membrane. The activities most often modified in pleiotropic nuclear respiratory-deficient mutants are the cytochromes aa_3 and b as well as the oligomycin-sensitive ATPase. It is particularly interesting that the exact same activities are also controled by the mitochondrial DNA. They are not detected in the cytoplasmic rho^- mutants (6, 7, 8) and contain components synthesized by the mitochondrial ribosomes (9 to 19). It is also very significant that several distinct nuclear genes produce pleiotropic mutations with apparently similar phenotypes (3, 4, 20, 21).

Unfortunately the most drastic pleiotropic pet mutants of *S. cerevisiae* are generally very unstable (1, 2, 5). In an overnight batch culture, about 40% of the pet cells might be spontaneously induced into cytoplasmic rho^- and thus be transformed in double mutants. Since both nuclear (pet) and cytoplasmic (rho^-) mutations affect the same mitochondrial activities, the biochemical study of such cultures might be very complicated.

An easy way out of this difficulty is the use of a "petite-negative" yeast species like *Schizosaccharomyces pombe*, where cytoplasmic rho^- mutations cannot be induced (22, 23, 24). This haplontic yeast divides by fission like a giant bacterium. Although it contains only four times more DNA than *Escherichia coli*, this well differentiated eukariot possesses at least six chromosomes, the genetic mapping of which is in progress (25, 26). The physiology of *S. pombe* is rather well known (27). It grows on glycerol as non-fermentable substrate (28) and its mitochondria are submitted to glucose repression (29). Although anaerobic growth is very restricted, *S. pombe* grows well aerobically on glucose in the absence of respiration (28). Nuclear respiratory-deficient mutants are easily obtained by a variety of mutagenic treatments (3, 4, 22, 23, 30, 31).

For instance, 5% of the survivors of nitrosoguanidine treatment were pet mutants able to grow on glucose but not on glycerol (Table I). In none of them was rho^- induction detected and all tested strains

TABLE I
FREQUENCY OF NUCLEAR RESPIRATORY-DEFICIENT MUTANTS IN SCHIZOSACCHAROMYCES POMBE TREATED BY N'-NITROSOGUANIDINE

Respiration	Cyt aa_3	Oligo sensitive ATPase	Dio-9 sensitive ATPase	Frequency among survivors
+	+	+	+	95%
-	-	+	+	3%
-	-	-	+	1.5%
-	-	-	-	0.5%

Nitrosoguanidine mutagenesis and characterization of the deficiencies were carried out as described in ref. 31.

showed a Mendelean 2:2 segregation for growth on glycerol. This absence of rho^- greatly facilitated the screening of the nuclear respiratory-deficient strains for growth on glycerol. The vast majority of the mutants had no cytochrome aa_3 absorption peak. Most of these were pleiotropic and showed multiple modifications of their absorption spectrum (3). About 40% of the cytochrome aa_3-deficient strains showed in addition modifications of their mitochondrial ATPase either by loss of oligomycin-sensitivity only or by total disappearance of the Dio-9 sensitive ATPase activity. Thus as much as about 2% of the survivors of the mutagenic treatment were deficient in at least two different proteic complexes of the inner mitochondrial membrane: the ATPase and cytochrome aa_3. This high frequency probably reflects the highly integrated nature of the mitochondrial membrane and of its biogenesis.

The cytochrome and ATPase deficiencies of one particularly stable mutant, M126, were analyzed in more detail. Despite numerous attempts, no reversion to the wild type could be induced with

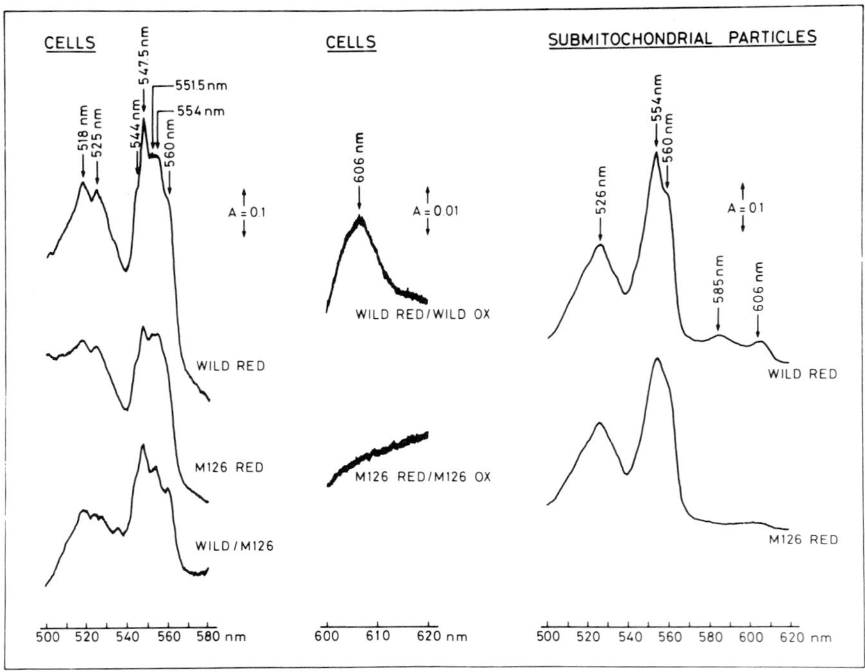

Fig. 1. Low temperature absorption spectra of Schizosaccharomyces pombe 972h$^-$ and the mutant M126 (reproduced by permission of the editors of the Journal of Biological Chemistry, ref. 4). All spectra were obtained from glucose-grown cells and carried out at liquid nitrogen temperature. On the left: spectra of cellular suspensions of 500 mg wet weight per ml reduced by dithionite; top: absolute spectrum of S. pombe M126. In the middle: differential spectra of cellular suspensions of 500 mg wet weight per ml reduced by dithionite versus oxidized by ferricyanide; top: S. pombe 972h$^-$; bottom: S. pombe M126 (14.6 mg protein per ml). RED and OX are abbreviations for reduced by dithionite and oxidized by ferricyanide.

or without mutagenic treatments. This strain obtained by X-irradiation is thought to be the result of a restricted deletion in only one or a few very closely linked nuclear genes (3, 4).

CYTOCHROME DEFICIENCIES IN THE MUTANT M126

The differential low temperature spectrum of reduced versus oxidized mutant cells demonstrates the absence of the 606 nm absorption peak of cytochrome aa_3 (Fig. 1). As expected, the mutant cytochrome c oxidase activity is markedly decreased to 5% that of the wild type (4). Cytochrome c is detected at 547 nm (α_1 band), 544 nm (α_2 band) and 519 nm (β band) in the absolute reduced spectrum of cell mutants. A pigment, possibly cytochromes c_1 or b_2, absorbing at 551 nm, is also seen in the same M126 spectrum. Another absorption peak is clearly observed at 554 nm in the mutant sonicated submitochondrial particles where cytochrome c and the 551 nm pigment are extracted. The 554 nm pigment (wild type) is reduced by succinate in the presence of antimycin A but not by ascorbate plus NNN'N' tetramethyl-p-phenylene diamine dihydrochloride; it is thus a cytochrome b (32). Cytochromes b absorbing around 560 nm are more difficult to analyze because of their low content in glucose-repressed cells. The mutant cells spectra show a shoulder at 558-560 nm most clearly observed in the stationary phase of growth where the 554 nm interference decreases. Compared to the absolute wild type spectrum, the differential spectra of the reduced wild type versus the reduced mutant cells shows a clear increase at 560 nm, suggesting a lower content in the mutant of at least one component of this absorption peak. It is difficult to distinguish the several forms of cytochrome b (32) in the mutant but it is clear that one component of complex II is deficient, as demonstrated by the marked decrease of the mutant succinate:cytochrome c reductase activity, while succinate dehydrogenase is unaffected (4). Growth of the wild type in the presence of chloramphenicol produces a spectrum idential to that

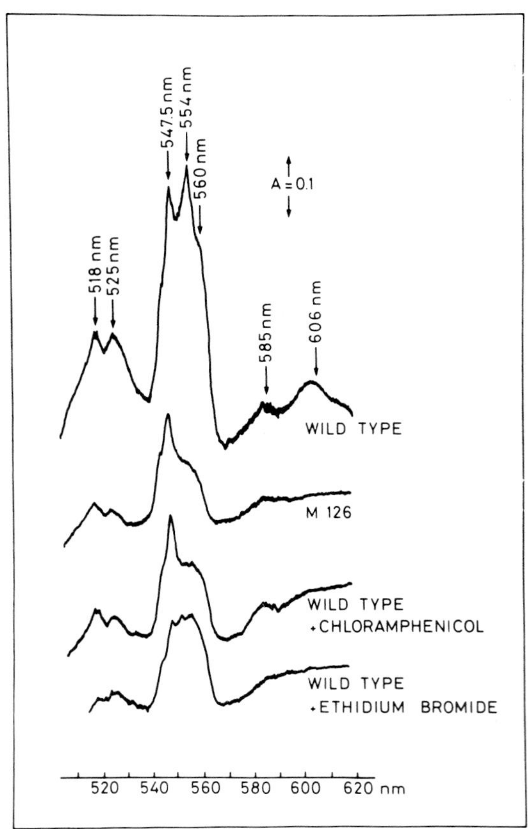

Fig. 2. Effects of ethidium bromide and chloramphenicol on cytochrome absorption peaks of *Schizosaccharomyces pombe* 972h$^-$ (wild type) compared to the mutant M126 (reproduced by permission of the editors of the Journal of Biological Chemistry, ref. 4). *Schizosaccharomyces pombe* M126 was grown for 20 hr in glucose medium, brought to pH 6.5. *Schizosaccharomyces pombe* 972h$^-$ was grown for 16 hr in glucose medium, pH 6.5, or for 24 hr in the same medium supplemented with 20 μg/ml of ethidium bromide or 4 mg/ml chloramphenicol. The differential low temperature spectra were carried out with cellular suspensions of 500 mg wet weight per ml. The references were oxidized by 0.3% H_2O_2, except for the ethidium bromide-treated cells, which were oxidized by ferricyanide. The samples were reduced by dithionite.

of the mutant (Figure 2). This observation suggests the decrease in the mutant M126 of a cytochrome b of long wavelength which has been shown to be inhibited by chloramphenicol in Neurospora (14, 33).

ATPase DEFICIENCIES IN THE MUTANT

The cellular homogenate was centrifuged at 100,000 × g for one hour and separated in a pellet and a soluble supernatant. In Fig. 3, the surface of each rectangle is proportional to the ATPAase units in each fraction obtained from 100 mg of homogenate proteins. No oligomycin-sensitive ATPase was found in the mutant. The mutant soluble Dio-9-sensitive ATPase

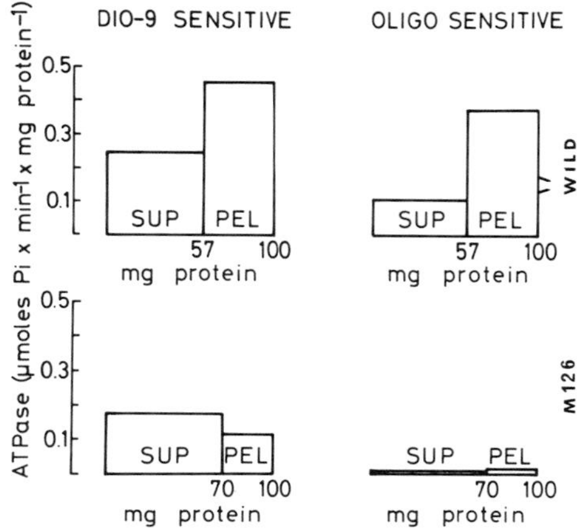

Fig. 3. Distribution of ATPase activities in soluble and particulate fractions of Schizosaccharomyces pombe 972h⁻ (wild) and mutant M126 (reproduced by permission of the editors of the Journal of Biological Chemistry, ref. 4). The homogenates obtained from glucose-grown cells as described in reference 4 were further fractionated into a supernatant and pellet by centrifugation for 1 hr at 100,000 × g. The abscissa reports the percent of homogenate protein recovered in each fraction. The rectangular surfaces are proportional to the ATPase units recovered in each fraction from 1 mg of homogenate protein.

TABLE II

PROPERTIES OF ATPase PURIFIED FROM SUBMITOCHONDRIAL PARTICLES OF SCHIZOSACCHAROMYCES POMBE WILD-TYPE AND MUTANT M126

	ATPase activity (nmoles Pi × min^{-1})		
	GLY wild	GLU wild	GLU wild
Complete	49	33	22
- $MgCl_2$	1	2	1
- $MgCl_2$ + 6 mM $MnCl_2$	28	19	13
- $MgCl_2$ + 6mM $CaCl_2$	13	10	5
- $MgCl_2$ + 6 mM $SrCl_2$	2	5	-
+ 10 mM $NaHCO_3$	57	41	26
Fresh enzyme	66	32	39
+ 20°C	69	32	39
+ 2°C	15	8	3
+ 2°C + 10% glycerol	59	29	35
+ 2°C + 10% ethanol	78	31	33
+ 2°C + 40% $(NH_4)_2SO_4$	48	20	17
Complete	57	33	22
+ 4 µg/ml oligomycin	63	36	20
+ 50 µM DCCD	63	34	20
+ 154 µM NaN_3	12	9	3
+ 25 µg/ml Dio-9	8	8	3
+ 150 µM synthalin	29	21	11
+ 155 µm TBTC	8	7	4
Complete	57	33	22
- PEP	32	25	7
- PEP + 6 mM ADP	23	12	8
Complete	43	-	22
+ 100 µl antibody	10	-	4

The enzyme was purified as described in reference 4 except that the glycerol gradient was omitted for the glucose-grown enzymes. The reaction mixture contains: (continued on opposite page) activity is not markedly modified, and the particulate pellet contains decreased but significant Dio-9-sensitive ATPase activity.

Identical distribution of the oligomycin and Dio-9-sensitive ATPase are observed in M126 and in the wild-type grown in the presence of chloramphenicol or ethidium bromide (4).

The mutant membrane-bound, Dio-9-sensitive ATPase was extracted by extended sonication at pH 9.0 of sub-mitochondrial particles and purified up to a specific activity of 80 μmoles ATP hydrolyzed per mg protein and per min. Table II shows that the responses of this purified mutant enzyme to cations, anions, cold treatment, ADP inhibition and Dio-9, synthalin, sodium azide, and tributyl-tin chloride were similar to those of the wild-type enzyme purified from glucose-grown or glycerol-grown cells. The mutant ATPase activity was also inhibited by an antibody obtained against a wild-type enzyme of specific activity of 120 isolated from glycerol-grown cells.

(Table II explanation continued)
3mM ATP, 20 mM Tris-HCl, pH 8.6, 6 mM $NaCl_2$, 4 mM phosphoenol-pyruvate, 15 units of pyruvate kinase, 10 μl ethanol and either 1.5 μg protein of glycerol-grown wild-type ATPase or 3.0 μg of glucose-grown wild-type of M126 ATPase. The rabbit antibody was obtained by injections of an ATPase preparation of 120 of specific activity purified from glycerol-grown wild-type cells. The temperature treatments are carried out by preincubation of the freshly unfrozen enzyme in Tris-acetate pH 7.5, 2 mM ATP, 1 mM EDTA and the indicated concentrations of glycerol or ammonium sulfate at 20°C or 20C for 6 hr. DCCD and TBTC are abbreviations for N,N',dicyclohexylacarbodiimide and tributyl-tin chloride. When phosphoenolpyruvate (PEP) is omitted, no pyruvate kinase is added to the reaction mixture. GLY and GLU are abbreviations for enzyme purified from strains grown on glycerol or glucose containing media.

Sodium dodecyl sulfate-polyacrylamide gel electrophoresis patterns show that the mutant enzyme contains the five mitochondrial ATPase components of 61,000, 58,000, 32,000, 14,000 and 8,000 molecular weight (Fig. 4). The relative proportion of these components is

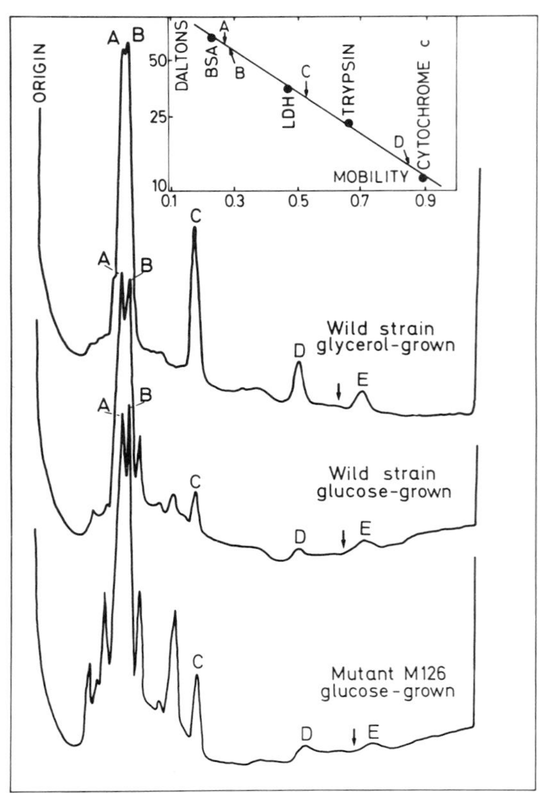

Fig. 4. Absorbance tracting of polyacrylamide gel electrophoresis of dissociated ATPase purified from Schizosaccharomyces pombe wild type and mutant M126 (reproduced by permission of the editors of the Journal of Biological Chemistry, ref. 4). The ATPase was purified from glycerol-grown cells of S. pombe 972h^{-} (middle trace) or S. pombe M126 (lower trace) as described in reference 4. The arrows indicate the migration of bromophenol blue. The letters A, B, C, D, E indicate the presumed ATPase subunits.

roughly similar in the glucose-grown wild-type and mutant enzymes but differs significantly from the glycerol-grown wild-type enzyme. In addition, bands of 40,000, 54,000 molecular weight and two bands of molecular weight higher than 65,000 are seen in the mutant. The significance of these additional bands is presently under study.

It is concluded that the membrane-bound ATPase of the mutant is very similar to the wild-type mitochondrial enzyme. The loss of oligomycin-sensitivity of the mutant is thus the result of modification of membrane factors. Since appreciable amounts of Dio-9-sensitive ATPase (inhibited by an antibody against the mitochondrial wild-type enzyme; data not shown) is located in the mutant post-ribosomal supernatant, it appears that one function of the deficient membrane factor(s) is to anchor the ATPase in the membrane. This conclusion is strengthened by the data of Table III, which shows that concentrations of Triton-X 100, which solubilizes up to 40% of the mutant submitochondrial particles, does not solubilize the Dio-9-sensitive membrane-bound ATPase. Under the same conditions, 58% of the proteins and 28% of the ATPase are solubilized in glucose-grown wild-type submitochondrial particles.

MITOCHONDRIAL PROTEIN SYNTHESIS IN THE MUTANT

All previously described deficiencies would be consistent with the existence of a deficient mitochondrial protein synthesis in the mutant. Chloramphenicol, acriflavin and ethidium bromide-sensitivities of cycloheximide-resistant anino-acid incorporation have, however, been detected *in vivo* in the mutant (4). The relative extent of these inhibitions (about 20%) is low because of high cycloheximide resistance of *S. pombe* but was similar to that of the wild-type. The absolute extent of these inhibitions is in the range expected for mitochondrial protein synthesis (about 4% of the total cellular amino-acid incorporation).

No firm conclusions about the intactness of the mitochondrial protein synthesis can, however, be drawn until the acriflavin and ethidium-bromide sensitive peptides of M126 have been identified and compared to the mitochondrial products of the wild-type.

CONCLUSIONS

The chromosomal mutation M126 mimics most of the deficiencies observed in the cytoplasmic rho^{-} mutation of _S. cerevisiae_ or produced by inhibition of the mitochondrial protein synthesis in the wild-type. The primordial product modified by the single-gene chromosomal mutation has not yet been identified. At least three possibilities remain open: 1) A nuclear-dependent deficiency of the mitochondrial protein synthesis which could produce chloramphenicol and ethidium bromide-sensitive but non-fully-functional peptides. 2) Deficiency of a nuclear-coded assembly factor specific for the products of the mitochondrial protein synthesis. 3) Deficiency of a nuclear-coded factor regulating the synthesis of nuclear or mitochondrial DNA-coded components of the oligomycin-sensitive ATPase, cytochrome oxidase and, possibly, cytochrome b complexes.

REFERENCES

1. SHERMAN, F., AND SLONIMSKI, P. P. (1964) _Biochim. Biophys. Acta_ _90_, 1.
2. SUBIK, J. KUZELA, S., KOLAROV, J., KOVAC, L., AND LACHOWICZ, T. M. (1970) _Biochim. Biophys. Acta_ _205_, 513.
3. GOFFEAU, A., COLSON, A. M., LANDRY, Y., AND FOURY, F. (1972) _Biochem. Biophys. Res. Commun._ _48_, 1448.
4. GOFFEAU, A., LANDRY, Y., FOURY, F., BRIQUET, M., AND COLSON, A. M. (1973) _J. Biol. Chem._ _248_, 7097.
5. EBNER, E., AND SCHATZ, G. (1973) _J. Biol. Chem._ _248_, 5379.
6. SLONIMSKI, P. P. (1953) _in_ La Formation des Enzymes Respiratoires chez la Levure, Masson,

Paris.

7. SCHATZ, G. (1968) J. Biol. Chem. 243, 2192.
8. PERLMAN, P. S., AND MAHLER, H. R. (1970) Bioenergetics 1, 119.
9. TZAGOLOFF, A. (1971) J. Biol. Chem. 246, 3050.
10. WEISS, H., SEBALD, W., AND BUCHER, TH. (1971) Eur. J. Biochem. 22, 19.
11. SCHWAB, A. J., SEBALD, W., AND WEISS, H. (1972) Eur. J. Biochem. 30, 511.
12. SEBALD, W., WEISS, H., AND KACKL, G. (1972) Eur. J. Biochem. 30, 413.
13. TZAGOLOFF, A., AND MEAGHER, P. (1972) J. Biol. Chem. 247, 594.
14. WEISS, H. (1972) Eur. J. Biochem. 30, 469.
15. WEISS, H., LORENZ, B., AND KLEINOW, W. (1972) FEBS Letters 25, 49.
16. MASON, T. L., AND SCHATZ, G. (1973) J. Biol. Chem. 248, 1355.
17. RUBIN, M., AND TZAGOLOFF, A. (1973) J. Biol. Chem. 248, 4275.
18. TZAGOLOFF, A., RUBIN, M. S., AND SIERRA, M. F. (1973) Biochim, Biophys. Acta 301, 71.
19. WEISS, H., SEBALD, W., SCHWAB, A.J., KLEINOW, W., AND LORENZ, B. (1973) Biochimie 55, 815.
20. EBNER, E., MENNUCCI, L., AND SCHATZ, G. (1973) J. Biol. Chem. 248, 5360.
21. GOFFEAU, A., COLSON, A. M., LANDRY, Y., FOURY, F., AND BRIQUET, M. (1974) Biochem. J., in press.
22. HESLOT, H., LOUIS, C., AND GOFFEAU, A. (1970) J. Bacteriol. 104, 482 (1970).
23. WOLF, K., SEBALD-ALTHAUS, M., SCHWEYEN, R. J., AND KAUDEWITZ, F. (1971) Molec. Gen. Genetics 110, 101.
24. SCHWAB, R., SEBALD, M., AND KAUDEWITZ, F. (1971) Molec. Gen. Genetics 110, 361.
25. LEUPOLD, U., (1970) Methods in Cell Physiology 4, 169.
26. DA CUNHA, M. (1970) Genet. Res. Cambridge 16, 127.
27. MITCHISON, J. M. (1970) Methods in Cell Physiology 4, 131.

28. HESLOT, H., GOFFEAU, A., AND LOUIS, C. (1970) J. Bacteriol. 104, 473.
29. FOURY, F., AND GOFFEAU, A. (1972) Biochem. Biophys. Res. Commun. 48, 153.
30. BACHOFEN, V., SCHWEYEN, R. J., WOLF, K., AND KAUDEWITZ, F. (1972) Z. Naturforsch. 27b, 252.
31. COLSON, A. M., COLSON, C., AND GOFFEAU, A. (1974) Methods in Enzymology 32, in press.
32. POOL, R. K., LLOYD, D., AND CHANCE, B. (1974) Biochem. J., in press.
33. VON JAGOW, G., AND KLINGENBERG, M. (1972) FEBS Letters 24, 278.

MECHANISM FOR THE BIOGENESIS OF MITOCHONDRIAL MEMBRANES IN YEAST

R. M. Janki, H. N. Aithal, E. R. Tustanoff, and A. J. S. Ball

INTRODUCTION

The data presented here are an extension of the work on the biogenesis and control of mitochondria in the facultative yeast, *Saccharomyces cerevisiae*, that has been carried out in our laboratory over the past decade. Early work in yeast mitochondriogenesis [cf. reviews: Mahler (1) and Linnane *et al.* (2)] was concerned with characterizing the effects of environment on the synthesis of components of these organelles. From these and subsequent reports, it became clear that yeast mitochondria contain a functional DNA-RNA protein synthesizing system which complements the major cyto-nuclear protein assembly apparatus which is encoded in nuclear DNA. The control of these nucleo-mitochondriogenic systems is primarily under the influence of the fermentative carbon source (i.e., catabolite repression) and/or the gaseous atmosphere via O_2 induction.

Although these studies have provided a plethora of information, our understanding of the control and mechanism of mitochondrial assembly by 1970 was still unclear. It is well known that the biosynthesis and subsequent incorporation of lipids into phospholipid-protein complexes play an important role in the formation and integrity of biological membranes. Yeast cells offer a useful system to study these constructions, since the relative structure and hence function of their mitochondrial membranes can be manipulated by changing the conditions of growth as well as their nutritional lipid supplementation. These altered membranes kinetically reflect the cell's new lipid environment (3,4). The isolation of unsaturated fatty acid yeast auxotrophs by Resnick (5) and their subsequent exploitation by Linnane's group (6) further demonstrated that alterations in the fatty acid moiety of membrane phospholipids lead to changes in many of the membrane-dependent functions.

Taking advantage of the fact that biological membranes undergo a cooperative thermotropic transition, Fox (7) and Raison et al. (8) employed a classical physico-chemical technique, Arrhenius kinetics, to investigate the lipid milieu which engulfs specific membrane enzymes. These temperature transitions reflect a transformation of the membrane lipid bilayer from a melted gel-crystalline ordered state to a gel-liquid disordered state and this transition temperature is controlled by the melting point of the fatty acids found in the phospholipid side chains. Since the enzymes associated with mitochondrial membranes undergo this phase transition, the perturbation associated with this change is assumed to induce a conformational change in the active center of the enzyme protein (3).

We have utilized this technique to investigate a possible mechanism for the biogenesis of mitochondrial membranes in yeast. Initial reports from our laboratory (4) showed that cytochrome c oxidase activity was lipid dependent and that the Arrhenius transition points were characteristic of the lipid supplement. Furthermore, newly synthesized enzyme activity was incorporated into newly formed mitochondrial membranes. We have expanded this line of investigation and will attempt here to relate how the synthesis of the inner and outer mitochondrial membranes is effected during the biogenesis of the whole organelle by measuring the temperature characteristics of marker enzymes of these two membranes (cytochrome c oxidase and oligomycin-sensitive ATPase - inner membrane; and kynurenine hydroxylase - outer membrane). In support of these experimental findings we will present possible models to explain the mechanism of the integration of these enzymes into the two mitochondrial membranes.

METHODS

Saccharomyces cerevisiae, strain 77 (our standard laboratory strain), was grown on either complete medium supplemented with ergosterol and Tween 80 (9) or on minimal medium supplemented with ergosterol and a series of 0.02% unsaturated fatty acids (oleic, linoleic, linolenic or elaidic) (10). For aerobic growth experiments 3% galactose was used as a carbon source and these cells were harvested an hour prior to the onset of their stationary phase. For anaerobic-aerobic transition experiments, cells were anaerobically grown on 3% glucose minimal medium supplemented with ergosterol and 0.02% linoleic acid in a commercial fermentor with constant nitrogen sparging. Cells were harvested prior to stationary phase, washed free of the anaerobic lipid supplement and then transferred to fresh minimal medium containing a new fatty acid and then induced with oxygen by vigorous aeration with air. Samples of these adapting cells were taken as indicated in the text and mitochondria prepared as described below. The unsaturated fatty acid auxotroph KD-20, which was kindly supplied by Dr. S. Fogel, University of California, was also grown on 3% glucose minimal medium supplemented with 0.15% unsaturated fatty acids under both anaerobic and aerobic conditions.

Isolation of Mitochondria

Harvested cells were once washed with ice-cold distilled water and twice with ice-cold 0.04 M phosphate buffer, pH 7.4, containing 0.1% bovine serum albumin, chloramphenicol (4 mg/ml) and cycloheximide (25 μg/ml). The resulting washed cells were suspended in 0.1 M sorbitol containing 0.1 M Tris-HCl buffer pH 8.0, 1 mM EDTA and 0.1% bovine serum albumin. After 30 second homogenization in a Braun shaker a mitochondrial fraction was obtained by differential centrifugation after the procedure of Henson *et al.* (11). The final mitochondrial pellet was suspended in 0.25 M sucrose containing 10 mM Tris-HCl buffer (pH 7.4) to yield a suspension containing 5-10 protein/ml.

Enzyme Assay

Oligomycin-sensitive ATPase was determined by the method of Tzagoloff (12), while kynurenine hydroxylase activities were monitored by the procedure outlined by Schott *et al.* (13). Temperatures were varied from 2° to 37°C during various incubation periods by use of a constant temperature refrigerated bath. Cytochrome oxidase activity was measured spectrophotometrically as described before using thermo-jacketed cuvettes and immersible thermocouples to monitor reaction temperatures (4).

Fatty Acid Analysis

Lipids were extracted from 0.2 to 0.5 ml of mitochondrial suspensions by the method of Folch *et al.* (14). Methyl esters of the extracted fatty acids were prepared by the boron trifluoride method and were resolved on a EGSS-X column using a Beckman GC-45 chromatograph and employing standard gas chromatograph procedures.

Respiration was monitored polarographically using a Clark-type oxygen electrode and protein concentration was determined using the procedure of Lowry *et al.* (15) using bovine serum albumin as a standard.

RESULTS AND DISCUSSION

Yeast grown anaerobically are incapable of synthesizing unsaturated fatty acids since molecular oxygen is required to carry out the enzymic desaturation reactions (16) and thus these lipids must be added to the anaerobic growth medium. Aerobically unsaturated fatty acid synthesis in yeast is repressed in the presence of an excess of unsaturated fatty acid in the growth medium with the result that these cells preferentially incorporate the lipid supplement. It is therefore possible to imprint both mitochondrial precursor membranes (i.e., promitochondria) anaerobically and oxidatively functioning mitochondrial membranes aerobically with particular unsaturated fatty acids.

In our previous reports (4,10) on the synthesis of cytochrome c oxidase using Arrhenius kinetics, whole cells were used. To allay misinterpretation of these data on the grounds that French pressure cell treatment caused severe fragmentation of all cellular membranes which may result in their possible interaction, we have used isolated mitochondria in all experiments reported here. Figure 1 illustrates the results of Arrhenius plots of the three mitochondrial marker enzymes. Cells for this experiment were grown aerobically on 3% galactose minimal media containing 0.02% lipid supplement of either elaidic, linoleic, or oleic acids. The results for cytochrome c oxidase (Fig. 1A) confirmed those obtained with whole cells (4,10). Kinetics for the oligomycin-sensitive ATPase (Fig. 1B) and kynurenine hydroxylase (Fig. 1C) duplicate those of the oxidase. Mitochondria from elaidic-grown yeast are characterized by the same transition temperature, 22° for all the three enzymes, while mitochondria from linoleic and oleic grown cells produce transition temperatures of 8° and 12° respectively for these enzymic activities.

Data on the synthesis of the outer membrane marker enzyme, kynurenine hydroxylase, are provided in Fig. 2. Cells were grown aerobically on 3% galactose complete medium containing Tween 80. The isolated mitochondria from late exponential cells were lipid depleted by acetone extraction (acetone-H_2O, 9:1) (17) and were shown to have only 25% of the activity of the control mitochondria. When this extract was added back to the lipid depleted preparation, kynurenine hydroxylase activity was restored to 60% of the control, clearly demonstrating that this enzyme is lipid dependent. Furthermore, this enzyme is not under mitochondrial protein genome control, since its activity is not affected when cells are grown in the presence of chloramphenicol, an inhibitor of mitochondrial protein translation. It would appear that this outer mitochondrial membrane enzyme is synthesized on cytoplasmic ribosomes (18).

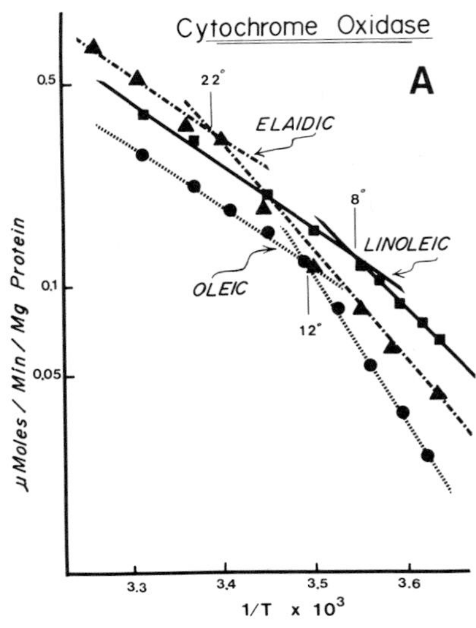

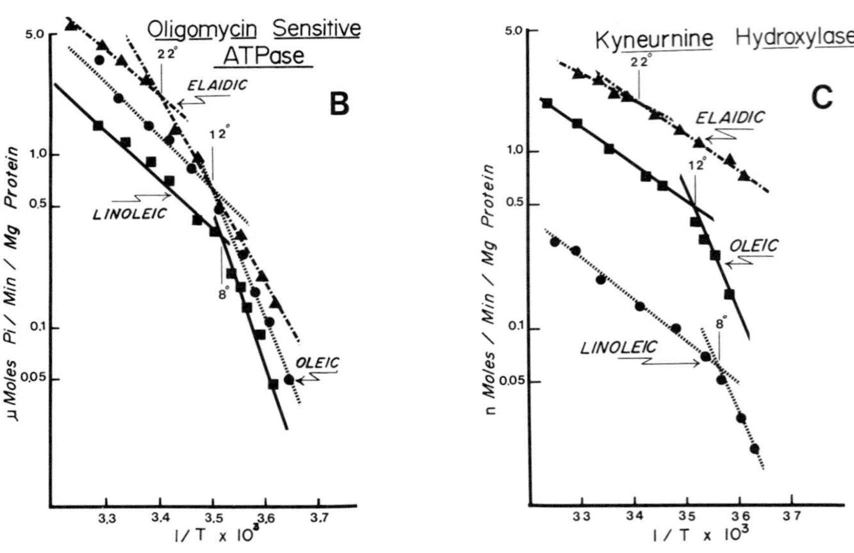

Fig. 1. (Legend on opposite page)

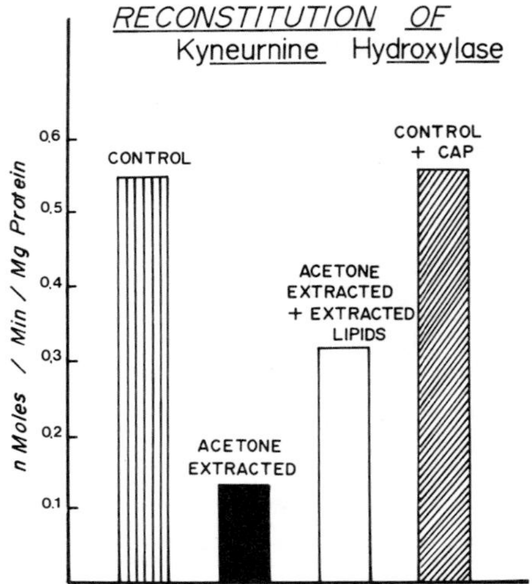

Fig. 2. Reconstitution of kynurenine hydroxylase. Yeast were grown aerobically on 3% galactose complete medium containing ergosterol and Tween 80. The isolated mitochondria were lipid depleted for 10 minutes as described in the text. In the reconstituted experiment the acetone extract was taken to dryness, resuspended in Tris-acetate buffer (pH 8.0, 0.05 M) and an aliquot equivalent in volume to the original sample added to the lipid-depleted mitochondria prior to assay. Inhibited cells were grown in the presence of 4 mg/ml D(-) chloramphenicol.

Fig. 1. Arrhenius plots of mitochondrial marker enzymes. Yeast were grown aerobically on 3% galactose minimal media containing 0.02% fatty acid supplement of either elaidic or linoleic acids, as described in the Methods section. Enzyme activities were determined on isolated mitochondria and are expressed as a logarithmic function at an absolute temperature (T).
A, Cytochrome c oxidase activity (EC 1.9.3.1);
B, Oligomycin-sensitive ATPase (EC 3.6.1.3); and C, Kynurenine hydroxylase (EC 1.14.1.2). The temperatures noted in the figure represent the transition point for each specific fatty acid.

TABLE I
EFFECT OF LIPID SUPPLEMENTATION ON THE FATTY ACID COMPOSITION OF MITOCHONDRIAL MEMBRANES

Lipid in growth medium	Percent fatty acid							
	$C_{12:0}$	$C_{14:0}$	$C_{16:0}$	$C_{16:1}$	$C_{18:0}$	$C_{18:1}$	$C_{18:2}$	$C_{18:3}$
Tween 80	1	2	13	33	3	33	trace	nil
Oleic	> 1	4	19	18	6	40	trace	nil
Linoleic	> 1	2	20	14	8	9	42	nil
Linolenic	> 1	3	21	10	7	7	trace	44
Elaidic	2	3	26	16	9	39	nil	nil

Yeast were grown aerobically on either 3% galactose complete medium containing ergosterol and Tween 80 or on a series of minimal media containing 0.02% fatty acids listed below. The cells were harvested in their late exponential phase and mitochondria were prepared as described in the Methods section. Fatty acid content of these organelles was determined by gas liquid chromatography as described in the Methods section.

The fatty acids are denoted by the convention: number of carbon atoms:number of unsaturated linkages.

Table I shows the effect of lipid supplementation on the fatty acid composition of mitochondrial membranes from yeast grown aerobically on 3% galactose complete - Tween 80 medium and a series of minimal media containing 0.02% oleic, linoleic, linolenic and elaidic acids respectively and harvested in their late exponential phase. When pure unsaturated acids are used, the supplement accounts for 40% of the total fatty acids found in these membranes and approximately 75% of the total unsaturated fatty acids. The Arrhenius transition temperature (T_t) changes proportionally with the melting point of the fatty acid supplement; gives the same temperature for all three enzymes (cf. Fig. 1); and the major portion of the fatty acid composition of this membrane is made up from the supplemented fatty acid (cf. Table I). It thus follows that it is reasonable to employ Arrhenius plots to investigate the nature of the lipid milieu which encompasses the different mitochondrial membrane-bound enzymes.

A question which arises in the study of biogenesis of mitochondria is "Are newly synthesized proteins integrated into new membrane or into preformed membranes?" To answer this, the following transfer experiments were carried out. Yeast were grown anaerobically to their late exponential phase on 3% glucose minimal medium supplemented with 0.02% linoleic acid. After harvesting, the residual linoleic acid was washed off the cell surfaces and the cells were transferred to new 0.02% elaidic minimal medium and adapted in air. Cells were sampled at specified times after the induction and mitochondria were prepared. The fatty acid content of these organelles was determined as well as activities of the three marker enzymes at temperatures from 2° to 37°. At zero time (cf. Table II) the major fatty acid is linoleic acid (18:2) and it makes up 35% of the total fatty acid of the organelle. After transfer to elaidic medium, the linoleic acid content falls to 20% and the proportion of elaidic acid (18:1) increases from 4 to 18% after 2 hours of aerobic adaptation. In the control experiment, an anaerobic linoleic acid to aerobic linoleic acid transfer, no changes are observed. The reduction of the 18:2 fatty acid and the subsequent increase of the 18:1 fatty acid in this transfer experiment must be due to the incorporation or exchange of the second fatty acid into the mitochondrial membrane.

TABLE II
FATTY ACID COMPOSITION OF MITOCHONDRIA OBTAINED FROM YEAST CELLS UNDERGOING AEROBIC ADAPTATION

Condition of growth	Percent fatty acid: Linoleic (N_2) → Elaidic (O_2)								Percent fatty acid: Linoleic (N_2) → Linoleic (O_2)							
	$C_{12:0}$	$C_{14:0}$	$C_{16:0}$	$C_{16:1}$	$C_{18:0}$	$C_{18:1}$	$C_{18:2}$	$C_{18:3}$	$C_{12:0}$	$C_{14:0}$	$C_{16:0}$	$C_{16:1}$	$C_{18:0}$	$C_{18:1}$	$C_{18:2}$	$C_{18:3}$
N_2	6.8	9.7	26.2	10.6	6.0	4.0	35.9	0	5.1	8.7	25.2	12.1	4.9	3.2	41.2	0
O_2 + ½ hr	3.0	9.4	20.8	15.4	5.9	13.3	24.9	0	2.5	5.6	27.7	16.7	6.0	2.4	40.7	0
O_2 + 1 hr	4.3	7.7	24.4	18.5	6.7	14.3	21.6	0	4.5	9.7	20.8	18.2	4.7	2.5	40.2	0
O_2 + 2 hr	4.3	5.1	21.6	24.7	5.8	17.7	20.6	0	3.3	5.4	23.0	16.6	4.2	3.0	42.7	0

Yeast were grown anaerobically on 3% glucose minimal medium supplemented with ergosterol and 0.02% linoleic acid to a point one hour prior to stationary phase and then adapted in minimal medium containing 0.02% elaidic acid as described in the Methods Section. Control cells were adpated in fresh 0.02% linoleic acid. Cells were harvested at times indicated and mitochondria were isolated. The fatty acid content of these organelles was determined by gas liquid chromatography.

The fatty acids are denoted by the convention: number of carbon atoms:number of unsaturated linkages.

Figure 3 illustrates the restructuring of the Arrhenius kinetics of membrane bound enzymes during the aerobic adaptation period. Each enzyme demonstrates its own characteristic changes. After 1/2 hr of exposure to air, cytochrome c oxidase (Fig. 3A) has two

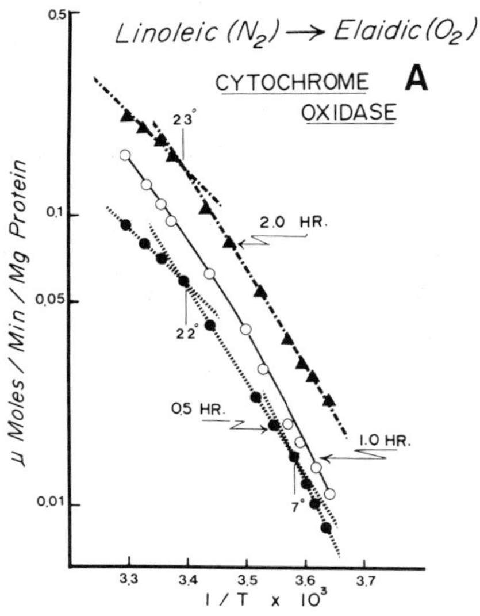

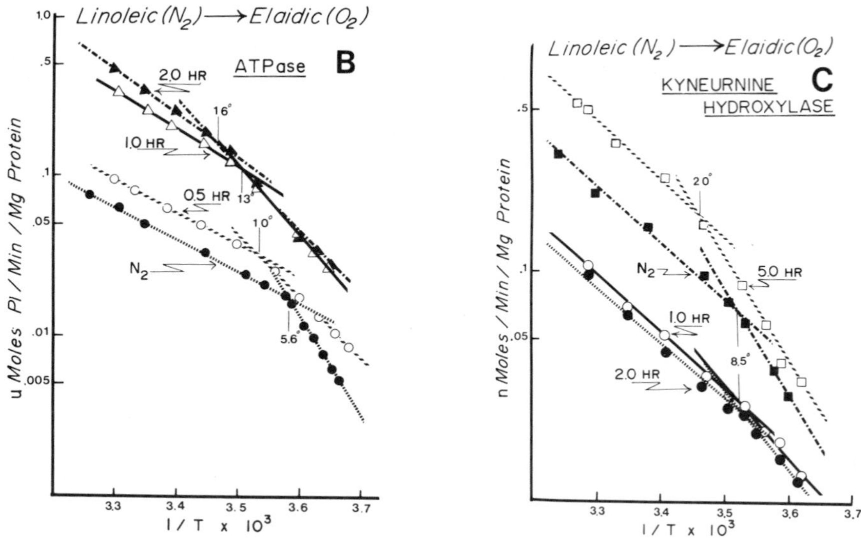

Fig. 3. (Legend on opposite page)

transition points, one associated with the anaerobic lipid environment (T_t = 7°) and the second with the new aerobic lipid, elaidic acid (T_t = 22°). This suggests that there are two discrete lipid domains containing this enzymic activity. After one hour it is difficult to decide between the two transitions, but after 2 hr the elaidic environment predominates in the membrane, giving one transition (T_t = 22°). The ATPase activity (Fig. 3B) presents entirely different kinetics. One always sees one transition point as a function of time, but the transition temperature progresses from an almost pure linoleic transition (T_t = 5.6°), through intermediate phases (T_t-0.5 hr = 10°; 1.0 hr = 13°; 2.0 hr = 16°) toward the elaidic transition temperature. The single constant transition argues for a homogeneous population. The gradual shift in the melting point (T_t) augurs for a mixed lipid environment (linoleic/elaidic), i.e., the transition point is proportional to the mixture of acids present. This implies that the ATPase creates a protein lipid environment which prevents the elaidic acid from segregating out into a pure phase as its freezing point is approached. During the first two hours of adaptation, the linoleic-elaidic acid transition has no effect on the kinetics of the outer membrane enzyme kynurenine hydroxylase despite the evidence that the membrane lipid environment is changing and the enzyme is still synthesized under these inductive conditions (Table II and Fig. 5C). However, not un-

---

Fig. 3. Arrhenius plots of mitochondrial enzymes obtained from yeast cells during aerobic adaptation. Yeast were grown anaerobically on 3% glucose minimal medium supplemented with ergosterol and 0.02% linoleic acid to a point one hour prior to stationary phase, harvested and treated as described in the text, and then adapted in new minimal medium containing 0.02% elaidic acid. Cells were harvested at times indicated and mitochondria were isolated and enzyme activities were monitored as described in Fig. 1. A, Cytochrome oxidase; B, ATPase; C, Kynurenine hydroxylase.

til 5.0 hrs after the transfer does the elaidic acid predominate the Arrhenius plot and no intermediate phases are observed. It may be noted that the anaerobic activity of this enzyme is greater than both the 1 and 2 hour aerobic activities but less than the 5 hour activity.

Figure 4 illustrates the temporal sequence of how some of the protein components are placed into these membranes. This figure shows data on the kinetics of total respiration of mitochondria which in turn reflects the kinetics of cytochrome oxidase synthesis. The control cells are grown anaerobically on 0.02% linoleic acid minimal media, transferred to

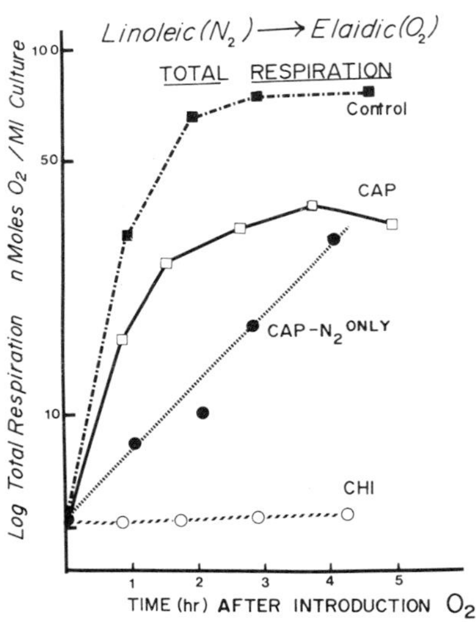

Fig. 4. Effect of protein translation inhibitors on aerobic adaptation in yeast. Yeast were grown anaerobically and adapted as described in Fig. 3. Cells adapted in the presence of inhibitors contained 4 mg/ml chloramphenicol and 25 μg/ml cycloheximide respectively. Chloramphenicol (4 mg/ml) was added to another lot of anaerobic cells two hours prior to harvest and these cells were washed free of inhibitor and adapted in 0.02% elaidic minimal medium.

elaidic acid medium and then induced with air at zero time. Mitochondria were prepared from cells sampled at hourly intervals. The respiration of these control mitochondria showed parabolic kinetics, suggesting some form of an auto-assembly process which is dependent upon anaerobically synthesized precursors. That this is so is demonstrated by the middle two curves. If cells are induced in the presence of chloramphenicol, partial inhibition is noted, this being due to curtailment of new precursor formation on mitochondrial ribosomes. Cytoplasmic ribosomes, however, continue to synthesize protein and then together with preformed chloramphenicol-sensitive proteins are able to provide a limited assembly of respiratory function (note the log scale). When cells are induced in the presence of cycloheximide no respiration is observed. This implies that no cycloheximide sensitive precursors accumulate under anaerobic conditions, for if they did, limited auto-assembly would be observed. Mason and Schatz (19) have reported similar results for cytochrome c oxidase and suggested that the assembly is the limiting step in the synthesis of the respiratory chain [cf. Ball and Tustanoff (20)]. When anaerobic cells are pre-incubated with chloramphenicol for 2 hrs prior to induction and then after washing in chloramphenicol-free medium are induced, the parabolic kinetics are abolished. Here the mitochondrial inhibitor prevented the anaerobic accumulation of oxidase precursors and we see a curve which reflects new synthesis from both pools.

The last question which affects membrane enzyme activity is lipid dependence. Unsaturated fatty acid auxotroph KD-20 was grown under anaerobic conditions in 3% glucose minimal medium containing an excess of oleic acid (0.15%) and after harvesting and washing was resuspended in fresh minimal medium with and without oleic acid (0.15%) prior to induction with air. Cells were harvested at precise time intervals after induction and mitochondria were isolated and enzyme activities monitored. Figure 5 demonstrates the effect of this type of lipid precursor starvation on the

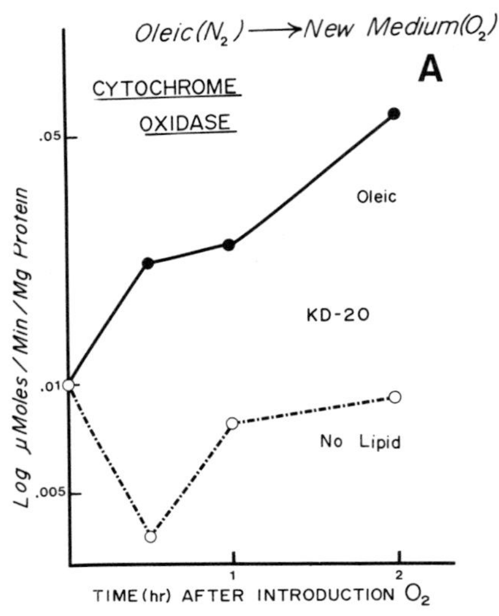

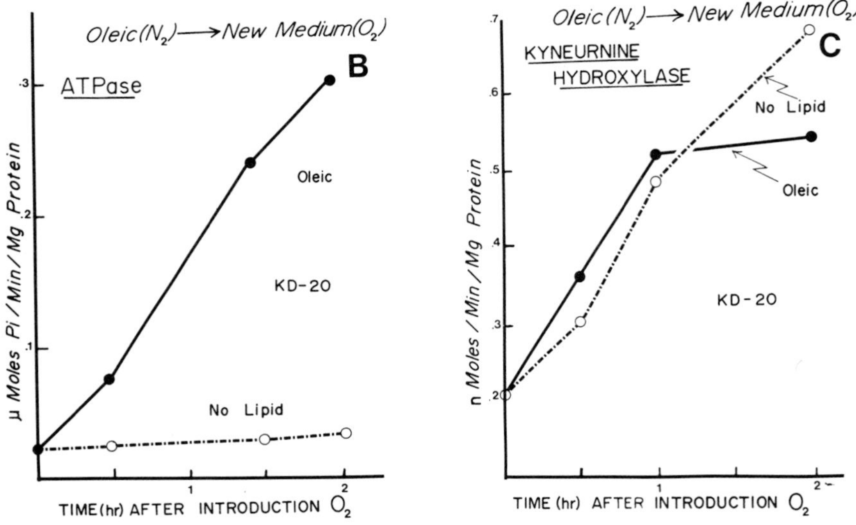

Fig. 5. (Legend on opposite page)

three membrane marker enzymes. Both cytochrome oxidase (Fig. 5A) and ATPase (Fig. 5B) have a free lipid precursor requirement whereas kynurenine hydroxylase (Fig. 5C) has none. The non-parabolic kinetics of ATPase control is to be noted. This implies that the component parts of the ATPase complex are coordinately synthesized in both anaerobic and aerobic cells in contrast to the cytochrome c oxidase complex.

PRESENTATION OF MODELS

The Arrhenius profiles used in this work strongly indicate a phase change in the lipid component of the membrane. Lipids derived from pure unsaturated fatty acids, such as elaidic and linoleic, should give very sharp melting points or phase transitions from the melted disordered to the frozen ordered state. Ideally, if an enzyme is surrounded by such a pure lipid environment, then the gel to liquid crystalline transition should be reflected in a change in activation energy, which seems to occur in mitochondrial membranes derived from yeast grown exclusively on one unsaturated fatty acid only (cf. Fig. 1 and Table I). Recently, Overath and Trauble (21,22), using a variety of physical techniques, have shown that transition temperatures determined for the β-galactoside transport system of E. coli correspond to the actual T_t of the membrane which supports that activity. All of these results from essentially mono-lipid membranes (cf. Fig. 1 and Ref. 20) are fairly unequivocal, but one must exercise considerable caution in analyzing

---

Fig. 5. Effect of lipid precursor starvation on mitochondrial enzyme activity in adapting yeast. KD-20 fatty acid auxotroph was grown in 3% glucose minimal medium containing 0.15% oleic acid and after harvesting and washing was adapted in fresh minimal medium in the presence and absence of 0.15% oleic acid. Enzyme activities were followed as described in the Methods section. A, Cytochrome oxidase; B, ATPase; C, Kynurenine hydroxylase.

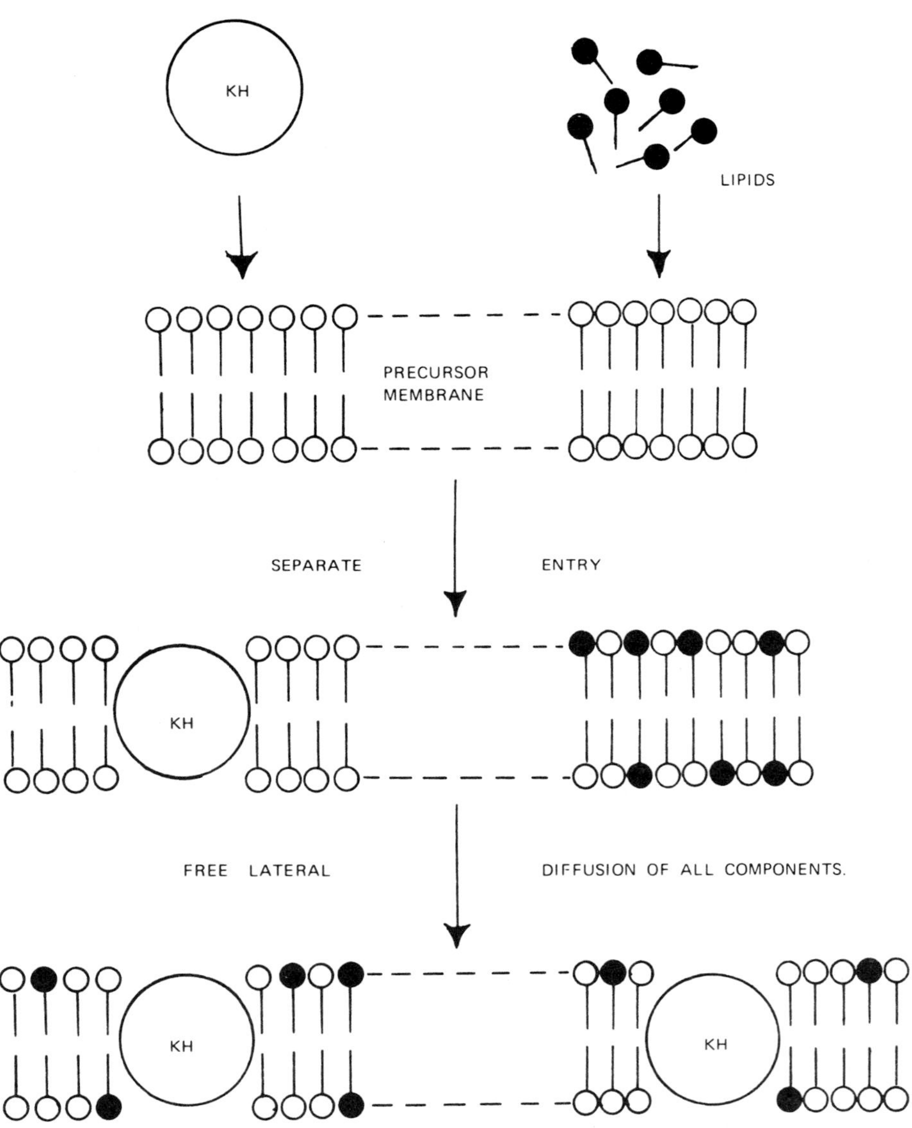

Fig. 6. (Legend opposite)

structures as encountered in cytochrome oxidase. Here we see a mixture of unsaturated fatty acids in the membrane as well as two distinct transition temperatures in the same Arrhenius plot. Phospholipids within a membrane are very mobile at physiological temperatures and even if lipids were bound to a protein before entering the membrane, one ought to expect that lateral diffusion and exchange would cause mixing almost immediately. Proteins will also diffuse in the membrane, although more slowly, and this should add to the mixing process. In addition, during an assay to obtain an Arrhenius profile one would expect the pure phospholipids which contain the same fatty acid side chain to freeze out, thus producing a mosaic membrane containing both melted and frozen zones (21). Bearing these points in mind it is still possible to construct two idealized membrane assembly models.

The first model (Fig. 6) requires that both the protein and the lipid be freely and independently soluble in the membrane; also that lateral diffusion is rapid and unhindered. In an Arrhenius profile of a mitochondrial membrane-bound enzyme the transition temperature strongly depends on the fatty acid associated with the phospholipids in the membrane. Moreover, if a situation exists where a mixed membrane is synthesized, indicating that the phospholipid moieties are composed of unsaturated fatty acids of two distinct types (i.e., linoleic or elaidic), one would expect the lipid zones to segregate out with the enzyme protein. These lipids would retreat from the frozen zones of the high melting point lipid into the still liquid zones of the low melting point components. Such an enzyme could therefore show a transition temperature characteristic of the lowest melting point phase (provided of course that this phase is large enough).

Kynurenine hydroxylase appears to fit such a model. It does not require free lipid for its acti-

---

Fig. 6. Model for kynurenine hydroxylase assembly into the mitochondrial membrane.

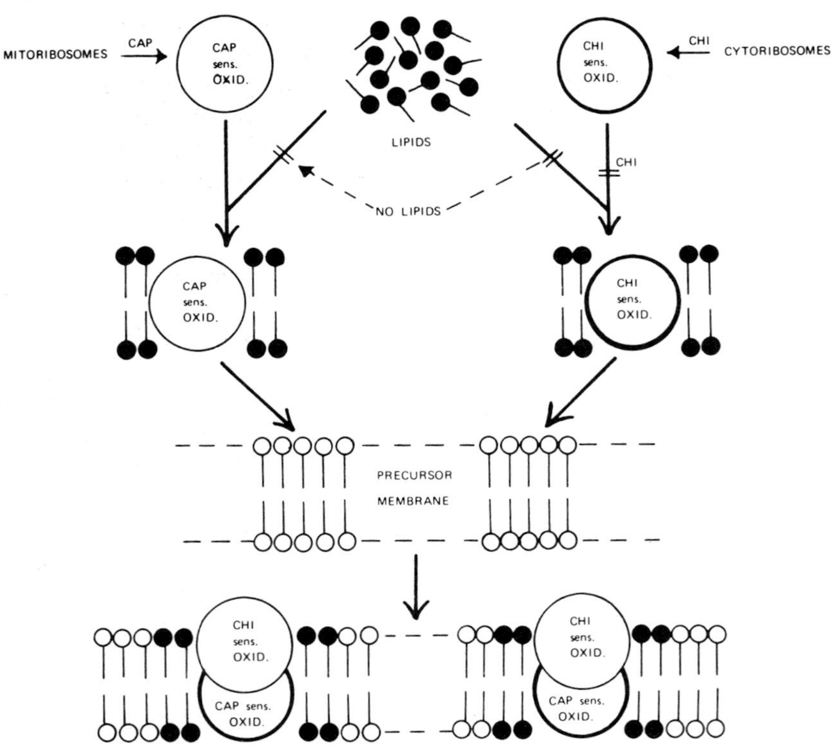

LIPID TIGHTLY BOUND, PRODUCES LIPID MOSAIC. TRANSITION TEMPERATURE CHARACTERISTIC OF ASSEMBLY LIPID. IN THE TRANSFER EXPERIMENTS AT +0.5 HRS EXPECT TWO KINDS OF OXIDASE ACTIVITY ASSEMBLED IN THE MEMBRANES.

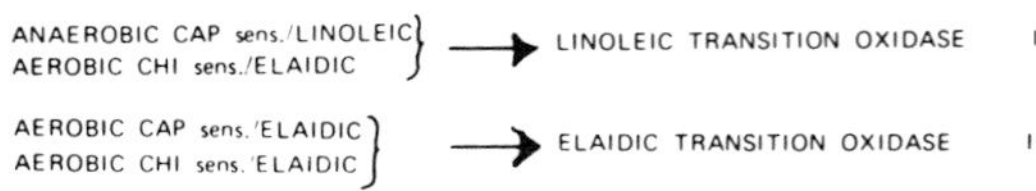

Fig. 7. Model for cytochrome c oxidase assembly into the mitochondrial membrane.

vation before entering the membrane (cf. Fig. 2 and 5C) and its transition profiles reflect the lowest melting point component (Fig. 3C) even under conditions where the membrane composition is changing (Table II) and more of the enzyme is being synthesized (Fig. 5C).

Under conditions of phospholipid synthesis one would expect considerable exchange between the old membrane lipids and the newly synthesized mitochondrial enzyme precursors. Moreover, under the induction conditions used in these experiments one would expect massive incorporation with very little exchange to be a prerequisite for the rapid development of mitochondrial structures. This is always observed (20, 23,24). It is therefore possible to conceive of another idealized model to demonstrate an assembly process where precursor polypeptides of the two enzymes of the inner mitochondrial membrane (cytochrome oxidase and ATPase) bind lipid moieties prior to the integration of the entire enzyme complex into the membrane (Figs. 7 and 8).

Ideally, the protein precursor binds free lipids irreversibly before entering the membrane. Once in the membrane, such a protein could act as a seed crystal, catalyzing the melting/freezing transition during the Arrhenius type assay. In a mixed membrane, containing similar enzyme molecules synthesized in the presence of two different lipids, one would expect to get one transition point for each pure phase available to the enzyme. For a linoleic-elaidic transfer as described in the experiment in Fig. 2, there should be two transition points as shown for cytochrome c oxidase.

The models for cytochrome c oxidase and ATPase (Fig. 7 and Fig. 8) are based on this model, with different degrees of reversible phospholipid binding occurring. We know from Schatz's work (25,26) that cytochrome c oxidase consists of a number of different precursors, some synthesized by the mitochondrion (chloramphenicol-sensitive) and some synthesized in the cytoplasm (cycloheximide-sensitive). The data

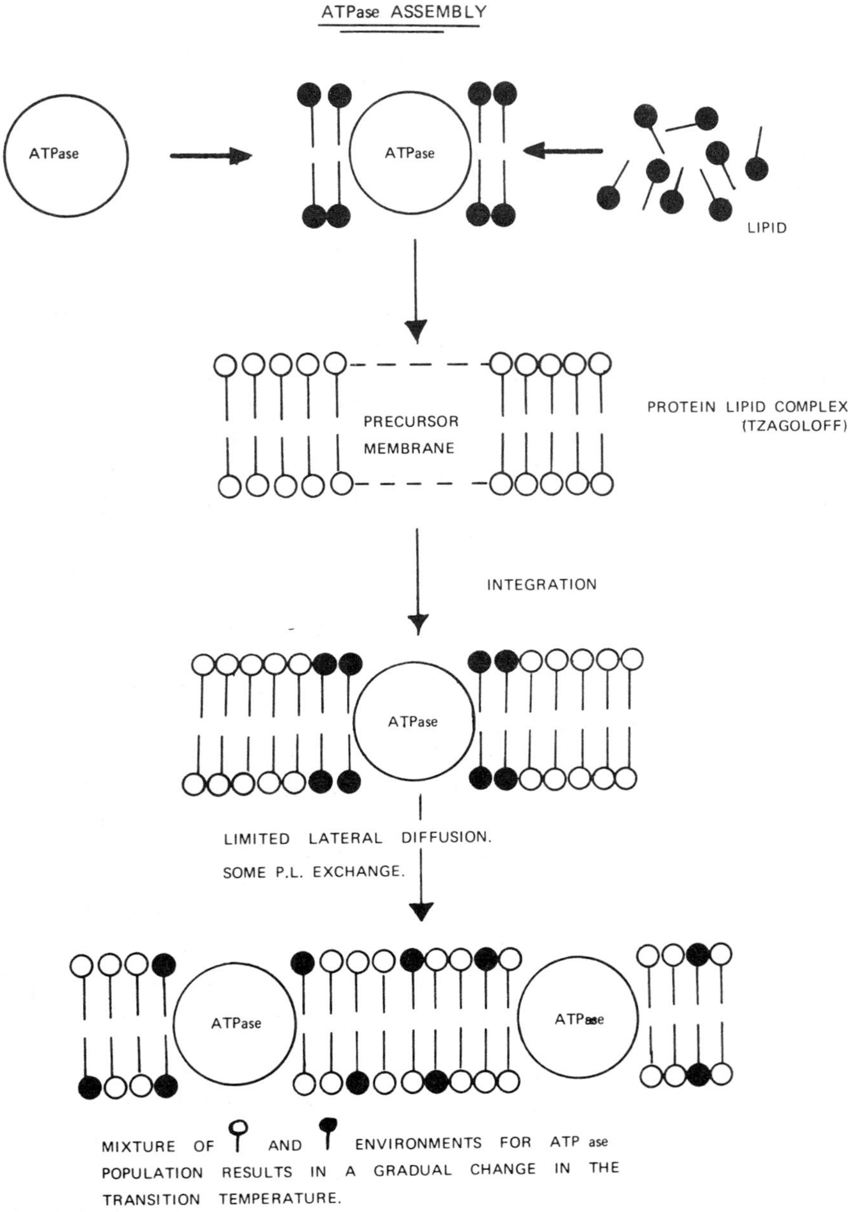

ATPase ASSEMBLY

ATPase

ATPase

LIPID

PRECURSOR
MEMBRANE

PROTEIN LIPID COMPLEX
(TZAGOLOFF)

INTEGRATION

ATPase

LIMITED LATERAL DIFFUSION.
SOME P.L. EXCHANGE.

ATPase

ATPase

MIXTURE OF AND ENVIRONMENTS FOR ATP ase
POPULATION RESULTS IN A GRADUAL CHANGE IN THE
TRANSITION TEMPERATURE.

Fig. 8. Model for oligomycin-sensitive ATPase assembly into the mitochondrial membrane.

from Fig. 4 suggests that only chloramphenicol sensitive mitochondrial precursors are synthesized anaerobically. The data from Figs. 4a and 5a show that the processes for both new synthesis and auto-assembly require free lipid precursors. This implies that at least one mitochondrial and one cytoplasmic precursor must bind lipid before assembly can take place.

Thus our model, Fig. 7, shows the independent synthesis of cytochrome c oxidase precursors, followed by tight binding of free lipid precursors (these lipoprotein complexes then dissolve into the membrane). Anaerobically, mitochondrial precursor-polypeptides of cytochrome oxidase should preferentially bind to the lipid moieties which contain linoleic acid as one of its major components. Thus, if there is little or no exchange of lipid between the protein and the membrane, two populations of chloramphenicol-sensitive precursors will exist soon after transfer (linoleic and elaidic bound mitochondrial precursors). These will bind newly synthesized cytoplasmic proteo-elaidic complexes to assemble both heterogeneous, linoleic-elaidic and homogeneous, elaidic-elaidic, cytochrome oxidase complexes. If the chloramphenicol sensitive precursor is concerned with the active site, and exchange (or diffusion) are slow, then a two-transition mosaic will result. This appears to be the case at 0.5 hours after induction. The situation at 1.0 hrs is confused and could be explained in terms of slow exchange, giving mixed lipid complexes in addition to the hetero- and homogeneous complexes proposed above. Also, during this experiment cytochrome oxidase activity increases rapidly, with the homogeneous elaidaic complex dominating the latter part of the experiment (elaidic transition at 2.0 hrs, cf. Fig. 3a and Fig. 4). A combination of these two effects could account for the 1.0 hr curve in Fig. 3a.

Our model for ATPase assembly (Fig. 8) is a variation on this theme. It is now known that oligomycin sensitive ATPase consists of both chloramphenicol-sensitive (mitochondrial) and cycloheximide-sensitive (cytoplasmic) protein components. Tzagaloff (27) in addition demonstrated auto-assembly using the inhibitor induced dissynchrony technique pioneered by Schatz (26).

However, there are no data available with respect to whether both types of precursors bind lipid before entering the membrane or whether only one kind of precursor is free lipid precursor dependent. The data of Fig. 5b could be accounted for by: chloramphenicol-sensitive protein + lipid → proteolipid and cycloheximide-sensitive protein + lipid → proteolipid or by a system requiring both steps (contrast Fig. 7). As a result, we have made the ATPase precursor protein anonymous in our scheme (Fig. 8). It should, however, be borne in mind that the ATPase, like cytochrome oxidase, is a multimeric enzyme and that the model is only an idealized interpretation.

The precursor(s) must bind lipid in this model (Fig. 8) before becoming part of the active membrane-bound complex. Once bound, there is a limited amount of lateral exchange of lipids which results in a mixed environment for ATPase during a linoleic-elaidic transition experiment. The gradual change in T_t (as demonstrated in Fig. 3b) and the gradual change in lipid composition (cf. Table II) imply that the lipid environment of the ATPase reflects the overall composition of the membrane, rather than reflecting the lowest melting point component (kynurenine hydroxylase) or the lipid present during synthesis (cytochrome oxidase, 0.5 hrs, Fig. 3a). This could result because of relatively tight lipid binding sites on the ATPase (which could prevent freezing out of pure elaidic phases) or it could result from ATPase having a reduced ability to migrate into melted zones due to protein-protein interactions in the inner mitochondrial membrane.

The distinct difference between Kynurenine hydroxylase and the other two enzymes with respect to lipid exchange and lateral mobility may be a function of the lipid/protein ratios of the inner and outer membranes. The outer membrane of mitochondria appears to be at least 50% lipid (1:1) whereas the inner membrane appears to be only about 20% lipid (4:1) (28). One would predict that lateral mobility in such mosaic membranes (29) would be altered in the manner suggested by our models and supported by the data of Figs. 2 and 5. The converse of course could also be true - that the kind of proteins present in the membrane probably dictate its structure. One would then argue that the strong protein-protein interactions exhibited by the inner membrane proteins (30) limit not only mobility and exchange but also tend to limit the amount of lipid bilayer in between the protein complexes.

Whichever of these speculations one favors, it seems clear that no single assembly process or model explains the synthesis of all enzyme-membrane complexes. The Arrhenius profiles obtained in transition-induction experiments like those reported above depend on the specific enzyme, the nature of the lipid supplements, and the temperature at which the experiment is carried out (7,21,31). Thus one must be careful in designing and interpreting analogous experiments in other systems.

The overall objective in these experiments was to gain an insight into the mode of mitochondrial biogenesis in yeast. Not unexpectedly, the evidence from the ATPase and cytochrome oxidase data support the idea that pre-formed anaerobic membranes (promitochondria) can be transformed into mitochondrial membranes via the integration of new proteins and lipids, whose synthesis is elicited by oxygen induction. From the cytochrome oxidase experiments one might hypothesize that the anaerobically synthesized chloramphenicol sensitive precursor(s) forms a template or organizes the membrane (20,32) in such a way as to facilitate the subsequent aerobic assembly process. Whether this is true for ATPase as well is not clear. That there is one mitochondrial organizer protein has been suggested on several occasions (19,31, 32) but results extant in the literature are equivocal. From the work on cytochrome c oxidase and ATPase subunits by Schatz (25,26) and Tzagoloff (27) it would seem that there are about 6 or 7 different mitochondrial chloramphenicol sensitive proteins. It seems quite probable that these constitute a class of "organizer proteins," which are coded for and synthesized in the mitochondrion and form a proteo-lipid template upon which the cytoplasmic components of the inner membrane are organized. Six or seven proteins, with recognition sites for particular catalytic proteins originating on cytoplasmic ribosomes, could account for the rapid, ordered and reproducible assembly of the three-dimensional proteo-lipid complex which constitutes a functional inner mitochondrial membrane.

REFERENCES

1. MAHLER, H. R. (1973) Critical Reviews in Biochemistry 1, 381.
2. LINNANE, A. W., HASLAM, J. M., LUKINS, H. B., AND NAGLEY, P. (1972) Ann. Rev. Microbiol. 26, 163.
3. RAISON, J. K. (1973) Bioenergetics 4, 285.
4. AINSWORTH, P. J., TUSTANOFF, E. R., AND BALL, A. J. S. (1972) Biochem. Biophys. Res. Commun. 47, 1299.
5. RESNICK, M. A., AND MORTIMER, R. K. (1966) J. Bacteriol. 92, 597.
6. PROUDLOCK, J. W., HASLAM, J. M., AND LINNANE, A. W. (1971) J. Bioenergetics 2, 327.
7. FOX, C. F. (1969) Proc. Nat. Acad. Sci. U.S.A. 63, 850.
8. RAISON, J. K., LYONS, J. M., AND THOMPSON, W. W. (1970) Arch. Biochem. Biophys. 142, 83.
9. TUSTANOFF, E. R., AND BARTLEY, W. (1964) Canad. J. Biochem. 42, 651.
10. AINSWORTH, P. J., BALL, A. J. S., JANKI, R. M., AND TUSTANOFF, E. R. (1974) J. Bioenergetics, in press.
11. HENSON, C. P., PERLMAN, P., WEBER, C. N., AND MAHLER, H. R. (1968) Biochemistry 7, 4445.
12. TZAGOLOFF, A. (1969) J. Biol. Chem. 244, 5020.
13. SCHOTT, H. H., ULLRICH, V., AND STAUDINGER, H. (1970) Z. Physiol. Chem. 351, 99.

14. FOLCH, J., LEES, M., AND STANLEY, G. H. S. (1957) J. Biol. Chem. 226, 497.
15. LOWRY, O. H., ROSEBROUGH, N. J., FARR, A. L., AND RANDALL, R. J. (1951) J. Biol. Chem. 193, 265.
16. BLOOMFIELD, D. K., AND BLOCH, K. (1960) J. Biol. Chem. 235, 337.
17. FLEISCHER, S., AND FLEISCHER, B. (1967) in Methods in Enzymology (Estabrook, R. W., and Pullman, M. E., eds.) vol. 10, p. 406, Academic Press, N.Y.
18. BANDLOW, W. (1972) Biochim. Biophys. Acta 282, 105.
19. MASON, T. L., AND SCHATZ, G. (1973) J. Biol. Chem. 248, 1355.
20. BALL, A. J. S., AND TUSTANOFF, E. R. (1971) in Autonomy and Biogenesis of Mitochondria and Chloroplasts (Boardman, N. K., Linnane, A. W., and Smillie, R. M., eds.) p. 466, North-Holland, Amsterdam.
21. OVERATH, P, AND TRAUBLE, H. (1973) Biochemistry 12, 2625.
22. TRAUBLE, H., AND OVERATH, P. (1973) Biochim. Biophys. Acta 307, 491.
23. SCHATZ, G., AND CRIDDLE, R. S. (1968) Biochemistry 8, 322.
24. PLATTNER, H., SALPETER, M., SALTZGABER, J., ROUSLIN, W., AND SCHATZ, G. (1971) in Autonomy and Biogenesis of Mitochondria and Chloroplasts (Boardman, N. K., Linnane, A. W., and Smillie, R. M., eds.) p. 175, North-Holland, Amsterdam.
25. MASON, T., EBNER, E., POYTON, R. O., SALTZGABER, J., WHARTON, D. C., MENNUCCI, L., AND SCHATZ, G. (1972) in Mitochondria: Biogenesis and Bioenergetics, FEBS, Proc. 8th Meeting, Vol. 8, p. 53, North-Holland, Amsterdam.
26. SCHATZ, G., GROOT, G. S. P., MASON, T., ROUSLIN, W., WHARTON, D. C., AND SALTZGABER, J. (1972) Fed. Proc. 31, 21.
27. TZAGOLOFF, A., RUBIN, M. S., AND SIERRA, M. F. (1973) Biochim. Biophys. Acta 301, 71.

28. PARSONS, D. F., WILLIAMS, G. R., THOMPSON, W., WILSON, D., AND CHANCE, B. (1967) in Round Table Discussion on Mitochondrial Structure and Compartmentation (Quagliariello, E., Papa, S., Slater, E. C., and Tager, J. M., eds.) p. 28, Adriatica Editrice, Bari.
29. GREEN, D. E., AND GOLDBERGER, R. F. (1967) Molecular Insights into the Living Process, p. 188, Academic Press, N.Y.
30. TSUKAGOSHI, N., AND FOX, C. F. (1973) Biochemistry 12, 2816.
31. EBNER, E., MENNUCCI, L., AND SCHATZ. G. (1973) J. Biol. Chem. 248, 5360.
32. BALL, A. J. S., AND TUSTANOFF, E. R. (1968) FEBS Letters 1, 255.

SYSTEMS APPROACH TO THE STUDY OF MITOCHONDRIOGENESIS

J. Jayaraman, K. Dharmalingam and N. Murugesh

INTRODUCTION

It would be redundant in a chapter like this to justify the choice of baker's yeast, *Saccharomyces cerevisiae*, as the experimental material for the study of mitochondriogenesis. Like several others, we have been interested in this problem for quite some time, but our approach was one of studying the interaction of various systems in this process. The breakdown of mitochondria under the influence of glucose (1-6) and its subsequent synthesis during the derepression phase was our experimental set up. The two parameters that have been employed to monitor the formation of mitochondria were (a) oxygen uptake capacity of the whole cells (or spheroplasts) and (b) cytochrome oxidase activity.

Part 1 of the study is concerned with the availability of phospholipids (particularly cardiolipin), ubiquinone and heme, in the yeast cells when new mitochondria are being made. Part 2 is more specifically concerned with the formation of the cytochrome oxidase enzyme and the relative contributions of mitochondria and cytoplasm to this event.

MATERIALS AND METHODS

Saccharomyces cerevisiae 3095 (diploid) was obtained from the National Collection of Industrial Microorganisms, National Chemical Laboratory, Poona, India. The maintenance and growth conditions have been described previously (3,7).

Abbreviations: CAP - chloramphenicol; CHI - cycloheximide; ALA - amino laevulinic acid; nDNA - nuclear DNA; mDNA - mitochondrial DNA.

Spheroplasts were prepared as described before with a few modifications (3). The snail gut enzyme used was obtained from locally available snail, *Pila virens*.

For isolation of particulate material, the spheroplasts were lysed by homogenizing in 0.1 M phosphate buffer for 30 seconds at medium speed in a Potter Elvehjem homogenizer. The 20,000 g pellet was obtained following the standard procedure.

The oxygen uptake capacity (3) and assay of enzymes, cytochrome oxidase (8), succinate-cyt.C reductase (9), alkaline phosphatase (10), acid phosphatase (11), protease (12), phospholipase A (13), phospholipase C (14), and phospholipase D (15) were carried out following the published procedures.

The extraction and estimation of phospholipids (16) and ubiquinone (17) have been described earlier. For isolation of cardiolipin, we used two columns. Initially the total phospholipids were chromatographed on a silicic acid column (18) and the fraction containing the acidic phospholipids was refractionated on Sephadex LH 20 to separate cardiolipin and phosphatidic acid (J. Jayaraman and N. Murugesh, unpublished results).

Estimation of ALA dehydratase was done according to Jayaraman *et al.* (19).

RESULTS

Availability of various factors needed for mitochondrial formation in yeast cells at the time of derepression

Proteins. Investigations from several laboratories have confirmed that proteins are contributed both by mitochondria themselves and the cytoplasm for the formation of the organelle. Data given in Table I, column 1, reiterate this fact under our conditions. Addition at the time of derepression of the cycloheximide or chloramphenicol, which inhibit protein synthesis by the cytoplasm and mitochondria respectively, prevented any recovery of oxygen uptake.

TABLE I
RESPONSE OF YEAST CELL TO GLUCOSE REPRESSION[a]

Status of cells	Oxygen uptake[b]	ALA dehydratase[c]	Ubiquinone[d]	Phospholipids[e]	Cardiolipin[f]
"0" time	-	-	31.2	67.2	150.0
Repressed (2-3 hr)	7.2	0.8	0	84.3	61.0
Derepressed (5-6 hr)	22.0	2.0	36.2	54.0	138.0
CAP treated	8.8	1.6	45.5	63.0	38.0
CHI treated	7.8	0.52	20.0	45.0	102.0
Glucose added	6.8	0.44	131.0	62.0	78.0

[a] Data has been pooled from several experiments. The general experimental procedure was as follows: Fully derepressed yeast cells were inoculated into a medium containing 1% glucose at a level of 250 mg wet wt/250 ml/flask. Between 2-1/2 - 3 hr, the cells are repressed. The derepressed values represent the time point of 5-1/2 - 6 hr. The additions were made at the repressed stage and the values given are the results at 6 hr. CHI was added at a level of 25 μg/ml. CAP 4 mg/ml and glucose 1%, final concentration.

[b] μmoles O_2 taken up/gm wet wt.

[c] nmoles porphobilinogen formed/mg protein.

[d] nmoles/gm wet wt.

[e] μg lipid phosphorus/gm wet wt.

[f] ng phosphorus/mg mitochondrial protein.

<u>Heme</u>. It has been suggested earlier by Bartley and his group (4,5,20) that loss of mitochondrial enzyme activity during glucose repression may be an example of inactivation repression; in other words, the proteins may be present but not the prosthetic groups. To check this possibility, we assayed the two enzymes, ALA synthetase and ALA dehydratase, under various conditions. ALA synthetase showed an oscillatory pattern and is not considered here. ALA dehydratase (Table I, column 2) was subject to glucose repression and thus may be a rate-limiting step in the biogenetic sequence.

<u>Ubiquinone</u>. Ubiquinone is a functional lipid component of mitochondria and as the results (column 3 of Table I) show it is subject to severe repression by glucose. To confirm this, we tried another series of experiments, where fresh glucose was added at the time when it was exhausted in the medium. This resulted in the anticipated block in the increase of oxygen uptake capacity, but ubiquinone synthesis continued. We interpret this result as indicating that although ubiquinone synthesis is glucose repressible, it may not be rate limiting.

Phospholipids. It was of obvious interest to study these compounds, since they contribute to the structural integrity of the organelle. Interestingly, it was found that all new synthesis of phospholipids took place before derepression started and there was only rearrangement of phospholipid classes during the period of investigation (see column 4 of Table I and reference 16). The behavior of cardiolipin, a specific phospholipid associated with mitochondria, however, was different. It did undergo repression by glucose (column 5 of Table I) but only a partial repression was noticed.

Do mitochondria and cytoplasm act in concert to synthesize cytochrome oxidase?

We attempted to extrapolate the above-mentioned results to the formation of a single function of mitochondria, rather than the whole, and chose cytochrome oxidase. The first question we addressed ourselves to was about the relative contributions of mitochondria and cytoplasm. For this purpose, we made use of the differential inhibitors cycloheximide and chloramphenicol. From Table I it can be seen that heme is probably donated by cytoplasm (ALA dehydratase inhibited by CHI) and cardiolipin probably by mitochondria (cardiolipin levels in CAP treated cells are very low). Then what of the proteins?

Use of Spheroplasts as experimental system. An inherent problem in studies using whole cells is the time spent in the preparation of intact mitochondria. The well accepted method of preparing spheroplasts takes about 2-3 hours and this makes it rather difficult to precisely define the physiological status of the cells (in our studies, the exact status of glucose repression). Looking for alternate methods, we found that spheroplasts themselves undergo repression-derepression of mitochondria in response to glucose, thus simulating the behavior of whole cells. The spheroplasts were maintained in a medium containing 1 M sorbitol as osmotic stabilizer and under those conditions they were stable up to 14 hours.

Action of chloramphenicol and cycloheximide. Addition of either chloramphenicol or cycloheximide to derepressing spheroplasts prevented derepression as monitored by the oxygen uptake. These inhibitory effects, however, were found to be reversible in nature. That is, spheroplasts washed free of the inhibitors showed increase in oxygen uptake. This reversibility was achieved as long as the incubation time of spheroplasts with inhibitors did not exceed about 4 hours.

Nature of the recovery process. The above mentioned results are summarized in Fig. 1. In this experiment, spheroplasts were exposed to glucose for a period of 210 minutes, and during this period oxygen uptake remained at the repressed level (58 nmoles/min/ml). By this time, the glucose in the medium gets exhausted and the spheroplasts start derepressing (Trace A). To one set of these derepressing spheroplasts, cycloheximide (25 g/ml) was added and incubated with shaking for 30 minutes. The oxygen uptake did not increase. At the end of this period, the spheroplasts were washed free of cycloheximide using sterile saline as washing medium. They were then resuspended in a similar medium without antibiotic and shaking continued. Aliquots were taken at different time points and oxygen uptake measured. Trace B indicates the results obtained. There was an initial rise, followed by a lag period (or small drop) and then another rise. Derepressing spheroplasts in another set of flasks were incubated with cycloheximide, washed at the end of 30 min, and then suspended in a similar medium containing chloramphenicol (4 mg/ml). Trace C indicates the results obtained. The initial rise was unaffected but the second rise in oxygen uptake was abolished.

Exactly similar results were obtained when cycloheximide and chloramphenicol were interchanged.

It occurred to us that these results could be explained as follows. During incubation with cycloheximide, there occurs in the cell an accumulation

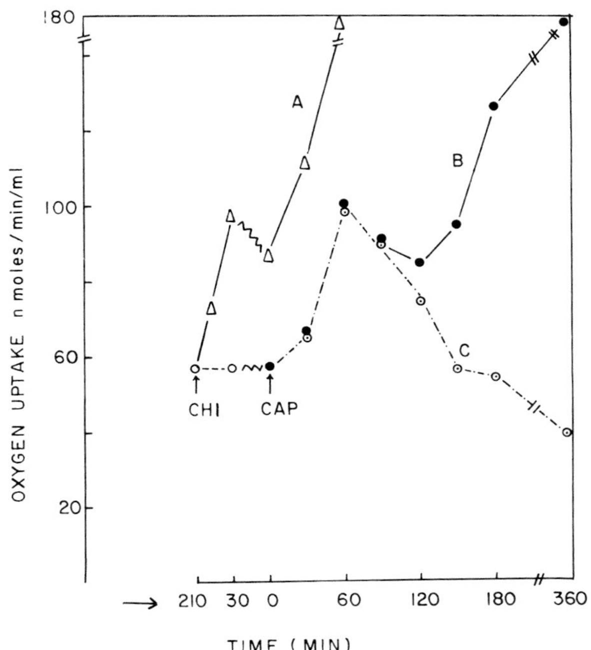

Fig. 1. Recovery of respiration after sequential addition of antibiotics. 200 mg wet wt. of spheroplasts in 50 ml medium were incubated with shaking at room temperature (28-30°C). 3 ml aliquots were taken for oxygen uptake measurements. Addition of antibiotics were made at time intervals specified in the figure. Δ——Δ, Control, receiving no antibiotic; o---o, Treated with antibiotics as indicated; ⊙—·—⊙, treated with CAP as indicated in the recovery period; ●——●, Control, receiving no antibiotic in the recovery period; ∿∿, Time elapsed during washing.

of the products made by mitochondria. On removing the inhibitor, the cytoplasmic products now start being synthesized and these immediately complement with the already accumulated products to exhibit their function. Once the accumulated material is exhausted, a lag period follows during which period both products are to be synthesized. It is only this latter synthesis that is affected by chloramphenicol and hence only the second rise is blocked. The reverse would be true when the antibiotics are interchanged.

In vitro complementation of cytochrome oxidase. If the above hypothesis, viz., mitochondria-made products accumulate in the presence of cycloheximide and cytoplasm-made products accumulate in the presence of chloramphenicol, is correct, then one should be able to complement them, to retrieve the function. This was what we attempted to do. Spheroplasts were incubated with cycloheximide for 30 min as described earlier. Another set of spheroplasts were incubated with chloramphenicol for the same period of time. At the end of this period, particulate matter was isolated from these and assayed for cytochrome oxidase activity. Incubation with antibiotics has reduced the level to about 30-50% in both cases. However, mixing these deficient particles gave a synergistic effect (Table II). This we believe is due to complementation of the partial products. In Table II

TABLE II
IN VITRO COMPLEMENTATION OF CYTOCHROME OXIDASE

Expt. no.	Sample	Specific activity of cytochrome c oxidase	% of control
308	Control	1.53	-
	a. CHI treated	0.70	46
	b. CAP treated	0.48	31
	c. a + b	1.00	75
351	Control	2.30	-
	a. CHI treated	0.90	39
	b. CAP treated	1.70	73
	c. Particulate fraction of a + b	1.80	80
	d. Particulate fraction of a + supernatant fraction of b	0.355	15

Note: Experimental conditions as described under Fig. 1. At the end of the incubation period the spheroplasts were washed, lysed, and particulate matter prepared separately. Equal amounts of protein from fractions a and b were mixed and incubated for 10 min at 30°C.

are also given results to show that partial products are membrane bound.

Vacuolar dynamics and mitochondriogenesis

During glucose repression, there occurs not only a decrease in electron transport enzyme activities, but also actual breakdown of the mitochondrial membrane (3,4,5). While looking for probably causative agents of this breakdown, we had earlier shown that there was an increase in phospholipase D activity in repressed yeast cells, in inverse relationship to oxygen uptake capacity (21). Extending this observation, we have now found that almost all hydrolytic enzymes show the same type of behavior (Table III). Although some technical problems are in the way of making positive statements, it can be said in general terms that these enzymes are localized in the vacuoles of yeast cells. Thus, it appears that the vacuolar dynamics determines in a large way the status and existence of mitochondria.

TABLE III
ACTIVITIES OF SOME HYDROLYTIC ENZYMES DURING REPRESSION AND DEREPRESSION

Enzymes	Repressed	Derepressed	Fold increase or decrease	Units
Alkaline phosphatase	180	170	-	μmoles p-nitrophenol formed/min/mg protein
Acid phosphatase	0.009	0.0006	- 15	units/min/mg protein
Protease	0.900	0.3000	- 3	mg BSA hydrolyzed/hr mg protein
Phospholipase A	0.500	0.3000	- 1.7	hamoglobin released, O.D.units/hr/mg protein
Phospholipase C	0.050	0.0060	- 9	μmoles Pi liberated/ 30 min/mg protein
Phospholipase D	3.700	0.6000	- 6	μmoles choline liberated/90 min/mg protein
Succinate cyt. C reductase	3.300	5.5000	+ 1.7	μmoles cyt.C red./min/ mg protein
Oxygen uptake	7.000	18.0000	+ 2.6	nmoles /min/mg wet wt. of cells

Note: Data has been collected from different experiments. The general experimental procedure is as described for Table I.

DISCUSSION

On glucose repression. Except the synthesis of phospholipids, all other parameters mentioned, synthesis of heme, proteins, ubiquinone and cardiolipin are subject to glucose repression. However, there is quantitative variation in their response to fresh glucose added at the time of derepression. Whether this is indicative of diverse control mechanisms operative at various levels is a subject of current investigation.

On the availability of precursors. The onset of derepression in the yeast cell is marked by multifarious concerted events, widely varying in nature. We have carried out a preliminary analysis of some of these events. Based on these results, we suggest the sequence of events as shown in Chart I.

CHART I
SUGGESTED SEQUENCE OF EVENTS LEADING TO MITOCHONDRIOGENESIS

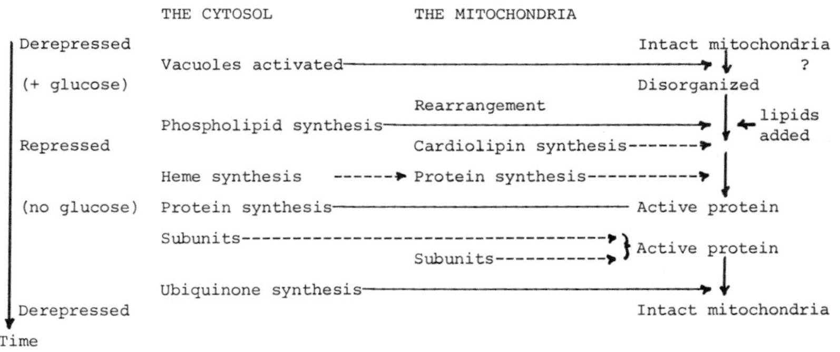

On the complementation of cytochrome oxidase. Results presented here demonstrate that partial products or subunits of the enzyme cytochrome oxidase are made both by mitochondria and cytoplasm and that they can be accumulated independently of each other. Further, by mixing they reassociate and give full enzymic activity. Very recently, we have been able to show a similar complementation of the enzyme ATPase also. It looks as though the assembly of the subunits is a physical phenomenon.

On a model for nuclear-mitochondrial interaction. Barath and Kuntzel (22) have recently suggested a model for explaining the interactions of nuclear and mitochondrial DNA functions. According to this, the nuclear DNA produces a repressor binding to mDNA and mDNA produces a repressor for nDNA. The controlled actions of these two repressors keep the balance between the proteins made in mitochondria and cytoplasm. Our experiments would support this model. Inhibiting cytoplasmic protein synthesis should inhibit the formation of mDNA repressor also. Hence the products made by mitochondria would tend to accumulate. The reverse would be true when mitochondrial protein synthesis is inhibited by chloramphenicol.

On the role of vacuoles. One of the effects of glucose repression seems to be the induction or activation of vacuolar hydrolytic enzymes, which are perhaps responsible for breaking down the organelles. Although the cause-and-effect relationship is yet to be established with certainty, one thing is clear. The status of vacuoles is an important environmental parameter in any study on mitochondriogenesis.

SUMMARY

Using derepressing yeast cells, the availability of several groups of mitochondrial components has been studied and a tentative sequence of events leading to mitochondriogenesis has been suggested. The spheroplasts have been shown to simulate whole yeast cells in undergoing a repression-derepression cycle actuated by glucose. Growth circumstances can be suitably modified to accumulate in spheroplasts, some of the subunits of cytochrome oxidase made in mitochondria. Similarly, subunits made in cytoplasm can also be accumulated. Mixing of these partial products results in the restoration of enzyme activity. The vacuolar dynamics in the cell has to be taken into account in any studies on mitochondriogenesis.*

REFERENCES

1. EPHRUSSI, B., SLONIMSKI, P. P., YOTSUYANAGI, Y., AND TRAVLITSKI, J. (1956) C. R. Trav. Lab. Carlsberg. Ser. Physiol. 26, 87.
2. UTTER, M. F., DUELL, E. A., AND BERNOFSKI, C. (1966) in Some Aspects of Yeast Metabolism (Mills, R. K., ed.), Oxford University Press.
3. JAYARAMAN, J., COTMAN, C., MAHLER, H. R., AND SHARP, C. W. (1966) Arch. Biochem. Biophys. 116, 224.
4. POLAKIS, E. S., BARTLEY, W., AND MEEK, G. A. (1965) Biochem. J. 197, 298.
5. CHAPMAN, C., AND BARTLEY, W. (1968) Biochem. J. 107, 455.
6. JAYARAMAN, J. (1969) J. Scient. Ind. Res. 28, 441.
7. DHARMALINGAM, K., AND JAYARAMAN, J. (1973) Arch. Biochem. Biophys. 157, 197.
8. WHARTON, D. C., AND TZAGOLOFF, A. (1967) in Methods in Enzymology (Estabrook, R. W., and Pullman, M. E., eds.), Vol. X, p. 245, Academic Press, New York.
9. TRISDALE, H. O. (1967) in Methods in Enzymology (Estabrook, R. W., and Pullman, M. E., eds.), Vol. X, p. 213, Academic Press, New York.
10. Worthington Enzyme Manual (1972) Worthington Biochemical Corporation, New Jersey, p. 71.
11. IKAWA, T. K., NISIZAWA, K., AND MIURA, T. (1964) Nature 203, 939.
12. MANDELSTAM, J., AND WAITES, W. M. (1968) Biochem. J. 109, 793.

---

*Note added in proof: It has been recently demonstrated in this laboratory that similar to cytochrome oxidase, the subunits of ATPase, made under nuclear and mitochondrial control, could be accumulated independently and these subunits on mixing in the proper ratio demonstrate complementation.

13. CONDREA, E., AND DEVRIES, A. (1964) Biochem. Biophys. Acta 84, 60.
14. OTTOLENGHI, A. C. (1969) in Methods in Enzymology (Lowenstein, J. M., ed.), Vol. XIV, p. 672, Academic Press, New York.
15. HAYAISHI, O. (1955) in Methods in Enzymology (Colowick, S. P., and Kaplan, N. O., eds.), Vol. 1, p. 672, Academic Press, New York.
16. JAYARAMAN, J., AND SASTRY, P. S. (1971) Indian J. Biochem. Biophys. 8, 278.
17. RAMAN, T. S., SHARMA, B. V. S., JAYARAMAN, J., AND RAMASARMA, T. (1965) Arch. Biochem. Biophys. 110, 75.
18. SWEELEY, C. C. (1969) in Methods in Enzymology (Lowenstein, J. M., ed.), Vol. XIV, p. 254, Academic Press, New York.
19. JAYARAMAN, J., PADMANABAN, G., MALATHI, K., AND SARMA, P. S. (1971) Biochem. J. 121, 531.
20. FERGUSON, J. J., BALL, M., AND HOLZER, H. (1967) Eur. J. Biochem. 1, 21.
21. DHARMALINGAM, K., AND JAYARAMAN, J. (1971) Biochem. Biophys. Res. Commun. 45, 1115.
22. BARATH, Z., AND KUNTZEL, H. (1972) Nature 240, 195.

RIBONUCLEOTIDE POLYMERIZING ACTIVITIES IN RAT LIVER MITOCHONDRIA

Cecilia Saccone, Carla De Giorgi, Palmiro Cantatore and Raffaele Gallerani

INTRODUCTION

In vitro and *in vivo* studies on mitochondrial transcription have clearly demonstrated that ribosomal RNA (rRNA) and several transfer RNA (tRNA) species are synthesized in the mitochondrion by using mitochondrial DNA as template (1). Furthermore, very recently, the first clear evidences for the presence of messenger RNAs (mRNAs) made on mitochondrial DNA have been presented (2). Like other mRNAs, mitochondrial mRNAs seem to contain a polyriboadenylate sequence at their 3' terminus implying that probably also a poly(A) polymerase activity is operating within the mitochondrion (3).

In this chapter we should like to present our studies on two enzyme activities present in rat liver mitochondria which could probably act in connection with each other. First activity is the DNA-dependent RNA polymerase, the enzyme probably responsible for the transcription of the mitochondrial genome which has been extensively characterized by us and rather recently also purified (4-7). Its properties will be discussed also in relation to those of the same enzymes so far isolated from mitochondria of other organisms. The second activity to which we should like to refer briefly is that responsible for the incorporation of labeled ATP into polyriboadenylate chains whose properties and function are now under investigation in our laboratory.

DNA-DEPENDENT RNA POLYMERASE

In Table I the major properties of the rat liver enzyme compared to those of other purified mitochondrial RNA polymerases from various organisms are shown.

TABLE I
PROPERTIES OF PURIFIED MITOCHONDRIAL RNA POLYMERASES

Source		Molecular weights Intact	Subunits	Optimal divalent cations conc.(mM) Mn^{++}	Mg^{++}	Inhib.by high ion. strength	Rifampicin sensit.	α-amanitin sensit.	Ref.
Yeast		200,000	59-63,000	1-3	20	+	+	-	(8)
Yeast	I	500,000	2 large	2	10 (broad)	+	-	-	(9)
	II	500,000	3 large	4	none	+	-	+	
	III	500,000	2 large	1	3	+	-	-	
Neurospora crassa		n.m.	64,000	1.6	30	n.m.	+	-	(10)
X. laevis ovaries		100-150,000	46,000	none	10	+	-	n.m.	(11)
Rat liver		n.m.	60-64,000	1	3	+	+	-	(7)

n.m. = not measured.

It can be seen that a DNA-dependent RNA polymerase so far has been isolated in a purified form only from the following organisms: Yeast (8-9), _Neurospora_ (10), _Xenopus_ ovaries (11), and rat liver (7). As far as the yeast is concerned controversial results have been obtained from two groups. Criddle _et al._ (9) claim that in yeast mitochondria probably more than one polymerase is present, some of them sharing common properties with the nuclear polymerases. On the other hand Scragg (8) finds a mitochondrial enzyme which very closely resembles the enzymes purified from other organisms, completely different from that described by Criddle _et al_. With regard to rat liver enzyme, partial purification was also obtained by Reid and Parsons (12).

From Table I it emerges clearly that all mitochondrial RNA polymerases, so far purified, with the one exception of Criddle's enzymes, display the following common properties:

(1) The enzymes are composed of a single polypeptide chain of relatively low molecular weight; about

60,000 daltons for Neurospora, yeast, and rat liver, about 46,000 for Xenopus. The polypeptide chains in low salt tend to form aggregates of higher molecular weight.

(2) The enzymes are inhibited by high ionic strength.

(3) The optimum of Mn^{++} concentration does not exceed 2-3 mM, whereas higher concentrations have an inhibitory effect.

(4) The enzymes are able to transcribe mitochondrial DNA, even if in several cases they display stronger specificity for other templates.

(5) The enzymes are insensitive to α-amanitin.

(6) The enzymes from rat liver, yeast and Neurospora seem to be rifampicin sensitive like bacterial polymerases.

The rifampicin sensitivity has been one of the major controversial points in the study of mitochondrial transcription mechanism. Table II summarizes results obtained in various laboratories on this enzyme property. For some organisms results obtained either

TABLE II
SENSITIVITY TO RIFAMICIN OR RIFAMPICIN OF DIFFERENT MITOCHONDRIAL RNA POLYMERASES

Source	% Inhibition		Reference
	Native enzyme	Purified enzyme	
Neurospora crassa	0	90 (6 μg/ml)	A.Kuntzel and K. P. Schafer (10)
" "	0	n.m.	E. Wintersberger (13)
Yeast	0	3-8 (20 μg/ml)	T. R. Eccleshall and R. S. Criddle (9)
"	0	n.m.	E. Wintersberger (13)
"	?	95 (10 μg/ml)	A. H. Scragg (8)
Rat liver	90 (10 μg/ml)	n.m.	Z. G. Shmerling (14)
" "	80 (50 μg/ml)	70 (20 μg/ml)	M. N. Gadaleta et al. (15), and R. Gallerani and C. Saccone (7)
" "	45 (10 μg/ml)	35 (10 μg/ml)	B. D. Reid and P. Parsons (12)
Rabbit liver	0	n.m.	E. Wintersberger (13)
Xenopus ovaries	n.m.	0 (50 μg/ml)	G. J. Wu and I. B. Dawid (11)
Sea urchin eggs	45 (20 μg/ml)	n.m.	P. Cantatore et al. (16)

using isolated organelles as enzyme source or the purified preparation are reported. The discrepancies can be summarized as follows: 1) In the case of yeast the rifampicin sensitivity has not been found by some authors either with purified (9) or with the native enzyme (13), whereas it has by others (8). 2) In the case of Neurospora, the native enzyme seems to be unaffected by rifampicin which instead completely inhibits the purified mitochondrial RNA polymerase preparation. 3) For rat liver, our group and others have been able to demonstrate inhibition by rifampicin using the native as well as the purified enzyme (14,15,7,12).

The discrepancies reported above can be explained in one of the following ways. When isolated organelles are used as enzyme source a permeability barrier at the level of the inner mitochondrial membrane may enable the drug to reach the DNA-enzyme complex. In this regard we have demonstrated (15) that when isolated rat liver organelles are used, pretreatment of mitochondria in order to alter their permeability barrier is necessary to demonstrate inhibition by rifampicin. In the same system, furthermore, the insensitivity of mitochondrial RNA synthesis to rifampicin may be due to the fact that in experimental conditions used by some authors, only the elongation of the already present RNA chains is occurring. Since it is known that rifampicin inhibits initiation of the chains, this could explain the lack of inhibition by rifampicin. On the other hand in the case of Neurospora Kuntzel hypothesizes that the native-enzyme contains a co-factor which protects it against rifampicin (10). The lack of inhibition by rifampicin of the purified enzymes can be due to the purification procedure used to isolate the enzyme, according to Scragg (8). The author states that after ammonium sulphate precipitation he was unable to detect RNA polymerase activity sensitive to rifampicin. On the other hand the same author has been able to demonstrate the binding of ^{3}H-rifampicin to the enzyme in crude extracts (8). In the case of rat liver enzyme, although the rifampicin sensitivity was preserved after ammonium sulphate precipitation, we often observed, in our first attempts to purify the enzyme, that DNA-dependence and rifampicin sensitivity were simultaneously lost during purification (17). This observation has been confirmed by Reid and Parsons (personal communication).

Taking into account the above-mentioned observations, we are tempted to suggest that the rifampicin sensitivity is a peculiar property of all mitochondrial polymerase enzymes. It is interesting to stress that the same discrepancies regarding rifampicin sensitivity have been reported for the DNA-dependent RNA polymerase of chloroplasts (18,19).

With regard to the problem of whether the purified enzyme is the same as that which in the native state transcribes mitochondrial DNA we can say that in our case the properties of rat liver mitochondrial RNA polymerase do not change during purification. Table III shows that the major peculiar properties of the enzyme are preserved along the purification procedure.

TABLE III
PROPERTIES OF RAT LIVER MITOCHONDRIAL RNA POLYMERASE IN THE NATIVE AND PURIFIED FORM

	Native	Purified
Optimum divalent cation concentration	2-3 mM Mg^{++}, 2 mM Mn^{++}	3 mM Mg^{++}, 1 mM Mn^{++}
% Inhibition by:		
rifampicin	80 (50 μg/ml)	70 (20 μg/ml)
α-amanitin	0 (0.2 μg/ml)	0 (0.2 μg/ml)
cordycepin	10 (20 μg/ml)	0 (60 μg/ml)
cycloheximide	33 (100 μg/ml)	63 (100 μg/ml)
ethidium bromide	70 (10 μg/ml)	23 (40 μg/ml)
ionic strength	50 (0.15)	50 (0.17)
Transcription product	14-8 S	19,14 and 4 S

Only ethidium bromide inhibition seems to be higher in the crude extract, showing that probably its activity is linked to the physical state of the DNA templates (7). Experiments on ethidium bromide inhibition using covalently closed circular DNA molecules from rat liver mitochondria are now in progress in our laboratory.

We can therefore conclude 1) that for rat liver the purified enzyme behaves like the native one, 2) that rat liver mitochondrial RNA polymerase resembles the mitochondrial polymerase enzymes from other organisms very closely, 3) that there are no reasons to believe that rat liver mitochondria contain more than one polymerase enzyme.

POLY(A)POLYMERASE

Direct and indirect lines of evidence clearly suggest that rat liver organelles possess in addition to a DNA-dependent RNA polymerase other ribonucleotides polymerizing activities and this fact in our opinion can explain most of the discrepancies reported in the literature concerning the properties of the mitochondrial RNA polymerase, especially in non-purified preparations. In our laboratory we have found in a solubilized extract from rat liver mitochondria an enzyme which incorporates labeled ATP into polyriboadenylate chains. This enzyme can be partially purified until the ammonium sulphate fraction with the procedure used to prepare the DNA-dependent RNA polymerase with few modifications (see legend of Fig. 1). Any attempt to further purify it was so far unsuccessful. Figure 1 shows the time course of poly (A) polymerase activity in the mitochondrial extract solubilized by Na deoxycholate (DOC extract) and in the ammonium sulphate fraction. The specific activity, very low in the DOC extract, increases considerably after ammonium sulphate precipitation and goes on almost linearly up to 30 minutes.

The presence of adenylate sequences in our product is based on the following properties: 1) the acid-insoluble product is hydrolyzed by 0.3 M KOH; 2) the acid-insoluble product is, on the other hand, resistant to RNase, and 3) is retained on the millipore filters in the presence of 0.5 M KCl which does not bind other ribonucleotide polymers.

Table IV shows some properties of the enzyme catalyzing the incorporation of ^{3}H-ATP. The enzyme is insensitive to the addition of DNA and almost unaffected

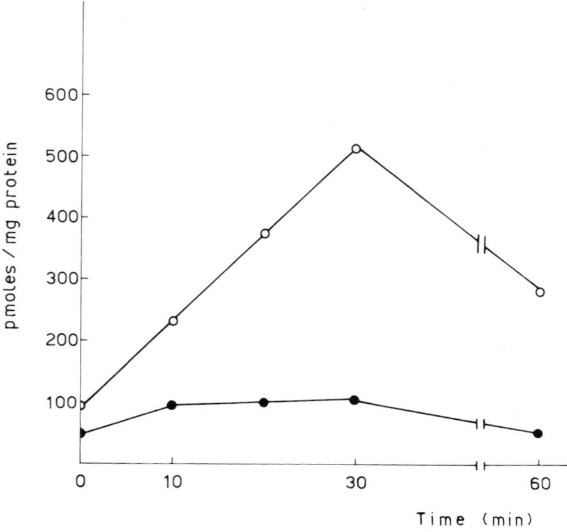

Fig. 1. Time course of poly(A)polymerase activity obtained by using DOC extract and ammonium sulphate fraction. The purification procedure was the same as that used for DNA dependent RNA polymerase from rat liver mitochondria (6) with the only exception that the DNase treatment was omitted. The incubation mixture contained 56 mM Tris-HCl pH 7.4, 64 mM KCl, 1.5 mM $MgCl_2$, 1.25 mM $MnCl_2$, 0.1 mM ^{3}H-ATP (specific activity 1 Ci/mmole). Final concentration of glycerol and DTT were 10% and 0.4 mM respectively in the DOC extract and 2.8% and 0.12 mM respectively in the ammonium sulphate fraction. At each time an aliquot of 75 ul was chilled in 1 ml of 5% ice-cold TCA containing 50 ug of albumin. Acid-precipitable material was collected on Whatman fiber-glass filter, washed by cold TCA solution, dried and counted in toluene scintillation fluid. ●——●, DOC extract; o——o, A.S. fraction.

by DNase and RNase. The lack of stimulation by poly(U) or RNA together with the RNase insensitivity seems to indicate that the enzyme does not require a primer as described for other similar activities. The enzyme was able to incorporate also ^{3}H-UTP and ^{3}H-CTP

TABLE IV
PROPERTIES OF THE POLY(A) SYNTHESIZING ACTIVITY FROM RAT LIVER MITOCHONDRIA

System	Enzyme activity (% of control)
Complete	100
+ GTP (0.1 mM)	100
+ CTP (0.1 mM)	52
+ UTP (0.1 mM)	70
+ Poly(U) (60 μg/ml)	94
+ *E. coli* DNA (100 μg/ml)	114
+ *E. coli* tRNA (200 μg/ml)	104
+ rat liver cit. RNA (200 μg/ml)	65
+ DNase (100 μg/ml)	78
+ RNase (100 μg/ml)	85

into acid insoluble material but at a much lower rate compared to ATP. The activity found by using ^3H-GTP was negligible. Figure 2 shows the Mg^{++} and Mn^{++} dependence of the enzyme. At the top (A and B) activities in the presence of increasing concentrations of either ion are reported. Down below (C and D) the effect of increasing concentrations of each ion has been measured in the presence of a fixed amount of the other. From these data it is possible to conclude that optimal activity is at about 3 mM Mg^{++} plus 2.4 mM Mn^{++} concentration. Figure 3 shows that the optimum of KCl concentration is about 64 mM while higher con-

Fig. 2 (opposite page, above). Mn^{++} and Mg^{++} dependence of the poly(A)polymerase from rat liver mitochondria. A, effect of increasing Mg^{++} concentration. B, effect of increasing Mn^{++} concentration. C, effect of increasing Mg^{++} concentration at 2.4 mM Mn^{++}. D, effect of increasing Mn^{++} concentration at 2.4 mM Mg^{++}. The incubation mixtures were as described in Fig. 1 except for changes in Mn^{++} and Mg^{++} concentrations. Other technical details are reported in legend of Fig. 1.

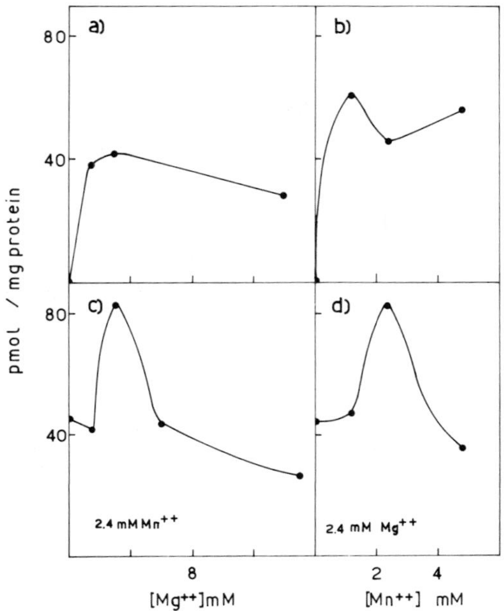

Fig. 2. (Legend on opposite page)

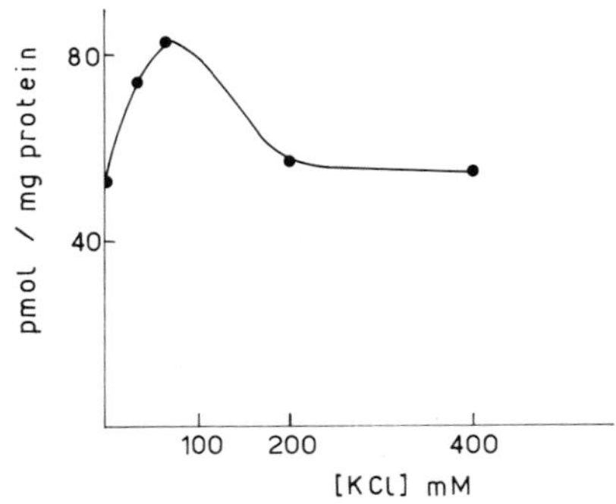

Fig. 3. Effect of increasing KCl concentration on rat liver mitochondrial poly(A)polymerase. The reactions were carried out as described in Fig. 1 in the presence of 2.4 mM Mn^{++} and 2.4 mM Mg^{++}.

centrations do not inhibit the enzyme as found for the mitochondrial RNA polymerase of the same organism (7).

Table V shows the effects of several inhibitors on poly (A) synthetase activity. The enzyme is inhibited by low concentrations of cordycepin which do not

TABLE V
EFFECT OF VARIOUS INHIBITORS ON POLY (A) SYNTHESIZING ACTIVITY FROM RAT LIVER MITOCHONDRIA

Conditions	Inhibitor concentration (μg/ml)	Synthesizing activity (% of control)
Control		100
+ cordycepin	20	63
	100	46
+ α-amanitin	0.33	71
+ actinomycin D	20	119
	100	100
	200	54
+ rifampicin	30	87
	70	85
+ rifampicin AF/013	100	77
+ ethidium bromide	5	64
	10	60
	50	46

affect mitochondrial RNA polymerase (7); it is practically insensitive to actinomycin D and rifampicin, inhibitors of the DNA-dependent RNA polymerase and sensitive to ethidium bromide. The sensitivity to ethidium bromide was the same when activity was measured in the absence or in the presence of mitochondrial DNA clearly showing that sensitivity was not due to a residual DNA-dependent activity. This led us to suggest that the drug may more directly inhibit the enzyme by a mechanism different from that described for DNA-dependent RNA polymerase.

The properties above reported differentiate this enzyme from the mitochondrial DNA-dependent RNA polymerase which on the other hand, in the purified form, displays only a negligible ability to polymerize ATP.

Jacob et al. (20) solubilized poly (A) polymerase activity from rat liver mitochondria which shows several properties similar to those of the enzyme here described.

What is the meaning of this and eventually other ribonucleotides polymerizing activities present in rat liver organelles? Although it is not possible at the moment to give a satisfactory answer to this question, it must be recalled that a poly (A) synthesizing activity can be correlated to the presence of poly (A) segments, either free or linked to RNA, which have been widely observed. As far as mitochondria are concerned, the presence of RNA species containing poly (A) has been already described by several authors (Table VI); in one case (HeLa cells) (9) it has been demonstrated that the poly (A) containing RNA species are made on mitochondrial DNA as template; in another case (yeast) a messenger activity of this fraction has been demonstrated (22).

TABLE VI
PRESENCE OF POLY (A) CONTAINING RNAs IN MITOCHONDRIA

Source	Length of poly (A) sequence (number of nucleotides)	Hybridizability with mit. DNA	Messenger activity	Reference
HeLa cells	70	+	n.m.	ATTARDI et al. (3)
HeLa cells	50-80	n.m.	n.m.	PERLMAN et al. (21)
Yeast	n.m.	n.m.	+	COOPER et al. (22)
Ehrlich ascites	150-180	n.m.	n.m.	AVADHANI et al. (23)

Further experiments will be necessary to clarify both the role of mitochondrial poly (A) and the function of the enzyme activity which catalyzes its synthesis.

CONCLUSION

We can summarize our studies in this field as follows. Rat liver mitochondria are able to synthesize RNA by an enzyme which is a typical DNA-dependent RNA polymerase. This enzyme can be extracted in a purified form and it displays properties differentiating it both from bacterial polymerases and from nuclear polymerases of eukariotic cells. Mitochondria are, however, also able to synthesize poly (A) and probably other ribo-homopolymers. A poly (A) polymerase activity has been partially purified and seems to be different from other similar activities described either in nuclei and cytoplasm of eukariotic cells or in bacteria. The role of this enzyme within the mitochondrion remains to be clarified.

REFERENCES

1. BORST, P. (1972) Ann. Rev. Biochem. 41, 333.
2. KROON, A. M., AND SACCONE, C. (1974) *in* The Biogenesis of Mitochondria, Academic Press, New York.
3. ATTARDI, G., CONSTANTINO, P., AND OJALA, D. (1974) *in* The Biogenesis of Mitochondria (Kroon, A. M., and Saccone, C., eds.) p. 9, Academic Press, New York.
4. SACCONE, C., GADALETA, M. N., AND GALLERANI, R. (1969) Eur. J. Biochem. 10, 61.
5. AAIJ, C., SACCONE, C., BORST, P., AND GADALETA, M. N. (1970) Biochim. Biophys. Acta 199, 373.
6. GALLERANI, R., SACCONE, C., CANTATORE, P., AND GADALETA, M. N. (1972) Fed. Eur. Biochem. Soc. Lett. 22, 37.
7. GALLERANI, R., AND SACCONE, C. (1974) *in* The Biogenesis of Mitochondria (Kroon, A. M., and Saccone, C., eds.) p. 59, Academic Press, New York.
8. SCRAGG, A. H. (1974) *in* The Biogenesis of Mitochondria (Kroon, A. M., and Saccone, C., eds.) p. 47, Academic Press, New York.
9. ECCLESHALL, T. R., AND CRIDDLE, R. S. (1974) *in* The Biogenesis of Mitochondria (Kroon, A. M., and Saccone, C., eds.) p. 31, Academic Press, New York.

10. KÜNTZEL, H., AND SCHÄFER, K. P. (1971) Nature New Biology 231, 265.
11. WU, G. J., AND DAWID, I. B. (1972) Biochemistry 11, 3589.
12. REID, B. D., AND PARSONS, P. (1971) Proc. Nat. Acad. Sci. USA 68, 2830.
13. WINTERSBERGER, E. (1972) Biochem. Biophys. Res. Commun. 48, 1287.
14. SHMERLING, Z. G. (1969) Biochem. Biophys. Res. Commun. 37, 965.
15. GADALETA, M. N., GRECO, M., AND SACCONE, C. (1970) Fed. Eur. Biochem. Soc. Lett. 10, 54.
16. CANTATORE, P., NICOTRA, A., LORIA, P., AND SACCONE, C. (1974) Cell Differentiation, in press.
17. GALLERANI, R., unpublished observations.
18. POLYA, G. M., AND JAGENDORF, A. T. (1971) Arch. Biochem. Biophys. 146, 635.
19. BOGORAD, L., MULLINIX, K. P., STRAIN, G. C., AND WOODCOCK, C. L. F. (1970) Abstr. Ann. Meeting Am. Soc. Cell Biol. p. 21a.
20. JACOB, S. T., AND SCHINDLER, D. S. (1972) Biochem. Biophys. Res. Commun. 48, 126.
21. PERLMAN, S., ABELSON, H. T., AND PENMAN, S. (1973) Proc. Nat. Acad. Sci. USA 70, 350.
22. COOPER, C. S., AND AVERS, C. J. (1974) in The Biogenesis of Mitochondria (Kroon, A. M., and Saccone, C., eds.) p. 289, Academic Press, New York.
23. AVADHANI, N. G., KUAN, M., VAN DER LIGH, P., AND RUTMAN, R. J. (1973) Biochem. Biophys. Res. Commun. 51, 1090.

TRANSPORT OF NUCLEAR DNA TO MICROSOMAL MEMBRANES

L. M. Narurkar, C. P. Giri, and M. V. Narurkar

The extra-mitochondrial cytoplasmic DNA has been a controversial topic in recent years. After Bach's suggestion of the presence of microsomal DNA in 1962 (1), it was only since 1969 that a number of reports on cytoplasmic DNA has been published (2-6). One of the first reports was from Bond *et al.* (2), who, using ^{3}H-thymidine as a DNA-label in mouse liver, demonstrated the presence of microsomal DNA and claimed that it had the highest specific activity as compared to that of the nuclear or the mitochondrial DNA. Further, based on their studies on thermal denaturation and reannealing characteristics they suggested that it was a unique entity. Almost simultaneous was an exhaustive report from Schneider and Kuff (3), who demonstrated the presence of microsomal DNA with a specific activity in between that of the nuclear and the mitochondrial DNA in mouse and rat liver, and that it closely resembled nuclear DNA in its melting behavior, banding in CsCl, sedimentation rate, electron microscopic appearance and hybridization with cytoplasmic RNA. These authors also showed that microsomes contained RNA polymerase activity which was, however, not associated with microsomal DNA, thus ruling out its possible transcribing function. Eugene Bell (4) in the same year published a stimulating article on the role of microsomal DNA during differentiation. Using cultured embryonic chick muscle cells in an *in vitro* system Bell suggested that the cytoplasmic DNA is informational and that it may act as an intermediary carrying genetic information from the nucleus to the cytoplasm. Fromson and Nemer (7), on the contrary, have challenged this "proposed conjunction of these cytoplasmic structures with DNA." Based on two methods for preparing cytoplasmic extracts from sea urchin embryos they claimed that DNA associated with the microsomes may be an artifact since it could not be detected when cells of sea urchin embryos were disrupted by passing them through a hypodermic needle. According to Williamson (8) the cytoplasmic DNA from cultures of embryonic mouse liver cells originates from nuclei, probably arising from lysis during primary cell culture, and contains all rather than a selection of information present in the cell genome, thus suggesting that this DNA does not have any informational role (9). Scolnick et al. (10) have postulated that this DNA is probably transcribed by an RNA dependent DNA polymerase from cytoplasmic RNA. In a brief survey of cultured cells from different species, Koch and Pfeil (11) have shown that the cytoplasmic DNA is not a common feature of all the cells and therefore the significance and function of this DNA is still an open question. In his recent report, based on specific inhibitors of synthesis of nuclear DNA and informational DNA, Bell (12) has shown that hydroxyurea inhibits the synthesis of the co-called I-DNA whereas 5-fluorodeoxyuridine and cytosine arabinoside inhibit the synthesis of nuclear DNA without affecting the synthesis of I-DNA. In yet another recent paper, Bell (13) has hypothesized that the DNA associated cytoplasmic particles constitute a class of organelles which function in transcription and possibly in translation as well.

The present chapter deals with the association of this DNA largely with the rough endoplasmic reticulum, its transport from nucleus to the membranes and finally its possible role in maintaining the integrity of the rough membranes. We have chosen 24 hr regenerating rat liver as a system to demonstrate the presence of microsomal DNA in relation to the nuclear and the mitochondrial DNA. For these experiments, rats 22 hr after partial hepatectomy were given intraperitoneally 0.7 mCi of ^{3}H-thymidine (6-9 Ci/mmole, BARC) and were killed two hr later. Results of these experiments are given in Table I. It can be seen that

TABLE I
^{3}H-THYMIDINE INCORPORATION INTO DNA IN CELL FRACTIONS OF NORMAL AND REGENERATING RAT LIVER

Cell fraction	Normal liver		Regenerating liver (24 hr post hepatectomy)	
	μg DNA/gm equi.liver	Sp. activity cpm/μg DNA	μg DNA/gm equi.liver	Sp. activity cpm/μg DNA
Homogenate	1682 (1580-1892)	26 (23-40)	1288 (800-1437)	2191 (1480-2559)
Nuclei*	-	55 (50-63)	-	2796 (1778-3289)
Mitochondria	13 (8.5-19.5)	142 (116-260)	26 (22-28)	732 (683-1192)
Microsomes	9 (6-14)	76 (70-89)	14 (12.5-21)	226 (155-350)

The values represent averages of at least 6 experiments. Figures in parentheses represent the range of values.

*Nuclei were purified according to the procedure of Blobel and Potter (19). Post-mitochondrial supernatant was spun at 105,000 ×g. The microsomal pellet was checked for mitochondrial contamination.

DNA was extracted from TCA insoluble precipitate with 10% NaCl-Tris 0.4 M (17), and estimated by modified indole method (18).

both in normal as well as in regenerating liver the presence of DNA in microsomes can be demonstrated. The microsomal DNA forms about 1% of the total nuclear and the mitochondrial DNA in regenerating liver. The specific activity of this DNA is closer to that of nuclear DNA in normal liver. In regenerating liver, however, this cytoplasmic DNA has the lowest specific activity.

In these experiments, due precautions were taken to avoid any possible contamination of the microsomal fraction with nuclei during cell disruption, by using only 8 strokes of a loosely fitting pestle (clearance 0.30 mm) for tissue homogenization. Nevertheless, experiments were carried out to rule out any possible nuclear contamination. Thus, purified labeled nuclei from 24 hr regenerating rat liver were added to freshly cut liver pieces from an unlabeled rat. This unlabeled liver tissue was then subjected to homogenization in a loosely fitting homogenizer and was then subjected to cellular subfractionation. The microsomes were then examined for any labeled nuclear contamination. Results of these experiments are given

TABLE II
CONTAMINATION OF MICROSOMAL FRACTION BY NUCLEI

	Total DNA (μg)	Total counts/ min.	Sp. Act. cpm/μg DNA	Contamination (percent of counts added)
A. Nuclei isolated from rats injected with ^{3}H-thymidine (0.75 mC/100 g body wt.)	302	568,640	1822	-
B. Microsomal fraction isolated from rat liver homogenized with labeled nuclei from A	30*	80*	2.6	0.014
C. Post microsomal super-natant from B	-	0	0	0

A 4 ml suspension of the purified labeled nuclei (302 μg DNA, 568,640 cpm) was mixed with freshly dissected liver pieces from normal rat, prior to homogenization.

*Figures represent values for total microsomal suspension.

in Table II. It is clear that out of 568,640 c.p.m. added, only about 80 c.p.m. were traced in the microsomal fraction giving a contamination level of 0.014%. Thus, the labeling of DNA in the microsomal fraction that is normally observed cannot be accounted for by nuclear contamination.

Results on the distribution of DNA in the microsomal subfractions are given in Table III. It can be seen that about 70% of the microsomal DNA is present in the rough endoplasmic reticulum while the smooth membranes and the free polysomes contain very small amounts. Further evidence about its location predominantly in the membrane component and not in the

TABLE III
DISTRIBUTION OF DNA IN MICROSOMAL
SUBFRACTIONS OF REGENERATING RAT LIVER

Cell subfraction	μg DNA/g eq.liver	Percent of microsomal DNA	Sp. Act. cpm/μg DNA
Microsomes	14 (12.5-21)	100	303
Rough endoplasmic reticulum (RER)	9.4 (7-12.6)	67	344
Smooth endoplasmic reticulum (SER)	2.1 (1.8-2.5)	15	-
Free polysomes	2.6 (2.2-3.4)	18.5	-

Subfractionation of microsomes into RER, SER, and FP was carried out by discontinuous sucrose gradient centrifugation according to Moyer et al. (20).

Figures in parentheses represent the range of values.

TABLE IV
EFFECT OF RNase AND DNase TREATMENT ON MEMBRANE DNA

Treatment	μg DNA/gm eq. liver	Sp. Act. cpm/μg DNA
Rough endoplasmic reticulum (RER)	10.50 (100%)	380
RER after RNase treatment, 50 μg/ml in 0.25 M EDTA	8.82 (84%)	364
RER after DNase treatment, 400 μg/ml, at 37° for 30 min	3.10 (30%)	224

bound polysomes is presented in Table IV. It can be seen that when the rough membranes are treated with DNase, thus stripping them of the attached ribosomes, most of the DNA remains associated with the membrane component. However, when the rough membranes are treated with DNase about 70% of the DNA is lost, thus confirming its presence in the membranes.

The cytoplasmic DNA is suggested to originate from the nucleus and the evidence in this respect is based on differential labeling, and early pulse to late pulse ratios after long exposures of *in vitro* cultures to radioactive precursors. Koch and Pfeil (11) have shown such a transport of nuclear DNA to the cytoplasm specifically with cultured cells of chick embryos. In the present experiments, a simple *in vitro* system, described by Ishikawa *et al.* (14), to demonstrate transport of mRNA from nucleus to the cytoplasm was employed with the exception of using ^{3}H-thymidine labeled purified nuclei from 24 hr regenerating rat liver and incubating these with the unlabeled post-mitochondrial supernatant. The counts recovered in microsomes after half an hour's incubation at 37° gave the measure of "transported DNA." Results of these experiments are given in Table V.

TABLE V
IN VITRO TRANSPORT OF NUCLEAR DNA* TO MICROSOMES

Incubation system	c.p.m. recovered in microsomes	Percent recovery
^{3}H-labeled nuclei* + unlabeled supt.	2,990	0.5%
(C)** Complete system	90,690	15%
(C) -ATP/ATP generating system	5,390	1%
(C) -GTP	23,050	3.8%

*^{3}H-thymidine labeled nuclei with 599,875 c.p.m. of acid precipitable counts were added to the unlabeled post-mitochondrial supernatant.

**Complete system consisted of ATP 3 mM, GTP 0.1 mM, creatine phosphate 4 mM, creatinkinase and reduced glutathione 2 mM in addition to labeled nuclei and unlabeled postmitochondrial supernatant from 24 hr regenerating rat liver.

Reaction mixture was incubated at 37° for 30 min, chilled immediately and spun at 5000 r.p.m. for 10 min. Post-nuclear supernatant was spun at 105,000×g for 2 hr to collect microsomes.

It is evident that when the complete system was added to the incubation mixture about 15% of the added counts were recovered in the microsomes. That the "transport" was not due to non-specific leakage was shown by the fact that only about 0.5% of the added counts could be recovered in the microsomal fraction when no other additions were made to the incubation mixture. Similarly, omission of ATP and ATP generating system gave a recovery of 1% of the added counts, and when GTP was excluded from the reaction mixture the recovery in microsomes was only about 4%. It is apparent, therefore, that nuclear DNA is "transported" to the microsomes and that the transport seems to be energy dependent. The distribution of "transported"

DNA in the microsomal subfractions is given in Table VI. It is evident that most of this "transported"

TABLE VI
DISTRIBUTION OF TRANSPORTED NUCLEAR DNA*
TO MICROSOMAL SUBFRACTIONS

Cell fractions	Recovery in fractions c.p.m.	Percent recovery
Microsomes	65,563	15%
Rough endoplasmic reticulum (RER)	22,334	5%
Smooth endoplasmic reticulum (SER)	3,181	0.7%
Free polysomes (FP)	3,869	0.9%
Supernatant	2,817	0.6%

*^{3}H-thymidine labeled nuclei with 448,690 c.p.m. of acid precipitable counts were added to the unlabeled post-mitochondrial supernatant. See Table V for details of the incubation system. After incubation at 37° for 30 min nuclei were spun down at 600×g. Supernatant was made to 1.35 M sucrose and spun over 2 M STKM for 3.5 hr at 275,000×g. Tubes were cut to recover smooth vesicles from top of 1.35 M sucrose, rough membranes at the interphase of 1.35 M/2 M sucrose and the pellet of free polysomes.

DNA is localized in the rough endoplasmic reticulum, the smooth membranes and the free polysomes accounting for about 10% and 13% respectively of the total microsomal activity. This distributional pattern is similar to that obtained for microsomal DNA in the intact animal (Table III). Effect of DNase on the transported DNA associated with microsomes is given in Table VII. It can be seen that about 14% of the added nuclear counts were recoverable in the microsomes. On treatment with DNase, about 95% of the

TABLE VII
EFFECT OF DNase ON TRANSPORTED DNA* IN MICROSOMES AND POST-MICROSOMAL SUPERNATANT

Cell fraction	Recovery in fractions c.p.m.	After treatment with DNase (c.p.m.)	% Inhibition
Microsomes	82,470 (14%)	4,320	95%
Postmicrosomal supernatant	3,674 (0.62%)	0	100

*^{3}H-Thymidine labeled nuclei with 589,071 c.p.m. of acid precipitable counts were added to the unlabeled post-mitochondrial supernatant. See Table V for details of the incubation system.

Figures in parentheses indicate percent recovery of added counts.

Treatment with DNase: 400 μg/ml in final concentration followed by incubation at 37° for 30 min.

activity was lost, confirming the presence of DNA in these microsomes. The post-microsomal supernatant which contained less than 1% of the added nuclear counts was also shown to be sensitive to DNase.

As mentioned before, the precise function of membrane DNA is an open question. It is, however, clear that this DNA forms an integral part of the rough endoplasmic reticulum. If this DNA does not have a transcribing function, does it have any other role? Its probable role as a constituent of the rough membranes was, therefore, examined. This was possible since an elegant *in vitro* reconstitution system (15) for rough membranes was available for such studies. This *in vitro* reconstitution in terms of polysome membrane interaction, according to Pitot (16), is a measure of the structural integrity of the rough membranes. According to this concept, the polysome-membrane interaction determines the stability of a translatable mRNA template required for the synthesis of a specific protein in the cytoplasm. Any change in the structural mosaic of the membrane would thus lead to a change in the template stability, resulting in an altered protein synthesis.

The reconstitution system essentially involves interacting ^{32}P-labeled liver polysomes with a separately isolated ribosome free (RNase stripped) preparation of unlabeled membranes in an <u>in vitro</u> system and then determining the percent binding of polysomes per unit membrane.

In order to study the possible role of membrane DNA in the structural integrity of the rough membranes, binding studies were carried out with the DNase stripped membranes. Results of these experiments are given in Table VIII. It can be seen that when ^{32}P-labeled polysomes were interacted with RNase

TABLE VIII
EFFECT OF DNase ON <u>IN VITRO</u> BINDING OF MEMBRANES WITH ^{32}P-LABELED POLYSOMES OF RAT LIVER

	Percent polysome-membrane binding
Rough membranes treated with RNase 50 μg/ml (stripped rough)	72%
Rough membranes treated with RNase followed by DNase 400 μg/ml	35%

Polysomal RNA (200 μg) was mixed with 1 mg of membrane protein on ice in a volume of 5 ml. The mixture was layered over 5 ml of 1.8 M sucrose and spun for 8 hr at 226,400×g.

Tubes were cut after the spin to recover reconstituted membranes at the interphase which were pelleted. Pellets of both the membranes and unattached polysomes were dissolved in formic acid and were counted after neutralization, in a liquid scintillation counter.

treated membranes from rat liver, the binding was about 72%. However, when the membranes were further treated with DNase and then subjected to interact with labeled polysomes, the binding was only to the extent of 35%, suggesting a 50% fall in the binding capacity of these membranes. Considering that only about 70% of the membrane DNA could be stripped by DNase treatment (Table IV), this difference of 50% in the binding of DNase treated membranes is, indeed, highly significant.

The fact that these DNase treated membranes were not able to bind the polysomes satisfactorily suggests the possible role of membrane DNA in directing polysome-membrane binding and hence maintaining the structural integrity of the membranes. In other words, this DNA may play a role in segregation of polysomes on the membranes. It would be difficult at this stage to state how exactly this DNA is involved in such an interaction. It would be pertinent to point out, however, that smooth membranes which do not contain sufficient DNA are not able to rigidly bind polysomes in an *in vitro* system (15). Thus, smooth membranes resemble the DNase stripped rough membranes in their binding characteristics. It is, therefore, tempting to speculate about the probable role of membrane DNA in the segregation and desegregation of polysomes and hence the conversion of rough membranes to the smooth ones.

REFERENCES

1. BACH, M. K. (1962) *Proc. Natl. Acad. Sci. U.S.* *48*, 1031.
2. BOND, H. E., COOPER, J. A. II, COURINGTON, D. P., AND WOOD, S. S. (1969) *Science* *165*, 705.
3. SCHNEIDER, W. C., AND KUFF, E. L. (1969) *J. Biol. Chem.* *244*, 4843.
4. BELL, E. (1969) *Nature* *224*, 326.
5. NAKAI, G. S., GUGANIG, M. E., KELLEY, R. O., AND LOFTFIELD, R. B. (1971) *Eur. J. Clin. Biol. Res.* *16*, 560.

6. HALL, M. R., MEINKE, W., GOLDSTEIN, D. A., AND LERNER, R. A. (1971) Nature New Biology 234, 227.
7. FROMSON, D., AND NEMER, M. (1970) Science 168, 266.
8. WILLIAMSON, R. (1970) J. Mol. Biol. 51, 157.
9. WILLIAMSON, R., McSHANE, T., GRUNSTEIN, M., AND FLAVELL, R. A. (1972) FEBS Letters 20, 108.
10. SCOLNICK, E. M., AARONSON, S. A., TODARO, G. J., AND PARKS, W. P. (1971) Nature 229, 318.
11. KOCH, J., AND PFEIL, H. V. (1972) FEBS Letters 24, 53.
12. BELL, E. (1972) Science 174, 603.
13. BELL, E., MERRILL, C., AND LAWRENCE, C. B. (1972) Eur. J. Biochem. 29, 444 (1972).
14. ISHIKAWA, K., UEKI, M., NAGAI, K., AND OGATA, K. (1972) Biochim. Biophys. Acta 259, 138.
15. SHIRES, T. K., NARURKAR, L., AND PITOT, H. C. (1971) Biochem. J. 125, 67.
16. PITOT, H. C. (1969) Arch. Path. 87, 212.
17. DAVIDSON, J. N., AND SMELLIE, R. M. S. (1952) Biochem. J. 52, 594.
18. HUBBARD, R. W., MATHEW, W. T., AND BUBOWIK, D. A. (1970) Anal. Biochem. 39, 190.
19. BLOBEL, G., AND POTTER, V. R. (1966) Science 154, 1662.
20. MOYER, G. H., MURRAY, R. K., KHAIRALLAH, C. H., SUSS, R., AND PITOT, H. C. (1970) Lab. Investig. 23, 108.

PROPERTIES AND FUNCTION OF CHLOROFORM-SOLUBLE PROTEINS FROM MITOCHONDRIAL AND OTHER MEMBRANES OF RAT LIVER

Bernhard Kadenbach and Paul Hadvary

Mitochondria possess a complete system for the synthesis of proteins, which differs in many respects from the corresponding cytoplasmic system. Mitochondrial DNA is small and circular and the ribosomes are smaller than cytoplasmic ribosomes. The product of the mitochondrial protein synthetic system is insoluble in water and consists of highly lipophilic proteins. Many attempts have been made to characterize these proteins. Up to now, however, no definite function of any particular protein synthesized in mitochondria could be demonstrated unequivocally. Some of the protein subunits of isolated cytochrome oxidase, cytochrome b, cytochrome c_1, and of the isolated ATP-ase complex of mitochondria have been shown to be synthesized in mitochondria (1-4). The function of an individual subunit of these enzyme complexes, however, is difficult to demonstrate, since separation of the complexes by sodium dodecyl sulfate gel-electrophoresis involves the destruction of their functional activities. From the seven components of isolated cytochrome oxidase the 3 larger protein subunits were shown by the groups of Mason and Schatz (1) and of Sebald _et al_. (2) to be synthesized in mitochondria. On the other hand, Komai and Capaldi (5) were able to isolate an active cytochrome oxidase, which does not contain the high molecular weight protein subunits synthesized by mitochondria.

These and other results suggest that mitochondria synthesize proteins, which have the function of organizing the steric arrangements of enzyme complexes within the inner mitochondrial membrane.

The lipophilic nature of the proteins synthesized in mitochondria has been a major impediment to the isolation and separation of these proteins. Therefore we have chosen the method described first by Folch-Pi and Lees (6), which extracts the so-called proteolipids together with the lipids from biological material by the use of chloroform-methanol (2:1). The proteolipids represent a small group of proteins, which are highly lipophilic and have been found in membranes of various cells. Our procedure involves a primary solution of the cell fractions in 5% neutral sodium dodecylsulfate buffer, and a subsequent washing with salt solution of the chloroform-methanol extract, which removes most of the methanol and some less lipophilic proteins.

We have applied this procedure to the isolation of proteins synthesized in mitochondria, using the label of ^{14}C-leucine incorporated in vitro as a marker for protein (7). As shown in Table I, approximately 23% of the incorporated radioactivity could be extracted with chloroform-methanol (2:1). If the previous dis-

TABLE I

EXTRACTION OF CHLOROFORM-SOLUBLE PROTEINS FROM MITOCHONDRIA LABELED IN VITRO WITH ^{14}C-LEUCINE

Treatment of Mitochondria		Radioactivity
Precipitation with TCA	Dissolution in SDS-buffer	% Chlor.-sol. protein
+	+	22.7
+	-	2.6
-	+	23.0
-	-	19.6

For details of the methods see ref. (7).

solution in sodium dodecyl-sulfate buffer was omitted, up to 19% of the incorporated radioactivity could be extracted by chloroform-methanol (2:1). The incorporation of ^{14}C-leucine into chloroform-soluble proteins is time dependent (Fig. 1). Other amino acids have also been incorporated into chloroform-soluble proteins (Table II). The incorporation of all amino

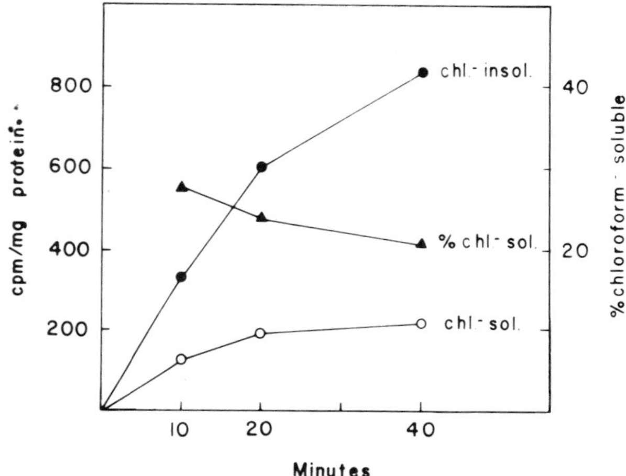

Fig. 1. Incorporation of ^{14}C-leucine into chloroform-soluble and insoluble proteins of isolated rat liver mitochondria. Taken from ref. (7).

TABLE II
INHIBITION OF INCORPORATION OF VARIOUS AMINO ACIDS INTO CHLOROFORM-SOLUBLE AND INSOLUBLE PROTEINS OF RAT LIVER MITOCHONDRIA BY CHLORAMPHENICOL

Amino acid	Chlor.-insol. protein -CA	+CA		Chlor.-sol. protein -CA	+CA	
	cpm/mg	% of control		cpm/mg	% of control	
Leu	852	80.5	10	439	34.9	8
Lys	228	47.6	21	92	13.9	15
Phe	453	64.0	14	179	29.1	16
Ser	365	60.5	16	254	122	48
Thr	337	50.0	15	154	21.8	14
Ile	1300	86.5	6	710	63.0	9

For details see ref. (7). CA = chloramphenicol.

acids is inhibited by chloramphenicol to the same extent as that into chloroform-insoluble proteins, except for serine, a portion of which is incorporated into phosphatidyl-serine. To demonstrate that chloroform-soluble proteins do not represent precursors of chloroform-insoluble proteins, synthesized by mitochondria, we have measured the change of radioactivity

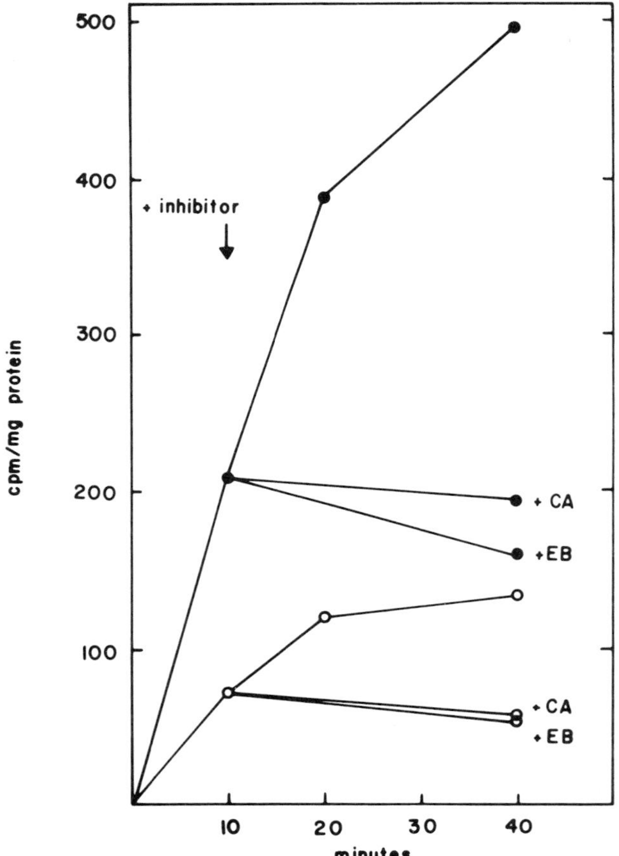

Fig. 2. Change of specific radioactivity of ^{14}C-leucine-labeled chloroform-soluble (o——o) and insoluble (●——●) proteins of rat liver mitochondria after addition of inhibitors. CA = chloramphenicol, EB = ethidium bromide. Taken from ref. (7).

of chloroform-soluble and insoluble proteins after a stop of incorporation by adding chloramphenicol or ethidium bromide (Fig. 2). The radioactivities of both chloroform-soluble and insoluble proteins decrease to the same extent indicating no transformation of one fraction into the other.

To analyze the molecular size of chloroform-soluble proteins, a sodium dodecylsulfate gel electrophoresis was performed from the chloroform-soluble and insoluble fraction of mitochondria labeled *in vitro*. As shown in Fig. 3, chloroform-soluble proteins have a smaller molecular weight than chloroform-insoluble proteins. The largest peak of radioactivity was found at a molecular weight of approximately 8000, but also

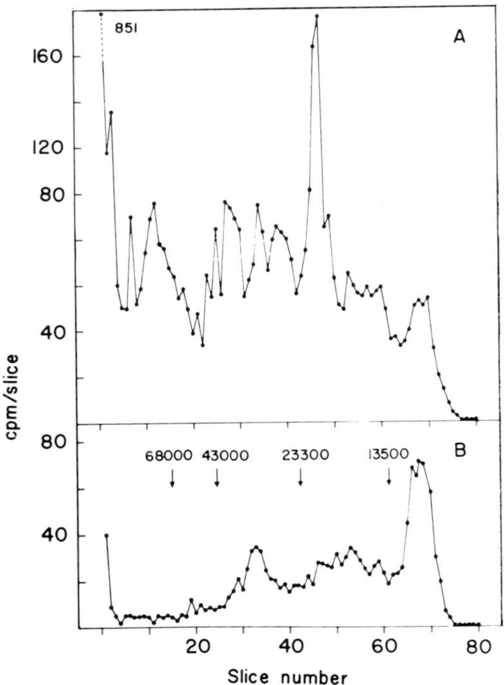

Fig. 3. Dodecylsulfate-polyacrylamide gel electrophoresis of chloroform-soluble (B) and insoluble (A) proteins of rat liver mitochondrial membranes labeled *in vitro*. Taken from ref. (10).

molecular weights up to 30,000 have been found. Since proteins with molecular weights around 8000 are also found in the fraction of chloroform-insoluble proteins, the solubility of proteins in chloroform is obviously caused by their structure rather than by their size.

Similar data on the incorporation of amino acids into chloroform-soluble proteins of mitochondria have been described by Murray and Linnane (8) and by Burke and Beattie (9).

Various attempts have been made to separate the proteins in the chloroform-extract from the bulk of lipids, particularly from phosphatides (10). A complete separation of the proteins from the phosphatides has been obtained by column chromatography on Sephadex LH-20 utilizing a mixture of chloroform-methanol-water-acetic acid (20:20:1:1) as the solvent. Proteins were identified by the radioactivity of ^{14}C-leucine incorporated *in vitro* into mitochondrial proteins. As shown in Fig. 4, the radioactivity appeared at the

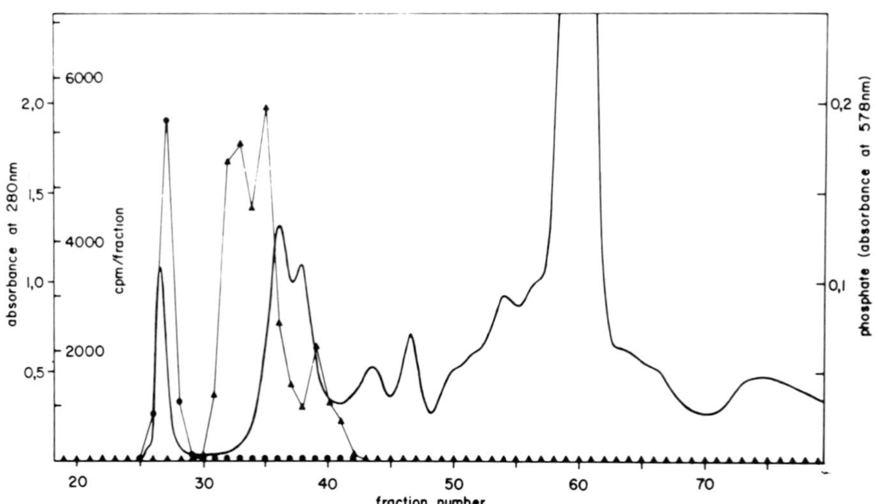

Fig. 4. Chromatography of the chloroform extract of mitochondria on a Sephadex LH-20 column. Taken from ref. (10). (——) Absorbance at 280 nm; (●——●) ^{14}C-radioactivity of fraction; (▲——▲) total phosphate.

front of the chromatogram, well separated from the phosphatides, estimated by phosphate determination. In a further experiment the fractions containing the protein peak were applied to a column of methylated Sephadex G-100 and eluted with chloroform-methanol (1:1) containing 10 mM acetic acid. Partial separation of the proteins has been obtained (Fig. 5), although no distinct peaks could be observed.

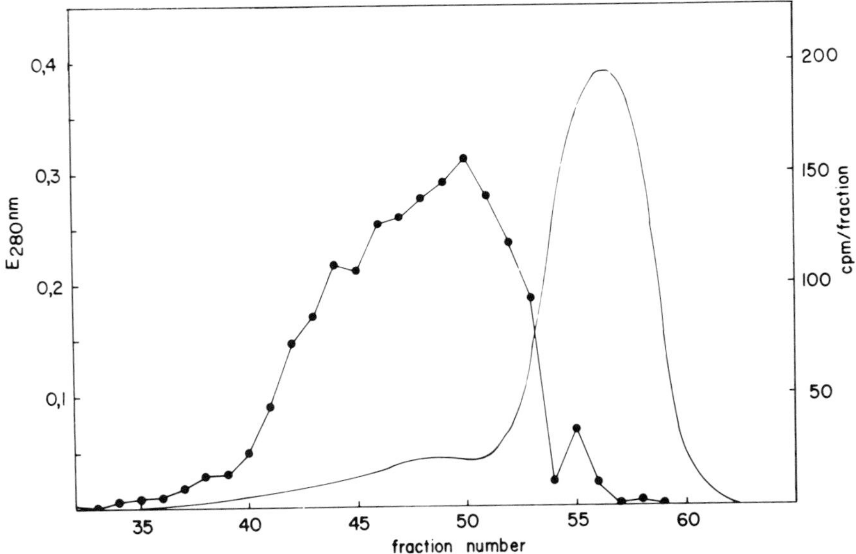

Fig. 5. Chromatography of phosphatide-free chloroform-soluble proteins on a methylated Sephadex G-100 column. (——) Absorbance at 280 nm; (●——●) ^{14}C-radioactivity of fraction. Taken from ref. (10).

If the phosphatide-free solution of proteins in chloroform was shaken with water, the radioactivity disappeared from the chloroform-phase and was recovered at the interphase. This denatured protein was completely insoluble in water, in 1N NaOH, in concentrated sodium dodecylsulfate-solutions and in various organic solvents. It could be solubilized in concentrated formic acid, but this solution was shown to contain aggregates of high molecular weight. From this behavior we concluded that water denatures irreversibly the chloroform-soluble proteins.

A better separation of the chloroform-soluble proteins was obtained by applying the chloroform-extract of mitochondria directly to methylated Sephadex G-100 column and eluting with chloroform-methanol-water-acetic acid (30:15:1:1). In Fig. 6, a parallelism of radioactivity and absorbance at 280 nm can be seen at the molecular weight range above 4000. The latter

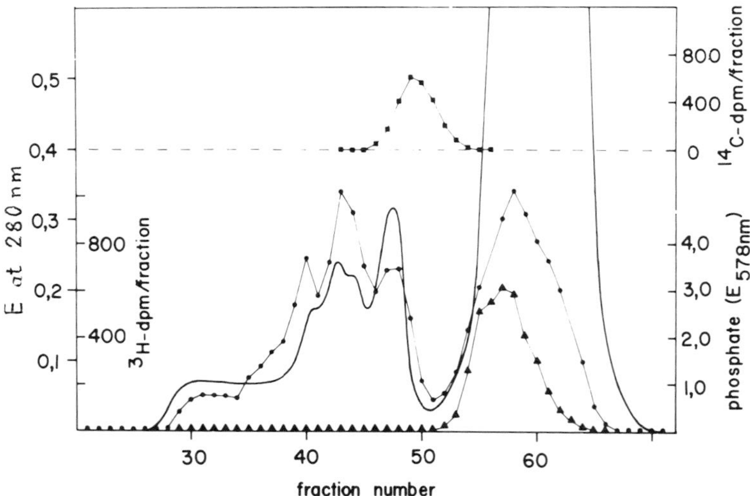

Fig. 6. Chromatography of the chloroform extract of mitochondria on a methylated Sephadex G-100 column. (——) Absorbance at 280 nm; (▲——▲) total phosphate; (●——●) ^{3}H-radioactivity of labeled proteins; (■——■) ^{14}C-radioactivity of polyethylenglycol-4000. Taken from ref. (10).

was measured by the radioactivity of ^{14}C-labeled polyethylenglycol-4000. Some radioactivity, however, was also found in low molecular weight fractions, where the phosphatides appear. This retention may be due to adsorption of the protein rather than due to molecular sieving.

We further investigated the occurrence of chloroform-soluble proteins in other fractions of rat liver cells. Identification of proteins was again performed by measuring the radioactivity of ^{14}C-leucine, which in this case has been incorporated *in vivo*.

1-^{14}C-leucine was injected into rats, and mitochondria, microsomes and the cytosol fraction were isolated 2 hours later. The chloroform-extracts of the three fractions were separated on Sephadex LH-20 columns as shown in Fig. 7. In all fractions a

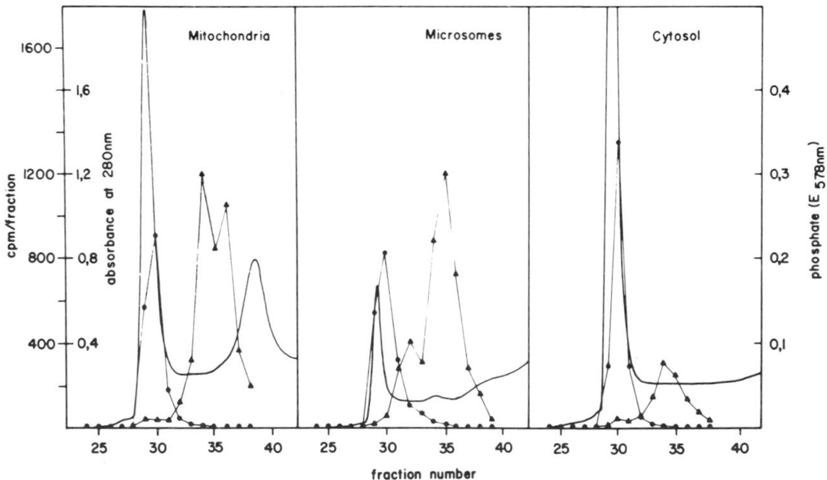

Fig. 7. Chromatography of chloroform extracts of various fractions from rat liver on Sephadex LH-30 columns. (——) Absorbance at 280 nm, (●——●) ^{14}C-radioactivity, (▲——▲) total phosphate. Taken from ref. (10).

radioactive peak was found in front of the chromatograms well separated from the phosphatides. The separated protein fractions were hydrolyzed and the amino acid composition determined by use of an automated amino acid analyzer. As follows from Fig. 8 the chloroform-soluble proteins contain more hydrophobic amino acids and much less acidic amino acids than the corresponding chloroform-insoluble proteins.

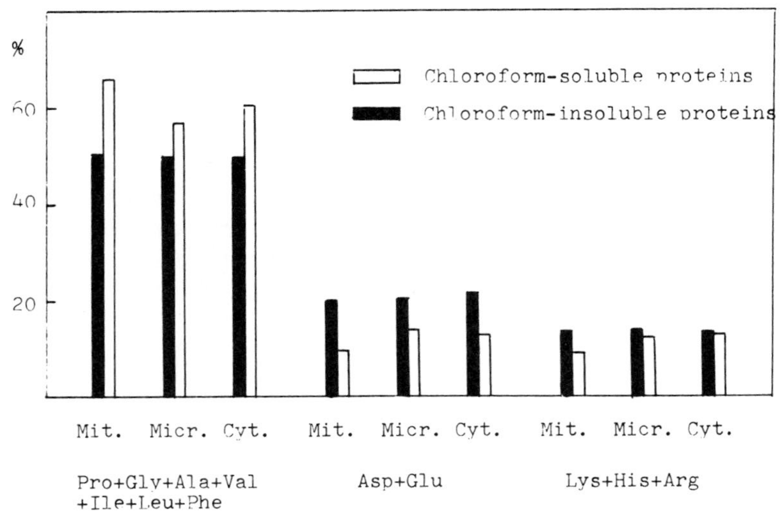

Fig. 8. Amino acid content of chloroform-soluble and insoluble proteins from various fractions of rat liver. Mit. = mitochondria; Micr. = microsomes; Cyt. = cytosol.

From the total amount of estimated amino acids the content of chloroform-soluble proteins in various fractions from rat liver was calculated as shown in Table III. In mitochondrial membranes, which almost

TABLE III
AMOUNT OF CHLOROFORM-SOLUBLE PROTEINS IN VARIOUS FRACTIONS FROM RAT LIVER

Fraction	Content of chl.-soluble proteins % from total protein
Mitochondria	0.5
Mitochondrial membranes	1.2
Mitochondrial matrix	0.025
Microsomes	2.7
Cytosol	0.15

For details see ref. (10).

exclusively contain the chloroform-soluble proteins of mitochondria, 1.2% of protein is soluble in chloroform. An even higher amount (2.7%) was found in microsomes. But surprisingly about 0.15% of the proteins of the cytosol fraction were found to be soluble in chloroform. This unexpected result may be explained by the assumption that these proteins are transported from their site of synthesis, the microsomes, to their site of action, the various membranes.

This assumption may be supported by the results shown in Table IV, where the specific radioactivities of chloroform-soluble and insoluble proteins of various fractions from rat liver are presented 2 hours after injection of 1-^{14}C-leucine, in the presence and absence of cycloheximide or chloramphenicol. Except

TABLE IV
EFFECT OF INHIBITORS ON THE SPECIFIC RADIOACTIVITY OF CHLOROFORM-SOLUBLE AND INSOLUBLE PROTEINS OF VARIOUS FRACTIONS FROM RAT LIVER 2 H AFTER INJECTION OF 1-^{14}C-LEUCINE

	Chloroform-insoluble proteins			Chloroform-soluble proteins		
	Control	+CHI	+CAP	Control	+CHI	+CAP
	cpm/mg	% from control		cpm/mg	% from control	
Mitochondria	900	7.2	95.6	4650	31.0	98.6
Microsomes	2090	6.2	89.3	2090	9.9	148
Cytosol	1060	2.3	109	13400	4.5	74.6

The data for chloroform-insoluble proteins were taken from ref. (10). The data for chloroform-soluble proteins have been recalculated according to a new amino acid analysis.

for microsomes, the specific radioactivity of chloroform-soluble proteins is higher than that of chloroform-insoluble proteins, indicating a higher turnover of these proteins in mitochondria and cytosol. The incorporation of 1-^{14}C-leucine into mitochondrial chloroform-soluble proteins in the presence of cycloheximide is 31% of the control, indicating that about 2/3 of mitochondrial chloroform-soluble proteins are synthesized on cytoplasmic ribosomes. On the other hand, chloramphenicol does not inhibit the incorporation of ^{14}C-leucine into mitochondrial chloroform-soluble proteins. This may be due to the combined effect of a stimulation by chloramphenicol of protein synthesis on cytoplasmic ribosomes, and an inhibition of incorporation into proteins, synthesized on mitochondrial ribosomes.

From the gel electrophoresis and from the separation studies on methylated Sephadex G-100 follows that chloroform-soluble proteins from mitochondria represent a mixture of different proteins. Therefore it may be suggested that the individual components have different functions. The lipophilic nature of these low molecular weight proteins suggests that some of them may act as carriers in membrane transport.

To test this possibility we estimated the binding of various anions, which are known to be transported through the mitochondrial membrane by specific carriers, to the chloroform extract of mitochondria (11). Table

TABLE V

COMPARISON OF THE UPTAKE OF VARIOUS LABELED ANIONS BY A CHLOROFORM-EXTRACT OF RAT LIVER MITOCHONDRIA

Labeled anion	Control extract	Mitochondrial extract	Difference
	(cpm in	chloroform-phase	pico moles)
^{32}P-phosphate	100	7700	31
^{14}C-malate	111	181	0.6
^{14}C-citrate	23	70	1.6
^{14}C-glutamate	108	109	0.0
^{3}H-ADP	255	501	0.9

Taken from ref. (11).

V indicates that among various anions, only ^{32}P-phosphate is taken up by the chloroform extract at a reasonable amount. Chromatography on Sephadex LH-20 showed that ^{32}P-phosphate is in fact bound by the proteins of the extract and not by the phosphatides (Fig.

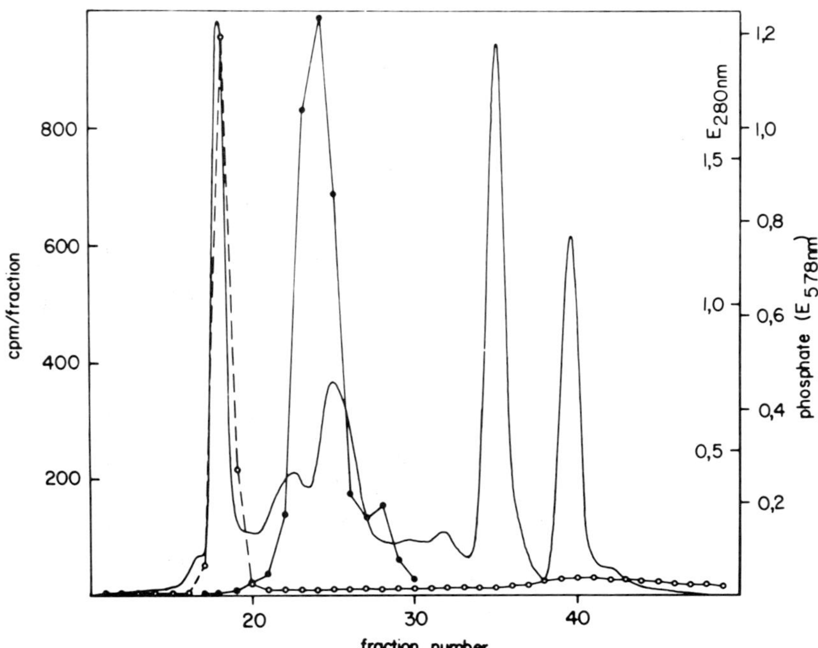

Fig. 9. Chromatography of a chloroform extract of rat liver mitochondria, labeled with ^{32}P-phosphate on a Sephadex LH-20 column. (——) Absorbance at 280 nm; (●——●) total phosphate; (o---o) ^{32}P-radioactivity. Taken from ref. (11).

9). Figure 10 shows that the binding of phosphate is inhibited by the SH-reagent ethacrynic acid. Mitochondria were incubated in the presence and absence of ^{14}C-labeled ethacrynic acid, extracted with chloroform-methanol and the extracts shaken with ^{32}P-phosphate. Figure 10 represents the two chromatograms

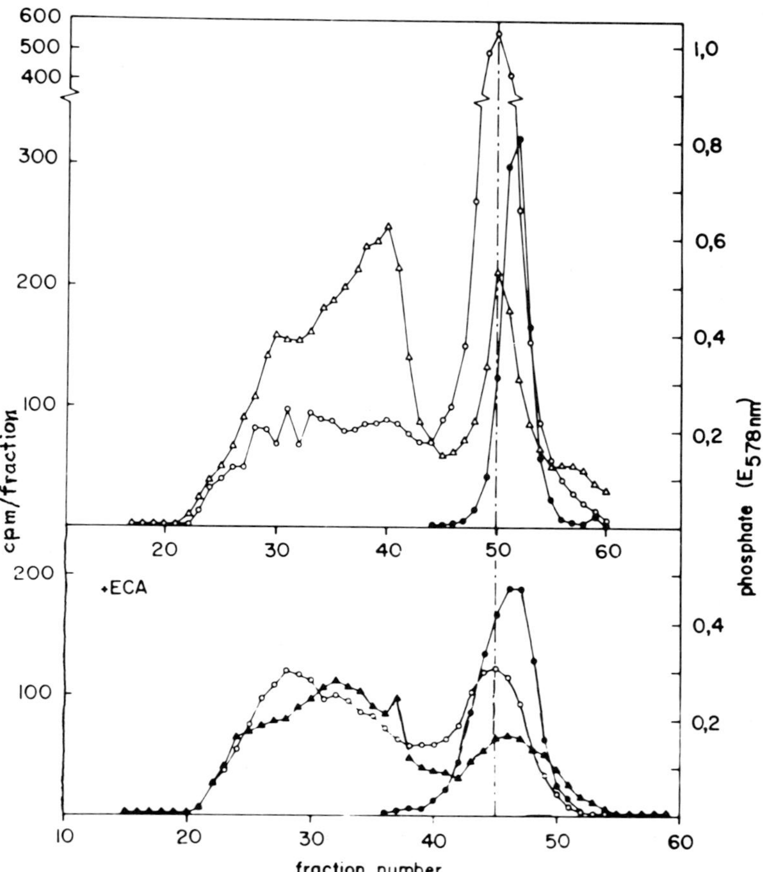

Fig. 10. Chromatography on methylated Sephadex G-100 of ^{32}P-phosphate-labeled chloroform extracts of mitochondria, preincubated in the presence and absence of ^{14}C-ethacrynate. (o——o) Radioactivity of ^{32}P-phosphate; (▲——▲) radioactivity of ^{14}C-ethacrynate; (Δ——Δ) radioactivity of ^{14}C-labeled protein from mitochondria, labeled <u>in vitro</u>; (●——●) total phosphate. Taken from ref. (11).

of the chloroform-extracts on methylated Sephadex G-100. The high molecular weight fraction of the two chromatograms were able to bind ^{32}P-phosphate to about the same extent. Just ahead of the phosphatide peak a protein fraction appeared - indicated by ^{14}C-radioactivity of in vitro labeled mitochondrial protein - which binds ^{32}P-phosphate to a high extent in the case of mitochondria incubated in the absence of ethacrynic acid. On the other hand, the extract of mitochondria, incubated with ^{14}C-ethacrynic acid binds ^{32}P-phosphate only to about 30% from the control. The radioactivity of ^{14}C-ethacrynic acid appeared in all fractions where protein was found. This may indicate that most of the chloroform-soluble proteins contain free SH-groups. These data suggest that a chloroform-soluble protein is involved in the transport of phosphate across the mitochondrial membrane.

Another function of chloroform-soluble proteins may be expected in the organization of biological membranes in general. Most membrane associated enzymes are directly or indirectly connected with transport processes. The transported substance may be an electron, a proton, a cation or anion or an undissociated substrate. From the low conductivity of biological membranes follows that the lipid bilayer of membranes is very tight. We therefore postulate that the chloroform-soluble proteins embedded within the lipid bilayer may represent the connecting, substrate specific unit between those proteins, localized on both sides of the membrane, which take up the substances from the aqueous phases. They may be understood as transporting units of membrane associated enzyme complexes.

Concerning the biogenesis of membranes, the chloroform-soluble proteins may represent the "crystallization center" of membrane associated enzyme complexes. They may be synthesized within the growing membrane as in the case of mitochondria or alternatively may be transported to the membrane together with a shield of lipids. The rapid labeling of chloroform-soluble proteins in the cytosol, which are associated with phosphatides, supports this assumption.

REFERENCES

1. MASON, T. L., AND SCHATZ, G. (1973) J. Biol. Chem. 248, 1355-1360.
2. SEBALD, W., WEISS, H., AND JACL, G. (1972) Europ. J. Biochem. 30, 413-417.
3. WEISS, H. (1972) Europ. J. Biochem. 30, 469-478.
4. TZAGOLOFF, A., AND MEAGHER, P. (1972) J. Biol. Chem. 247, 594-603.
5. KOMAI, H., AND CAPALDI, R. A. (1973) FEBS Lett. 30, 273-276.
6. FOLCH-PI, J., AND LEES, M. (1951) J. Biol. Chem. 191, 807-871.
7. KADENBACH, B., AND HADVARY, P. (1973) Europ. J. Biochem. 32, 343-349.
8. MURRAY, D. R., AND LINNANE, A. W. (1972) Biochem. Biophys. Res. Comm. 49, 855-862.
9. BURKE, J. P., AND BEATTIE, D. S. (1973) Biochem. Biophys. Res. Comm. 51, 349-356.
10. HADVARY, P., AND KADENBACH, B. (1973) Europ. J. Biochem. 39, 11-20.
11. KADENBACH, B., AND HADVARY, P. (1973) Europ. J. Biochem. 39, 21-26.

PHYSIOLOGICAL MODIFICATION OF ENERGY COUPLING IN *EUGLENA GRACILIS* IN RESPONSE TO EFFECTORS OF OXIDATIVE PHOSPHORYLATION

Joseph S. Kahn

Euglena gracilis is a facultative photoautotroph which will rely on photosynthesis as a major energy source only when utilizable organic energy sources are unavailable (1,2). While it is easy to obtain obligate organotrophic mutants of euglena, since the chloroplast-forming ability is readily lost, genetic modification of the respiratory mechanism has met with little success (3). We have attempted to modify the respiratory energy metabolism of wild-type *Euglena gracilis* by changes in the growth medium and growth conditions. Among other things, we tried to grow cells in the presence of 2,4-dinitrophenol (DNP). We found an unusual, reversible adaptation of these cells, which makes them resistant to a whole spectrum of uncouplers and inhibitors of oxidative phosphorylation. The fact that these chemically diverse effectors enter the cells without appreciably affecting cell growth -- provided the cells were adapted to one of the effectors -- necessitates the postulation of a drastically altered mode of energy coupling or of mitochondrial membrane permeability in these cells. The universality of the reactions involved in oxidative phosphorylation in mitochondria as usually seen makes one, as a rule, reluctant to postulate the existence of other energy coupling pathways, not involving the classical mitochondrial system.

When a clonal culture of euglena was grown in a medium containing 10^{-5} M DNP, they bleached, ceased to divide, appeared non-motile, oxygen consumption decreased greatly, and at least one-half, sometimes as many as 90% (based on an agar plate count), of the cells died (4). After a lag of 140-160 hours in this quiescent stage, the culture suddenly turned green again, the cells became motile again, and cell division commenced at a rate close to that of a normal culture (Fig. 1). These cells had become adapted to

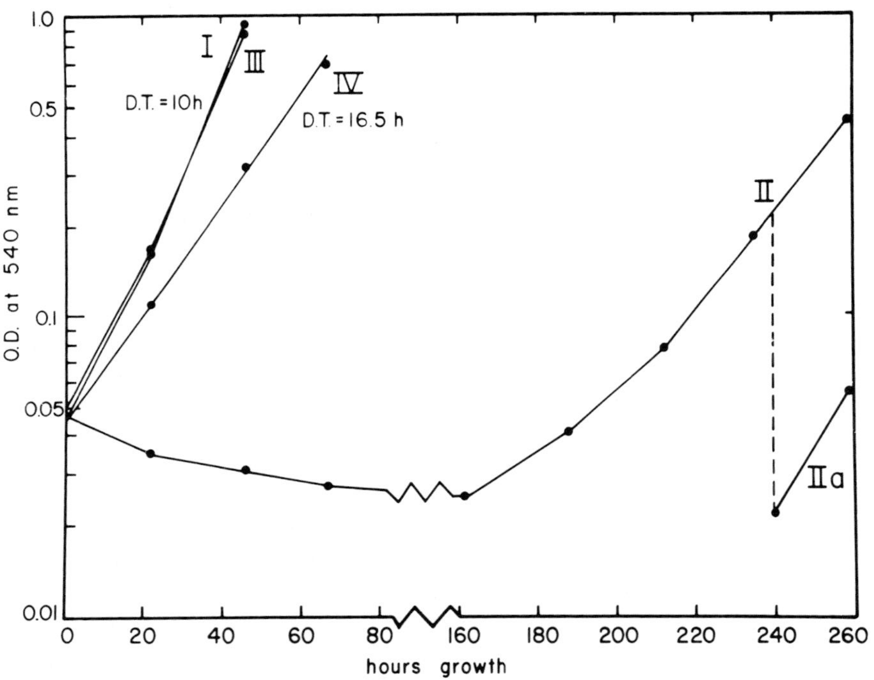

Fig. 1. Adaptation of Euglena gracilis to growth in media containing DNP. D.T. - doubling time. I, normal cells. II, normal cells + 10^{-5} M DNP. IIa, cells from II reinoculated into fresh medium containing DNP. III, adapted cells. IV, adapted cells + 10^{-5} M DNP.

growing in the presence of DNP, and when transferred to fresh medium containing DNP, continued to divide without any lag period. The adapted cells retained their resistance to DNP as long as DNP was present in the medium; but if grown for 2-4 doubling times (20-40 hours) on a medium lacking DNP, they completely lost this adaptation (Fig. 2). When grown on organotrophic medium, the adapted cells could grow in the dark in the presence of DNP, indicating that photosynthesis was not the alternative energy source in these cells.

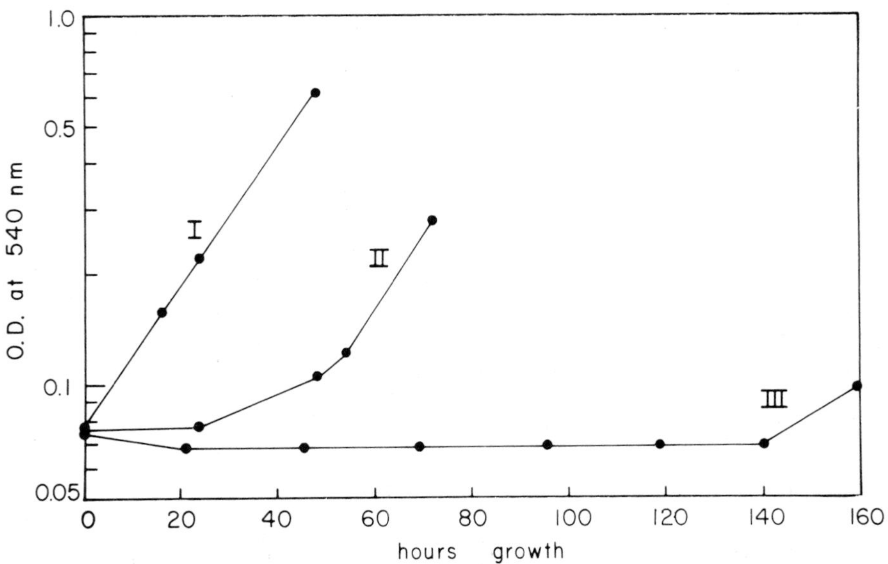

Fig. 2. Loss of resistance to DNP by adapted cells of Euglena gracilis. I, adapted cells grown for 8 hours in the absence of DNP prior to its addition at time 0. II, adapted cells grown for 16 hours in the absence of DNP prior to its addition at time 0. III, adapted cells grown for 40 hours in the absence of DNP prior to its addition at time 0.

The adapted cells did not metabolize DNP, since nearly 100% could be recovered at the end of the growth period. In addition, when DNP-containing

TABLE I

CELL SIZE AND COMPOSITION OF NORMAL AND DNP-ADAPTED *EUGLENA GRACILIS*

Cell Type	O.D. / 10^6 cells/ml	Chlorophyll	Protein	Carbohydrate	Lipid
		mg/10^6 cells			
Phototrophic					
Normal	0.57	0.028	0.47	0.103	0.284
Adapted	0.94	0.037	0.59	0.148	0.417
Organotrophic					
Normal	0.65	0.004	0.35	0.103	0.251
Adapted	1.05	0.009	0.53	0.213	0.462

Optical density was measured in 1/2" round cuvettes in a B & L Spectronic 20 colorimeter, and correlated to cell number with a Neubauer-type hemocytometer.

TABLE II

PHOTOSYNTHETIC ACTIVITY OF CHLOROPLASTS FROM NORMAL AND DNP-ADAPTED CELLS OF *EUGLENA GRACILIS*

Cell Type	Photophosphorylation		Ferricyanide Reduction	NADP Reduction
	Pyocyanin	Ferricyanide		
	μmole/mg Chlorophyll/hr			
Phototrophic				
Normal	56.2	24.9	119.3	19.1
Adapted	25.9	11.1	66.0	10.9
Organotrophic				
Normal	166	61.1	121.5	31.5
Adapted	145	60.6	138.5	27.0

medium from 4-day-old cultures was cleared by high-speed centrifugation and reinoculated with adapted and unadapted cells, the adapted cells grew normally while the unadapted ones required the normal 6-day lag period before commencing to grow.

The adapted cells were somewhat larger and more elongated than the normal cells, and light scattering per cell was higher. The differences in composition between normal and adapted cells (Table I) reflected the difference in size between them and was possibly a function of the difference in doubling time.

In photoautotrophically grown cells, all photosynthetic reactions measured were lower in adapted cells. The difference was small in organotrophically grown cells, but the very low chlorophyll content of the latter indicates photosynthesis to play only a minor role in their metabolism (Table II). Assuming comparable activity of chloroplasts *in vivo* as *in vitro*, photosynthesis can account only for a few percent of the energy requirement of organotrophically grown cells.

In order to determine whether the DNP actually affects cell energy coupling, we investigated the respiratory control of the mitochondria indirectly, in whole cells, using the inhibition of respiration by oligomycin and its stimulation by DNP as a measure. In normal, unadapted cells, respiration was greatly inhibited by oligomycin; and this inhibition was overcome and respiration was stimulated by the addition of DNP. In adapted cells, neither inhibition by oligomycin nor stimulation by DNP could be shown (Table III). Respiration in the adapted cells appeared to be loosely coupled, showing no respiratory control. Respiration in both kinds of cells was totally inhibited by 10^{-3} M KCN. It became clear that DNP must be able to enter the cell, and we looked for the conditions and extent of this uptake. Only when grown at a pH close to or below the pK_a of DNP (3.96) did the uncoupler penetrate into the cell and inhibit the growth of unadapted cells (Fig. 3). When grown at pH 5.4, unadapted cells grew normally in the presence of DNP and did not acquire any adapta-

TABLE III

RESPIRATION BY NORMAL AND DNP-ADAPTED CELLS OF EUGLENA GRACILIS

Cell Type	Oxygen Consumption nmoles O_2/min/10^6 Cells		
	Basal	+ 25 µM Oligomycin	+ 10^{-5} M DNP
Normal	2.13	----	4.10
Normal	2.30	1.10	3.84
Adapted	4.39	4.42	4.72
Adapted	4.58	----	4.58

Oxygen was measured polarographically in a medium containing: $MgCl_2$ - .005 M; KCl - .01 M; NaCl - .03 M; KH_2PO_4 - .001 M; Ethanol - .05M; and about 10^6 cells/ml. Final pH 4.0, temp. 21°. Additions were made consecutively.

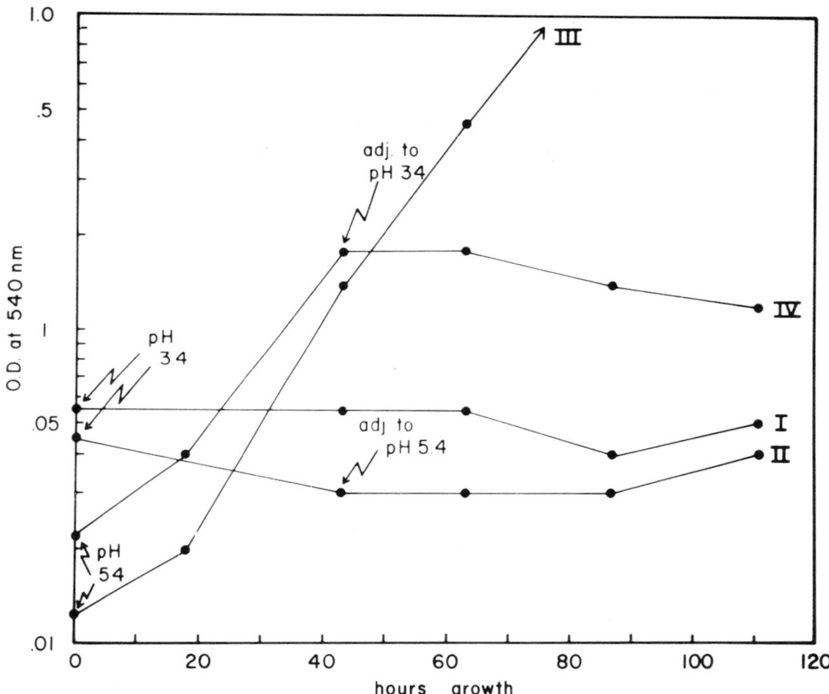

Fig. 3. Effect of pH of the medium on the growth of Euglena gracilis in the presence of 10^{-5} M DNP. D.T. - doubling time. I, normal cells in pH 3.4 medium. II, normal cells in pH 3.4 medium at arrow pH adjusted to 5.4. III, normal cells in pH 5.4 medium. IV, normal cells in pH 5.4 medium at arrow pH adjusted to 3.4.

tion; when the pH was lowered to 3.4, they became immediately inhibited. Conversely, when grown at pH 3.4 (below the pK_a) in DNP, the cells did not commence growing when the pH was raised to 5.4 (above the pK_a) (Fig. 3). Apparently the DNP, once accumulated, could not leak out from the cells.

We assayed adapted cells, grown in the presence of DNP at pH 3.4 and 5.4, for accumulation of the uncoupler. In cells grown at pH 5.4, <10 nmole DNP/gm fresh weight of cells was found. By contrast, in cells grown at pH 3.4, DNP accumulation ranging from 100 to 300 nmoles/gm fresh weight was found. Since the external concentration of DNP was only 1×10^{-5}M, this represents appreciable accumulation and cannot be accounted for by occlusion. In light of the fact that when cells were centrifuged at speeds higher than 1,000 × g they lost most of their DNP into the supernatant (centrifugation of adapted cells at that speed severely reduces cell viability), it appeared that the accumulated DNP was in a soluble form and neither precipitated nor bound to the membranes.

The question was raised whether the resistance to DNP involved a specific bypass of a DNP-sensitive site, or whether it involved a more generalized modification of energy coupling, involving possibly a different set of reactions for ATP synthesis. Using a number of the commonly used effectors of oxidative phosphorylation, we found that cells adapted to DNP were resistant to the inhibitory effect of carbonylcyanide, *m*-chlorophenylhydrazone (CICCP) (Fig. 4), as well as the aliphatic fluorocarbon uncoupler "1799" (1,7 hexafluoro-2,6 dihydroxy-2, 6 *bis*(trifluoromethyl)-4 heptanone). Neither the salicylanilide derivative S-13 nor gramicidin or chlorpromazine had any effect on the growth of euglena, and did not elicit DNP resistance either. The DNP-resistant cells were also resistant to oligomycin as well as to tri-*n*-butylchlorotin, both inhibitors of oxidative phosphorylation (Fig. 5). In the case of tributylchlorotin, secondary toxic effects slowed the growth even of adapted cells and caused them to bleach, although reversibly.

Dicyclohexylcarbodiimide inhibited both kinds of cells equally, and no resistance to it could be developed.

Atractyloside had no effect on growth of either kind of cell and did not induce the development of DNP resistance. Atractyloside, however, penetrated the cell membrane since other metabolic changes (drastic

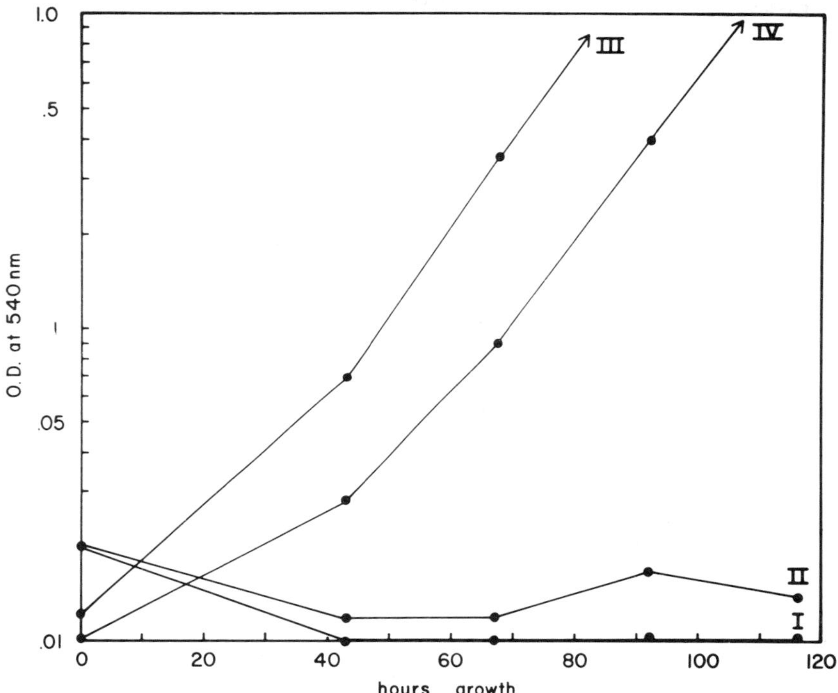

Fig. 4. Resistance of DNP-adapted cells to carbonyl-cyanide, m-chlorophenyl-hydrazone (ClCCP). D.T. - doubling time. I, normal cells + 10^{-5} M ClCCP. II, normal cells + 10^{-5} M ClCCP + 10^{-5} M DNP. III, DNP-adapted cells + 10^{-5} M ClCCP. IV, DNP-adapted cells + 10^{-5} M ClCCP and 10^{-5} M DNP.

changes in the level of some glycolytic enzymes) could be detected in cells grown in its presence.

Normal cells in ClCCP-containing medium showed identical symptoms to those in DNP-containing medium, i.e., bleaching, partial mortality, and a lag period of 140-160 hours until growth commenced (Fig. 6). When these ClCCP-adapted cells were transferred to a DNP-containing medium, they grew normally, like DNP-adapted cells (Fig. 6).

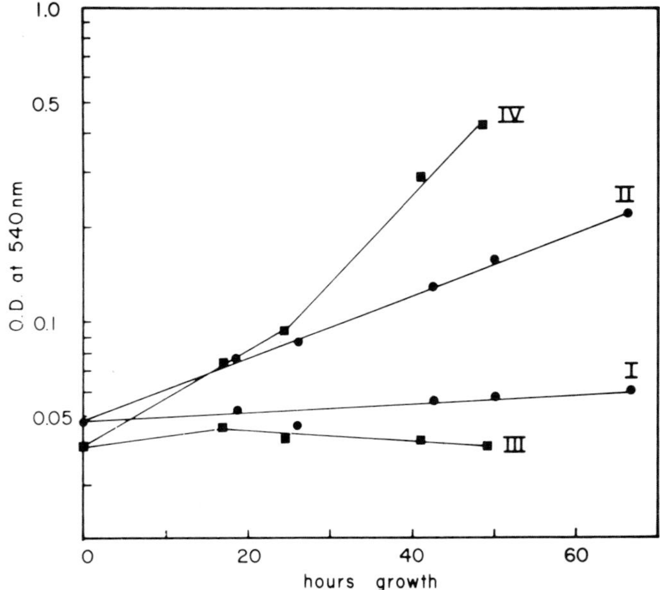

Fig. 5. Growth of normal and DNP-adapted cells in inhibitors of oxidative phosphorylation. I, normal cells + 2×10^{-6} M tri-n-butylchlorotin. II, DNP-adapted cells + 2×10^{-6} M tributylchlorotin. III, normal cells + 10^{-5} M oligomycin. IV, DNP-adapted cells + 10^{-5} M oligomycin.

Normal cells, growing on ethanol as the sole carbon source, appear to have a highly efficient metabolism, using about 1.2 gm of ethanol for every gm of dry weight produced. This highly efficient conversion of ethanol to biomass means that the TCA cycle plays only a minor role in the energetics of these cells. Adapted cells showed a somewhat lower efficiency, using about 1.4 gm ethanol/gm dry weight produced.

The short-term phosphate esterification by whole cells was measured with the aid of $^{32}P_i$. Adapted cells, although having a slightly lower rate of phosphate esterification, showed only a small response to

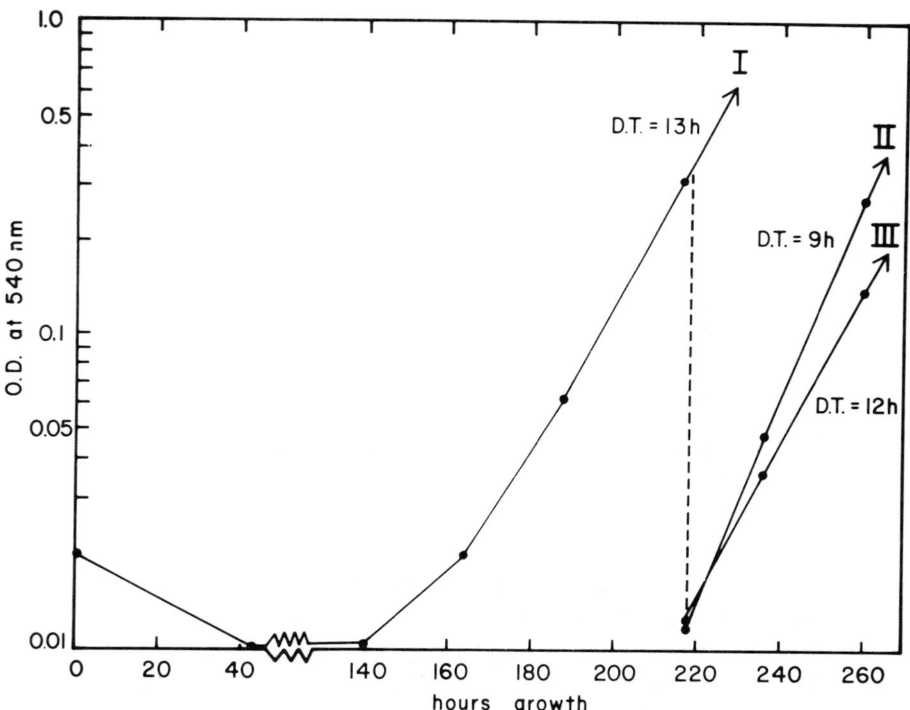

Fig. 6. Adaptation of Euglena gracilis to ClCCP and their subsequent resistance to DNP. I, normal cells + 10^{-5} M ClCCP. II, cells from I reinoculated into uncoupler-free medium. III, cells from I reinoculated into medium containing 10^{-5} M DNP.

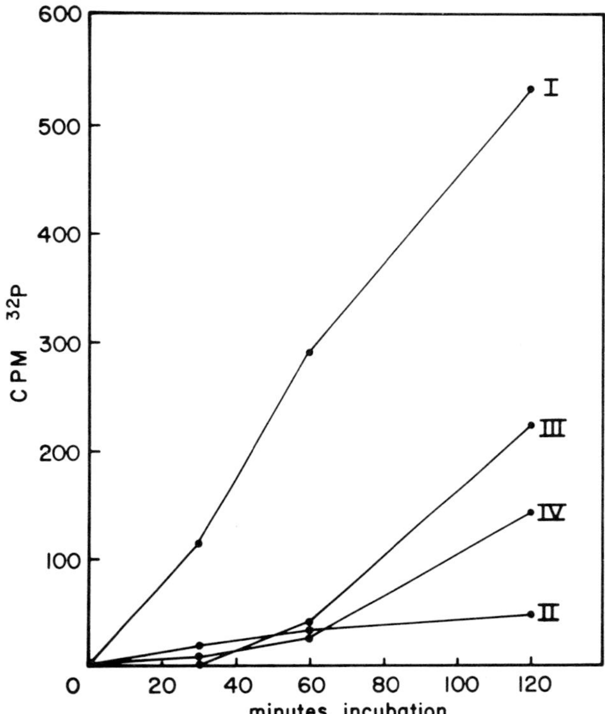

Fig. 7. Phosphate esterification by normal and DNP-adapted cells of <u>Euglena gracilis</u>. Cells grown in low phosphate medium were resuspended in normal medium, pH 3.4, in the presence of $^{32}P_i$. I, normal cells. II, normal cells + 10^{-5} M DNP. III, adapted cells. IV, adapted cells + 10^{-5} M DNP.

the addition of DNP, while the addition to normal cells totally inhibited phosphate esterification (Fig. 7). The one-hour lag in esterification by adapted cells could be due to either a larger internal pool coupled with the lower rate of utilization, or a slower exchange of intracellular inorganic phosphate with the external medium.

We checked a large array of activities of mitochondria from normal and adapted cells. No significant differences could be detected in the rate of respiration and the P/O ratios with various substrates, in the sensitivity to DNP, ClCCP, oligomycin, arsenate, valinomycin, atractyloside, azide, and cyanide, in the ATPase activity -- both DNP stimulated and oligomycin inhibited, and in the optimal pH for phosphorylation. The microsomal fraction showed some NADPH-dependent phosphorylation, which was DNP insensitive. However, there was no difference between normal and adapted cells. No significant phosphorylation could be detected in the cytoplasm with either NADH or NADPH as substrates. No acetyl phosphate formation from ethanol could be detected either.

The resistance to DNP which can be induced in euglena is clearly a physiological adaptation and does not represent a mutant selection. Not only was the adaptation readily reversible, but repeated adaptations and de-adaptations did not decrease the length of the lag period before growth commenced, as would be expected if mutant selection was involved. It represents the phenotypic expression of genetic information already present in the unadapted organism. The long lag period is very unusual (5,6), especially in light of the rapid loss of adaptation; and it requires either drastic modification of the energy coupling pathway or the formation of a completely new mitochondrial membrane (7), impermeable to most uncouplers and inhibitors of oxidative phosphorylation. The absence of any discernible differences in activities between cellular fractions of normal and adapted cells may indicate that the modified or alternative pathway is very labile and is lost during isolation, either due to denaturation or due to the fact that the modification involves some compartmentation within the cell.

The accumulation of DNP by the cells is probably a result of the pH gradient across the cell membrane and not due to a carrier mechanism. The normal pH in the cells is 6.5-6.9 at which DNP is completely ionized. Only the undissociated uncoupler penetrates and inhibits the cell, and once inside becomes ionized and thus unable to leak out. In the case of ClCCP, similar results were obtained and no inhibition observed if the medium pH was appreciably above 5.95, the pK_a of ClCCP. The same held true for the uncoupler "1799" whose pK_a is ≃ 6.1 (P. Heytler, private communication).

There exists the possibility that the changes associated with the adaptation do not involve any drastic metabolic changes but just reflect a general "tightening" of energy utilization--reducing energy use for purposes not essential under the laboratory growth conditions. We have observed a drastic decrease in motility, a much weaker cell wall (easier breakage by grinding or by centrifugation), and a greater sensitivity to extremes of pH and to anaerobiosis. The reduced growth rate and reduced phosphate esterification would then represent an accommodation to limited energy supply. It should be stressed that even adapted cells will not grow readily in higher concentrations of DNP (> 10^{-5} M), and will cease growing completely - although remaining metabolically active - at a concentration of 10^{-4}M. The same holds true for oligomycin, tributylchlorotin, and "1799". If, however, adaptation involves only such an accommodation to an "energy crisis", the question remains why it is such a traumatic event, with the bulk of the cells dying and 160 hours required for adaptation.

As stated in the beginning, the universality of the reactions involved in oxidative phosphorylation in mitochondria as usually seen makes one, as a rule, reluctant to postulate the existence of alternative pathways of ATP synthesis coupled to electron transport. However, the data presented here, and in particular the acquired insensitivity to a wide variety of chemically diverse effectors of oxidative phosphorylation, point strongly to the possibility of a drastically modified pathway for energy coupling to supply the cells' metabolic needs.

REFERENCES

1. GNANAM, A., AND KAHN, J. S. (1967) Biochim. Biophys. Acta 142, 475.
2. COOK, J. R. (1968) in The Biology of Euglena I (Buetow, D. E., ed.) p. 243, Academic Press, New York.
3. SCHIFF, J. A., LYMAN, H., AND RUSSEL, G. K. (1971) in Methods in Enzymology (San Pietro, A., ed.) Vol. 23, p. 143, Academic Press, New York.
4. KAHN, J. S. (1973) Arch. Biochem. Biophys. 159, 646.
5. SHARPLESS, T. K., AND BUTOW, R. A. (1970) J. Biol. Chem. 245, 58.
6. McCASHLAND, B. W., AND ANDERSEN, W. F. (1963) Growth 27, 47.
7. CALVAYRAC, R., VAN LENTE, F., AND BUTOW, R. A. (1971) Science 173, 252.

PART II

ARCHITECTURE OF MEMBRANES

The term "membrane architecture" is immediately misleading, in that it assumes the existence of some structural organization of membranes that is stable in time. If there is one thing we have learned in recent years, it is that membrane components are astonishingly mobile, and that, at best, we can describe only a time-averaged membrane structure. Even that description, however, omits much of what is important about membrane function. In fact, we have arrived at almost an "uncertainty principle" of membranology - we can know only what some of the molecules are doing some of the time.

A few relative certainties do exist: membranes contain mostly lipid and protein, and lipid spends much of its time in the form of a lipid bilayer. Some protein species are embedded in the bilayer, where they interact with the hydrophobic moieties of the lipid molecules, and other species are primarily associated with inner or outer surfaces of membranes. Even these generalizations have many exceptions, depending on which membrane one chooses to study, and thus emerges a second principle of membranology: do not generalize.

What, then, can we learn about membrane structure, bearing in mind the two principles above? The major questions, at least, have been specifically identified:

1. What is the physical condition of the bulk phase of membrane lipid? Is it fluid, or solid, or something in between? Are there changes in such physical properties as one alters temperature, or changes functional

state? Are different lipid molecules in different phases?

2. How do lipids interact with protein polypeptide chains? (We know that many membrane-bound enzymes are dependent on lipid, but we know little about whether this is a function of lipid hydrocarbon chain properties or head groups.)

3. Does membrane protein undergo conformational changes related to function? Do membrane proteins have rotational or translational diffusion?

4. How are proteins inserted into membranes during membrane biogenesis? (Much of the overall functioning of a given membrane must depend on a fairly precise ordering of specific proteins on membranes, but we have little insight into what mechanisms govern this important process.)

Every investigator working in the membrane area undoubtedly would add to this list of questions, and the chapters that follow, in fact, will reflect the diversity of questions among them. Fundamental understanding of membrane structure may still lie beyond our present grasp, but (or for this very reason) those of us conducting research on this topic are in common agreement that membranology is a very exciting area in which to work.

MOLECULAR ARCHITECTURE OF ENERGY TRANSDUCING MEMBRANES RELATED TO FUNCTION

Lester Packer

INTRODUCTION

One of the main functions of biomembranes is the partitioning of metabolism. This is particularly evident in the energy dependent systems for ion transport. An understanding of the molecular basis of the functions of membranes requires knowledge of the orientation of lipids and proteins in the membrane. Experiments will be reported herein using several membrane systems which provide clues to the understanding of this problem. Illustrations will be given for mitochondria and chloroplasts where very elaborate systems for energy generation exist.

Unfortunately there are only a few ways to study membranes directly. Among these are electron microscopy and x-ray defraction techniques. Most other techniques which probe membrane structure are averaging techniques. For this reason, despite the fact that the techniques of electron microscopy are still somewhat crude, we have emphasized their use in combination with other approaches to study membrane structure function relationships. The following questions will be considered.

I. What details of the architecture of membranes can be revealed by freeze fracture electron microscopy?

II. How is the orientation of protein and lipid in the membrane changed in relation to function?

III. How can changes in the structure of membranes seen by freeze fracture electron microscopy be made quantitative?

MEMBRANE ARCHITECTURE AS REVEALED BY FREEZE FRACTURE ELECTRON MICROSCOPY

An interpretation of the technique of specimen preparation for freeze fracture microscopy for a hypothetical membrane containing a lipid bilayer interspersed with protein is shown in Fig. 1. Two types of protein associations are shown, surface associated proteins, where surface-protein and protein-protein interactions occur, and proteins with a portion protruding into the hydrophobic center of the membrane,

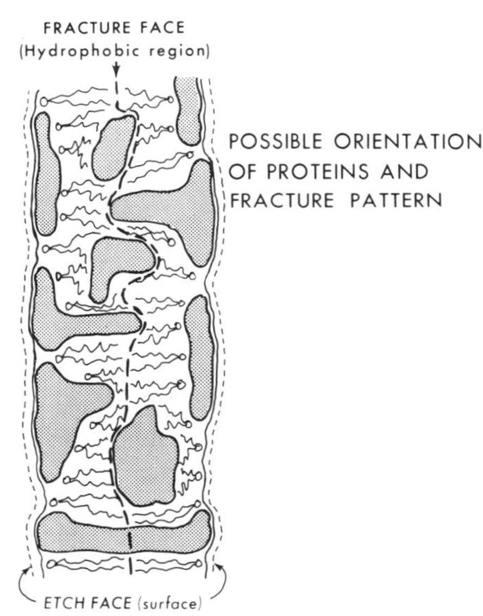

Fig. 1. Interpretation of structures seen in freeze fractured and etched membranes. The dotted line indicates that the fracture face occurs down the hydrophobic center of the membrane. In agreement with the interpretation of studies with complimentary replicas, the fracture face reveals integral components which are in the membrane interior. In freeze etching, the specimen is temporarily warmed to -100°C, causing the ice to be sublimed away under the vacuum, thus removing surface water, enabling one to detect the presence of surface components.

where lipoprotein interactions occur. In freeze fracture electron microscopy, a cleavage of the membrane occurs through the hydrophobic center to expose protein and lipids in two half-membrane faces (1). The components of the cleavage plane are replicated by a thin layer of platinum followed by a supporting thicker layer of carbon. Subsequently, the replica is dissolved away from the biological specimen and examined in a transmission electron microscope. Membrane structure can be examined with freeze fracture electron microscopy *in vivo* and *in vitro*. Figure 2 shows a typical example of a low magnification electron micrograph of a rat liver hepatocyte. The

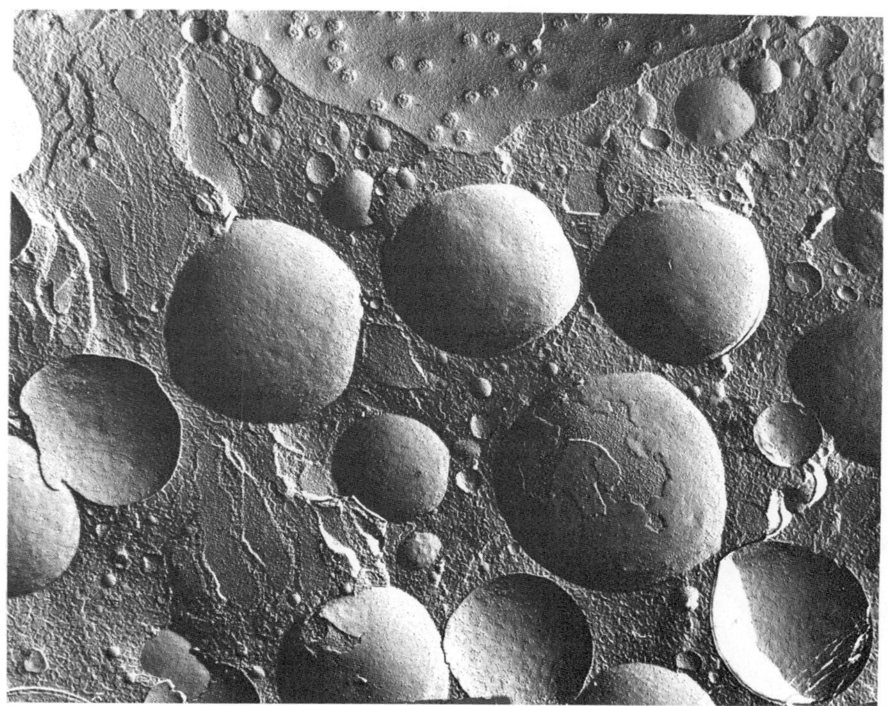

Fig. 2. Rat liver hepatocyte, low magnification freeze fracture electron micrograph.

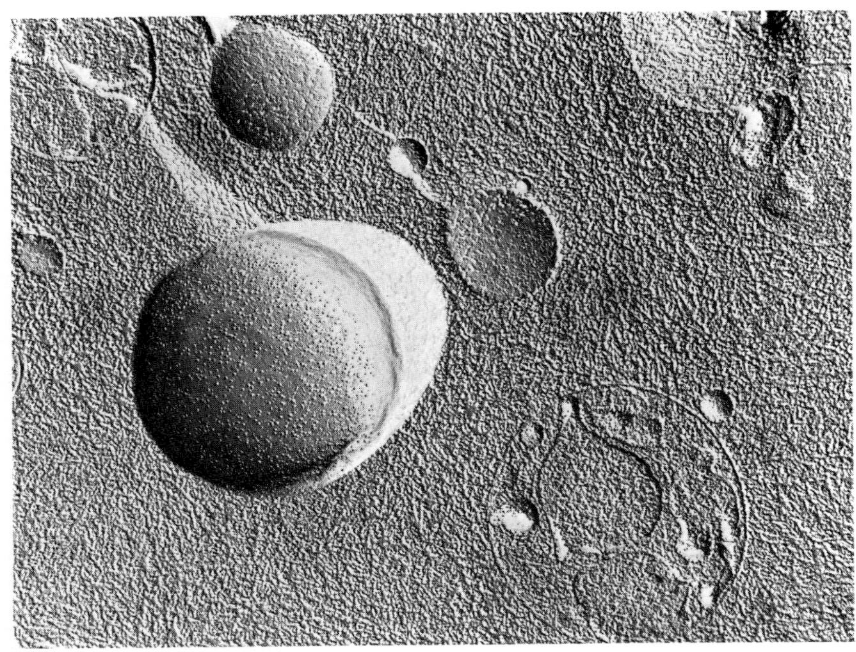

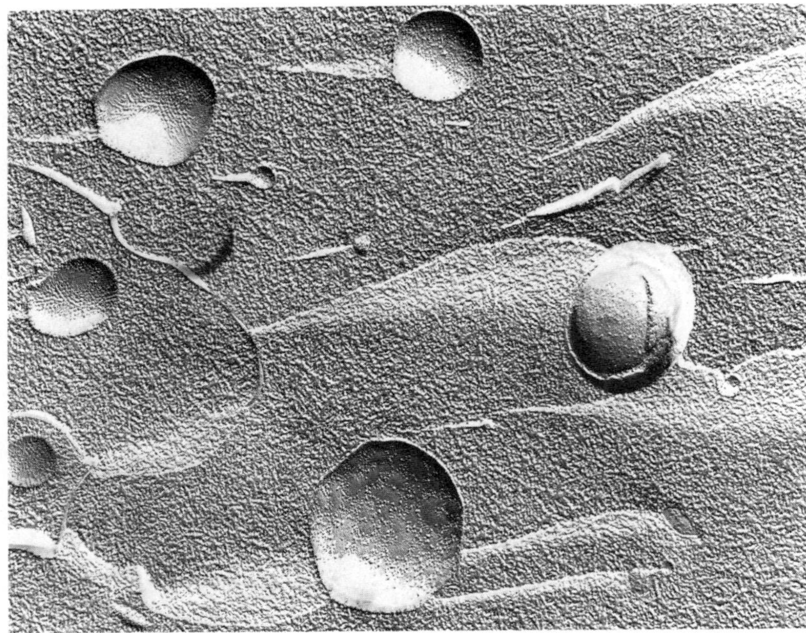

Fig. 3 (legend on opposite page) A, top; B, bottom.

fracture face of the nuclear membrane is recognized by the presence of numerous pore structures. Many cross-fractured mitochondria are seen in the vicinity of the nucleus. In some instances it is possible to recognize replicas of double membrane fracture faces identifying the organelles as mitochondria, since, apart from the nucleus, mitochondria are the only other cell organelles with double membranes. In the lower left-hand corner, a mitochondrion has been cross-fractured; this type of membrane profile shows details similar to that seen by conventional staining electron microscopy. The unique feature of the freeze fracture technique is the ability to discern in the fracture face of the membrane the disposition of particle (proteins or lipoproteins) and non-particle (lipid domains) areas in the half membrane.

Experiments with isolated mitochondria membranes were undertaken to compare the organization of concave and convex fracture faces seen *in vivo* and *in vitro*. In examining inner and outer membranes, results were obtained that were substantially in agreement on the organization of concave and convex fracture faces seen *in vivo* (2,3). Furthermore, a similar distribution of membrane particles (i.e. proteins and lipids) were seen when rat liver, heart and wild type and petit yeast mitochondria (4) were compared. Figure 3 A-B shows typical examples of the appearance of mitochondria isolated from yeast in cross-fracture and fracture face; the distribution of particles is similar in concave and convex fracture faces to that observed *in vivo*. From studies with isolated organelles, inner and outer membranes (5), and *in vivo* studies, an interpretation of the mitochondrial structure can be deduced,

---

Fig. 3. Freeze fracture electron micrographs of mitochondria isolated from *Saccharomyces cerevisiae*.
A. Mitochondria isolated from wild type cells showing concave fracture faces in cross section. 72°K.
B. Mitochondria isolated from petit mutant mitochondria showing concave and convex fracture faces and cross sections. 72°K.

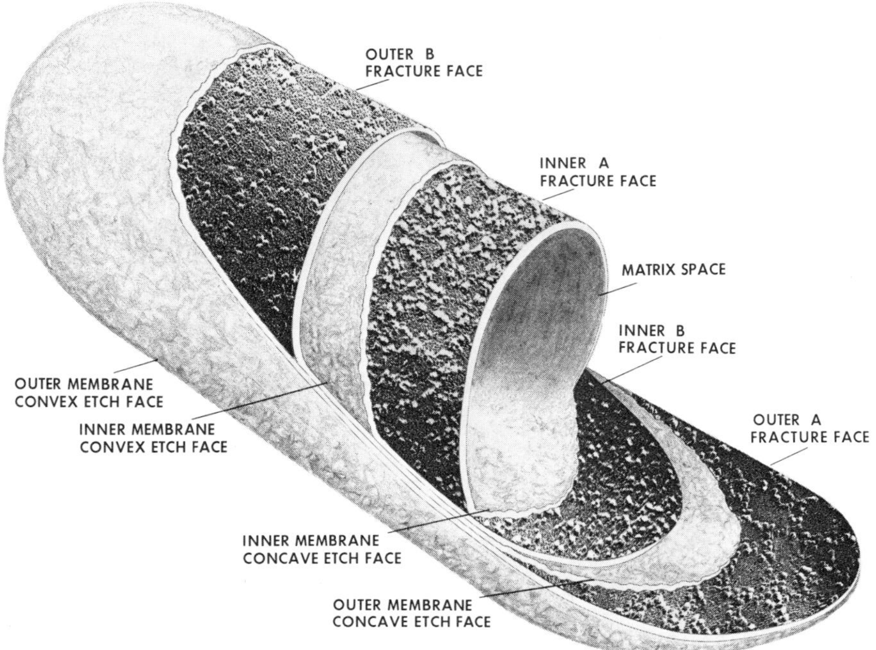

Fig. 4. Interpretation of the organization of mitochondrial membrane as revealed by examination of concave and convex fracture faces _in vivo_ and _in vitro_.

as illustrated in Fig. 4 (3). Both the outer and inner mitochondrial membranes have asymmetric particle distributions. Most of the particles are on the sides of the "half-membrane" facing the matrix, either the cytoplastic matrix for the outer membrane, or the inner mitochondrial matrix for the inner membrane. There are approximately 6,000 particles/μ^2 of inner membrane and about half that many particles in the outer membrane. The sides of the two membranes which face the intermembrane space contain the fewest number of particles. This suggests that strong hydrophobic attraction exists between the two membranes. This finding is consistent with the known close apposition _in vivo_ of outer and inner mitochondrial membranes. Indeed, time lapse studies _in vivo_ show that these two membranes move in unison and closely adhere to one-another. Only in certain pathological or physiological states does the separation between the inner and outer mitochondrial become evident in vivo (6). Hence, knowledge of the disposition of protein and lipid components in the membrane permits predictions to be made concerning the forces that operate to determine the morphology of mitochondria.

HOW DOES THE STRUCTURE OF THESE MEMBRANES CHANGE DURING FUNCTION?

To answer this question, we have studied the energy (substrate plus oxygen) or light dependent movement of water and ions by mitochondria and chloroplasts.

Under energized conditions the inner membranes of mitochondria can fold as the inner compartment expands (7). We have observed dispersal of particles in both the inner and outer mitochondrial membranes. In particular the A face, which contains most of its particles in aggregated clusters, shows a 10-15% increase in distance between particle clusters coinciding with the expansion of the inner membrane during the swelling of mitochondria when ions are accumulated. The most striking changes, however, are seen in the outer membrane where the cluster-like arrangement of particles characteristic of the outer membrane A face undergos increases in diameter and increasing distance between the particle network. Since these changes in molecular orientation are more marked in the outer than in the inner membrane, it suggests that expansion of the inner compartment and unfolding of the inner membrane has applied pressure upon the outer membrane. This tension shows up as a distortion in the lateral dispersal of protein within the membrane. These results also suggest that the outer membranes may break because of inner membrane expansion. Indeed, these results are consistent with the findings in vitro of high permeability of the outer mitochondrial membrane to relatively large molecules, e.g., as cytochrome C. This is somewhat of a puzzle because the high lipid protein ratio of the outer membrane leads one to predict that this membrane should be more impermeable than the inner membrane. A similar result could come from fragmentation of outer membranes during expansion of the inner membrane arising during the isolation procedure, or if the cell contains a few large mitochondria which fragment during isolation. These considerations are pertinent to the questions concerning the number of mitochondria in a cell and mixing of gene pools which has been raised in some of the chapters in this book concerned with the mitochondrial biogenesis.

HOW IS THE ROLE OF LIPID AND PROTEIN ORIENTATION IN MEMBRANE STRUCTURE RELATED TO ENERGY COUPLING?

To demonstrate the involvement of membrane lipid and protein organization to energy coupling we have pursued several approaches.

Lipid Depletion Studies

In these studies the progressive removal of lipid from inner mitochondrial membranes has been investigated together with Dr. C. Mehard of this laboratory and Dr. W. Zahler in S. Fleischer's laboratory (8). Phospholipase A_2 treatment is used to release fatty acids and the released fatty acids are removed by binding to bovine serum albumin according to a method developed in Fleischer's laboratory (9). Figures 5 A, B and 6 demonstrate the effects of such treatment in an actual experiment and in diagramatic fashion on freeze fracture replicas of mitochondrial inner membranes. The step-wise removal of lipid at first causes the smooth areas that surround the particles to become rough. Eventually, when perhaps half of the lipid is removed, the particles scatter. This dispersal of particles continues, until only an amorphous appearance of particles remains. These studies convincingly demonstrate that the particles seen in freeze fracture replicas are proteins or lipoproteins and that the smooth areas that surround them are the bulk lipid domains. Thus, it can be considered that the nature and amount of bulk lipid and tightly bound "boundary"

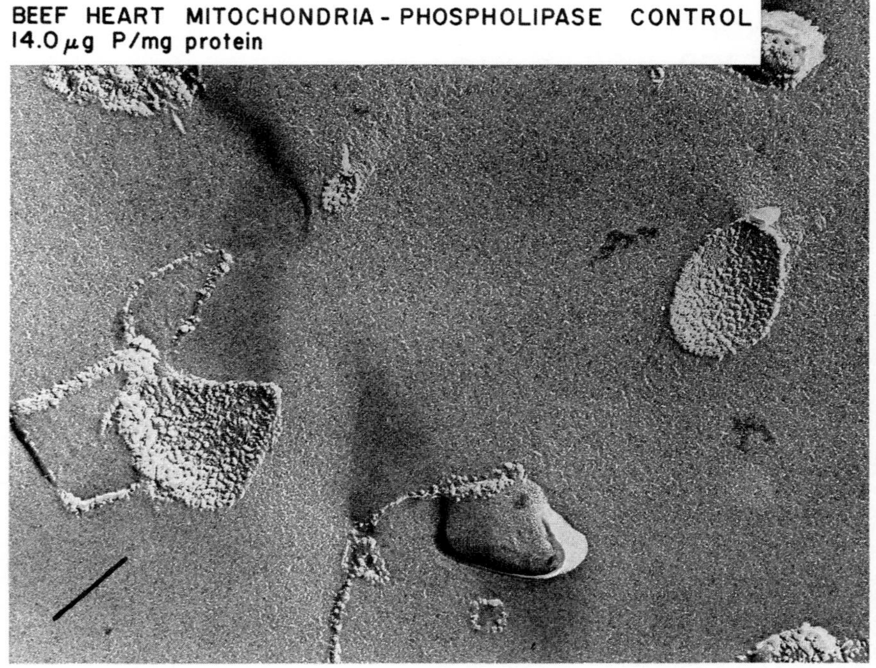

Fig. 5. Effect of lipid depletion by phospholipase treatment on structure of inner mitochondrial membranes seen by freeze fracture electron microscopy. A (above), control; B (following page), treated with phospholipase.

lipid (10) have a role in the lateral and perpendicular positions of proteins in the membrane.

Dietary Manipulation of Membrane Lipid

These studies have been carried out with rat liver mitochondria in collaboration with Dr. M.A. William's laboratory, isolated from animals grown on diets deficient or supplemented with unsaturated fatty acids (11). Freeze fracture studies on mitochondrial membranes from animals supplemented with unsaturated fatty acids show that the particles in the two half membranes that arise from fracturing are more asymmetrical than in the membranes containing lower amounts

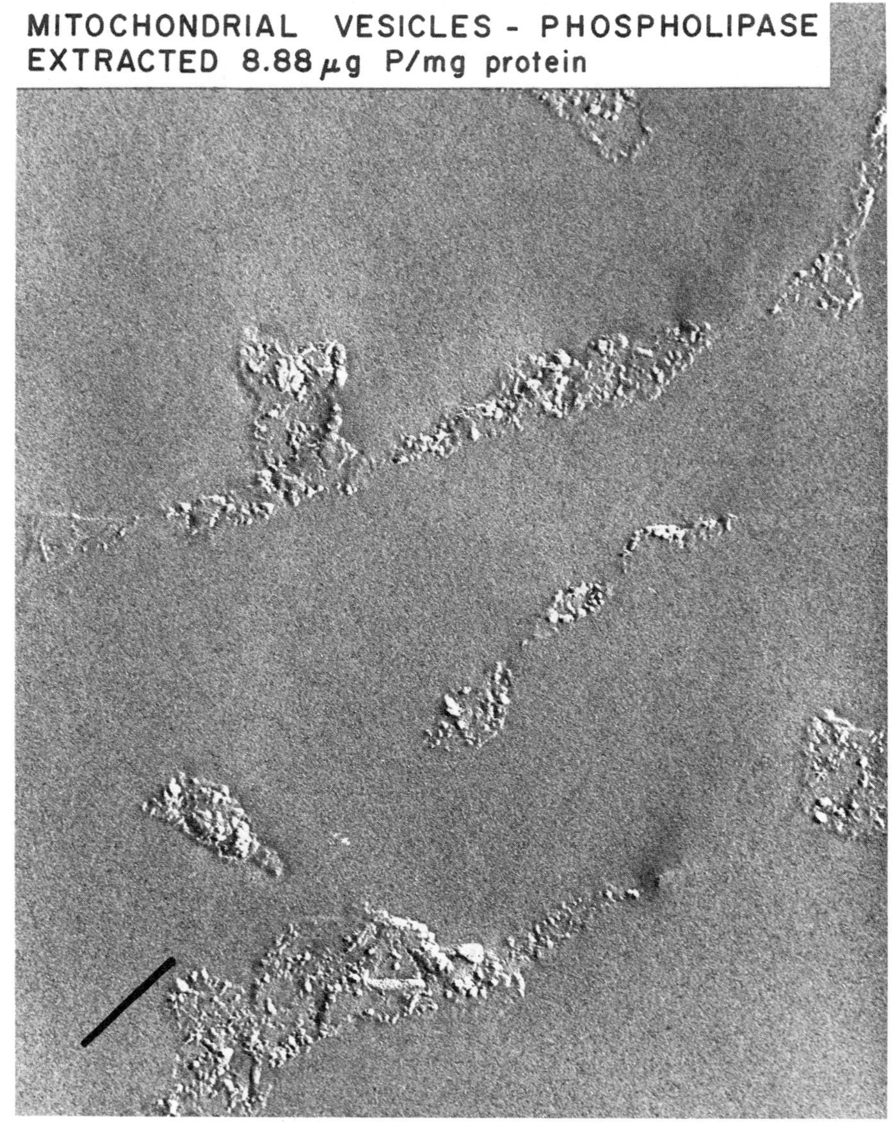

Fig. 5B (legend on previous page).

ARCHITECTURE OF ENERGY TRANSDUCING MEMBRANES

MITOCHONDRIAL INNER MEMBRANE VESICLES

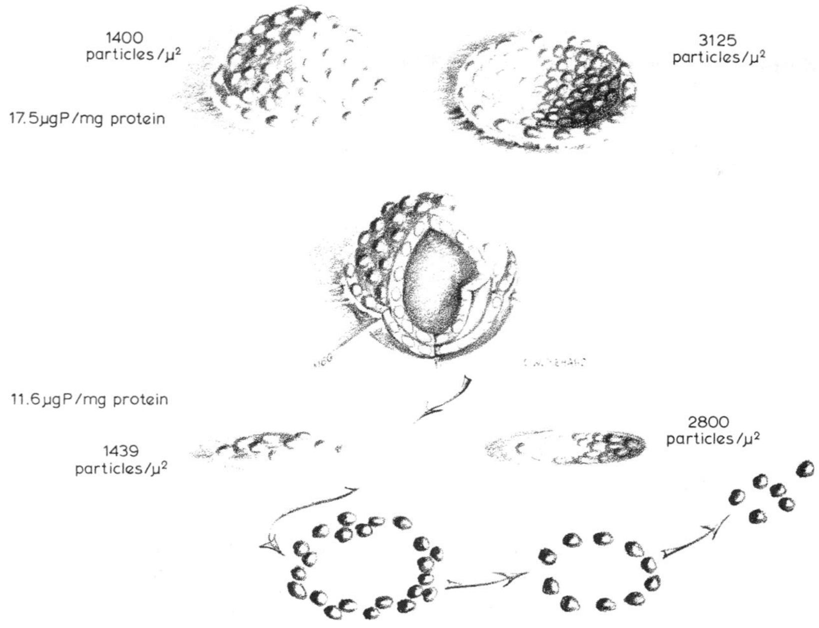

Fig. 6. Diagrammatic illustration of the effects of lipid depletion on the asymmetric distribution of particles in inner mitochondrial membrane vesicles (after C. Mehard). The diagram shows in step-wise depletion of lipids from 17.5 mg of phospholipid phosphorous/mg protein to 11.6 mg results in elimination of the smooth areas surrounding the particles. Further depletion of the lipid results in dispersal of the particles.

of unsaturated fatty acids (3, 12). These results are consistent with the studies in Mycoplasma membranes where changes in the fatty acid composition of cell membrane lipids have revealed similar results (13). Perhaps the increased fluidity conferred by unsaturated fatty acids accounts for this asymmetric splitting or partitioning of the membrane proteins. These results are relevant to mechanisms of the assembly and biogenesis of membranes.

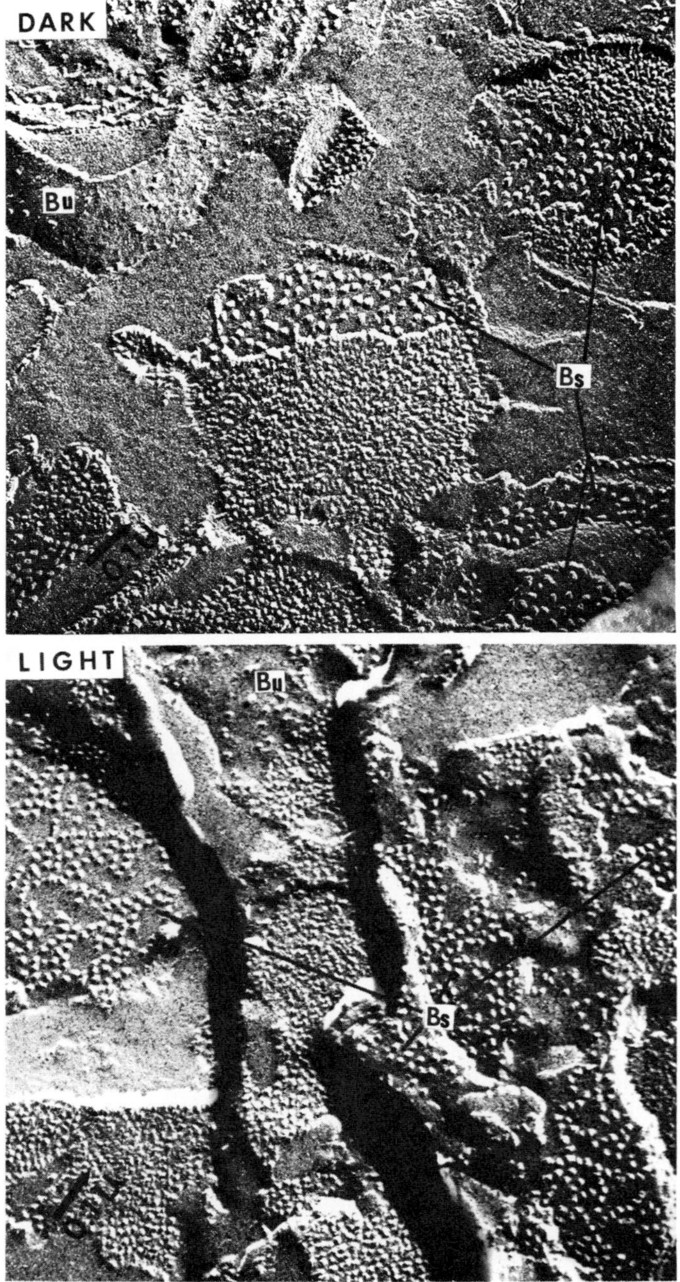

Fig. 7 (legend on opposite page).

In companion studies (11) it has been found that higher unsaturated fatty acid levels permit mitochondria to undergo more rapid oscillation of energy dependent ion transport and associated expansion and contraction of the mitochondrial inner membrane. Thus our studies indicate a greater fluidity and/or flexibility of the membrane and increased efficiency of energy coupling in membranes supplemented with unsaturated fatty acids.

Correlation of Results of Freeze Fracture Microscopy With Changes in Partitioning of Hydrocarbon Spin Labels - Electron Spin Resonance Studies

The results of freeze fracture electron microscopy showing changes in dispersal and clustering of proteins with concomitant changes in size of the lipid domains, can be confirmed and extended by examining the partitioning of hydrocarbon spin label molecules in the membranes hydrophobic core during changes in metabolic state.

An example of how the combined use of freeze fracture electron microscopy spin labelling techniques, with functional studies, can be successfully used as a strategy to elucidate the role of lipids in energy coupling can be shown in studies with isolated chloroplasts (14) or mitochondria (15). Isolated spinach chloroplast thylakoid membranes have been shown by Wang and Packer and Torres-Pereira *et al*. (14) to develop marked changes in the orientation of particles seen in the large particle containing fracture face upon illumination of chloroplasts in a medium containing weak acid anions and an electron cofactor. Figure 7 shows a typical experiment where the clustering of

---

Fig. 7. Freeze-fracture faces of dark and illuminated type II spinach chloroplasts. Chloroplasts were suspended at 0.1 m sodium acetate, 50 mg phenozine methosulfate, at pH 6.8. Samples were taken from freeze-fracture electron microscopy, in the dark and after 5 minutes illumination. Note changes in the distribution of particles in the large particle containing fracture face (B_S) which is from a region in the streaked grana areas (14).

the large particles occurs upon illumination of the samples. An interpretation of the nature of these changes is that lateral and perpendicular movement of membrane particles obtains in response to illumination (14-17).

However, a difficulty of the freeze fracture technique is the inability to quantitate the changes observed. In order to circumvent this, we have recently undertaken, in collaboration with Drs. R. Melhorn, A. Elsaeter, and D. Branton (18) to develop a powerful and unique computer based technique to analyze quantitatively the distribution of particles in freeze-fracture replicas. This technique involves the use of scanning machines first developed at the Lawrence Berkeley Laboratory for the analysis of photographic images of the tracks made in bubble chambers, in association with high energy physics experimentation. We have modified the film holders of this apparatus to accommodate electron microscope plates. The plates are displayed at a ten-fold magnification on a table. A pointer is then moved to locate the position of particles; the depression of a footswitch enters information on the location of the particle in time and space in the computer. Using this approach representation of the distribution of particles in dark and illuminated chloroplast membranes at a magnification of 800,000 diameters can be presented, as shown respectively in Figs. 8-9. Using this method for collection and representation of data we are currently using the Monte Carlo techniques to determine whether the particle distributions observed are random or non-random.

In a typical experiment, the computer is asked to examine a series of areas progressively increasing in size, to determine whether or not particles are present or absent, and in this way, to define a smooth area parameter. A surprising result which has emerged from analysis of many micrographs is that changes in the smooth area parameter are much more striking than the clustering of the proteins. The latter often appears to occur in non-random patterns. Hence, it is likely that the lipid domains may be more significant than

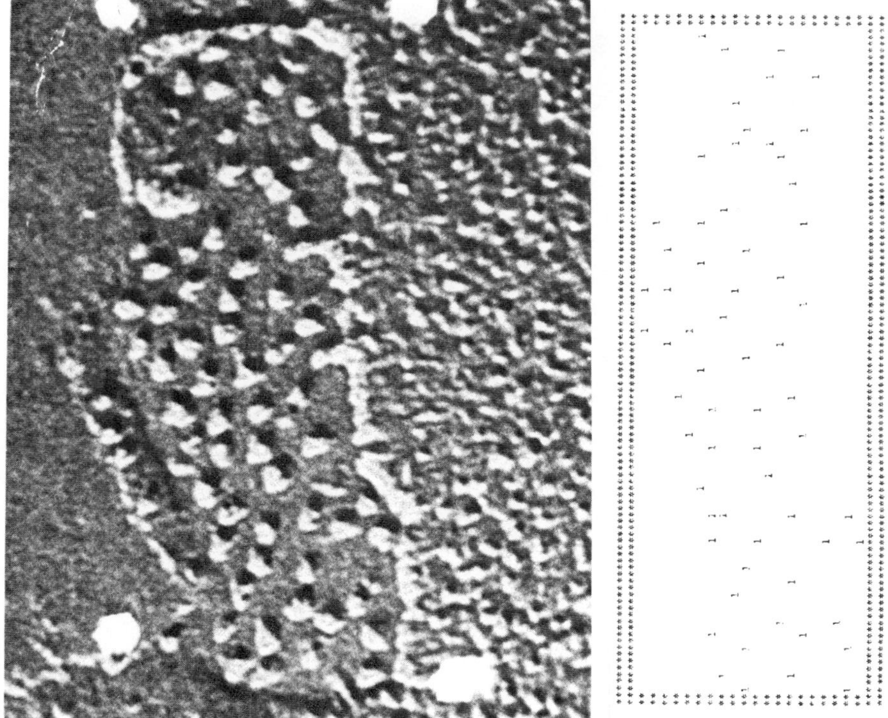

Fig. 8. Computer simulation of freeze fracture pattern seen in the large particle containing fracture face in dark incubated spinach chloroplasts.

previously considered in terms of understanding how lipid and protein organization is involved in the coupling of energy between the photosystems.

Using the hydrocarbon spin label 6N11 as a probe, Torres-Pereira _et al_. (15) found an increase in the partitioning of this spin label into hydrophobic domains in chloroplasts illuminated under the conditions which reveal changes in particle clustering and lipid

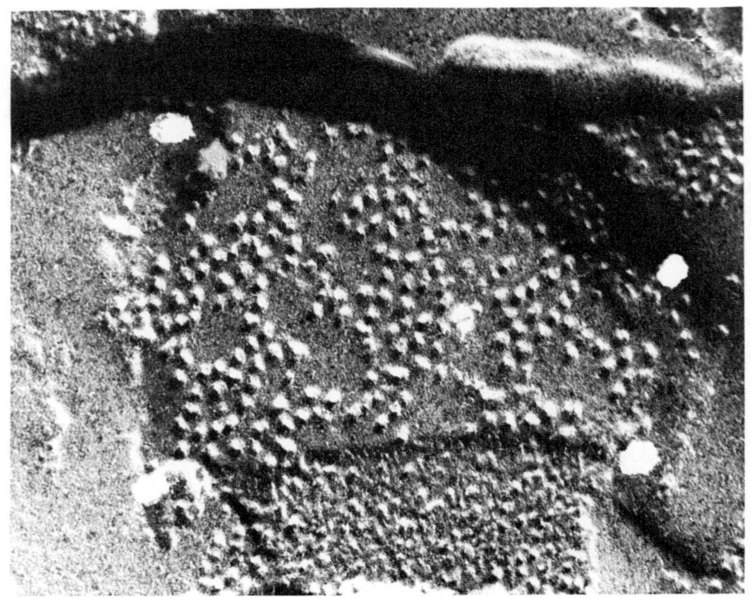

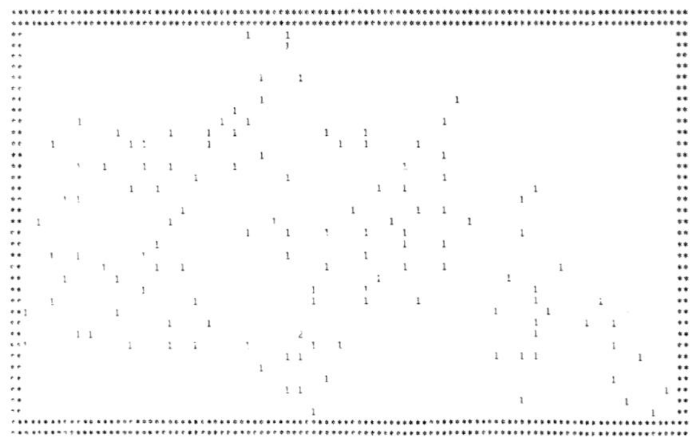

Fig. 9. Computer simulation of freeze-fracture pattern of large particles containing fracture face seen in illuminated chloroplasts. Note clustering of particles and increase in the size of the lipid domains as compared with dark control shown in Fig. 8.

domains shown by freeze fracture techniques. Typical results are shown in Fig. 10 and Table I, where the charges in ESR Spectra between dark and illuminated chloroplasts, and the calculated charges in partitioning are given. These results reinforce the interpretation that striking changes in bulk lipid domains accompany illumination of chloroplast thylakoid membranes.

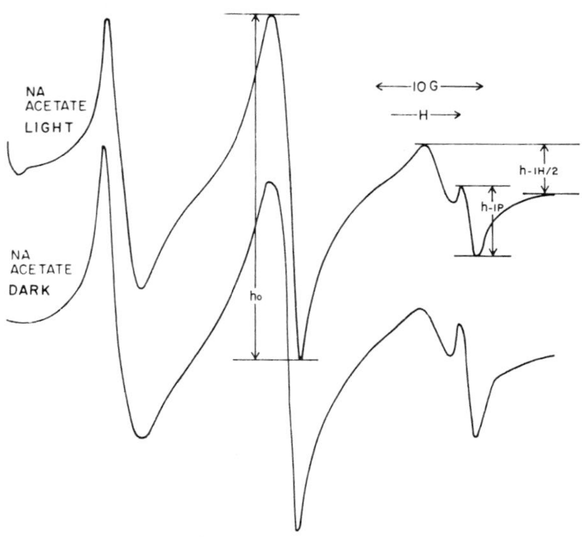

Fig. 10. Electron spin resonance spectra 3×10^{-5} M 6N11 spin label dispersed in chloroplasts fixed with glutaraldehyde while suspended in sodium acetate medium. Upper and lower spectra resulted from fixation carried out under illumination and in the dark, respectively. In this case and in Table I, the spin label was introduced after fixation, but similar results are obtained when it is introduced before fixation.

TABLE I

LIGHT-INDUCED CHANGES IN CHLOROPLAST VOLUME AND IN THE PARTITIONING OF 6N11 IN CHLOROPLAST MEMBRANES

Chloroplast volume			Packed volume (%)			
Sample	Chlorophyll (μg/ml)	6N11 (μM)	NaAc dark	NaAc light	NaCl dark	NaCl light
1	700	15	3.5	1.5	3.7	4.8
2	1600	30	11.5	3.5	13.5	17.0
3	1800	30	12.0	3.2	---	---

6N11 partitioning			h-1 HC/h-1P			
Sample	Chlorophyll (μg/ml)	6N11 (μM)	NaAc dark	NaAc light	NaCl dark	NaCl light
1	700	15	0.480 ± 0.010	0.540 ± 0.020	0.525 ± 0.045	0.420 ± 0.030
2	1600	30	0.430 ± 0.020	0.485 ± 0.015	0.565 ± 0.005	0.485 ± 0.025
3	1800	30	0.515 ± 0.035	0.575 ± 0.035	---	---

CONCLUSIONS

A brief survey of the history of modern molecular biology has demonstrated that knowledge of the structure of proteins and nucleic acids has led to functional predictions that could not have been made without precise knowledge of these structures. Precise details of the structure of biomembranes is only now being gained. Such knowledge is essential to predicing function and to comprehending the role of membranes in energetic processes of ion transport and in asymmetric partitioning of metabolism, and in understanding how membranes are formed and function in differentiation.

The few examples given in this chapter indicate that no single technique can be used to answer questions of the importance of the molecular orientation of lipid and protein components in the membrane to function. Among the direct techniques, the electron microscope, while still crude, is still the best technique. In particular, freeze-fracture electron microscopy, especially in conjunction with our new computer-based technique for quantitation, offers perhaps at the present time the best method currently available for scrutinizing the internal structure of membranes. Use of this approach is particularly enhanced when carried out in combination with functional studies and the use of probes such as spin labels which sense the changes in other environments within the membrane. It is apparent that only by a combination of a number of different techniques can a more comprehensive understanding be generated concerning the dynamic nature of the orientation of membrane lipids and proteins, and the importance of the mobility of lipids and proteins in membranes to function.

REFERENCES

1. BRANTON, D., AND DEAMER, D. W. (1972) Protoplasmalogia--Membrane Structure, Springer Verlag, Wien-New York.
2. WRIGGLESWORTH, J. M., PACKER, L., AND BRANTON, D. (1970) Biochim. Biophys. Acta 205, 125-135.

3. PACKER, L. (1972) J. Bioenerg. 3,, 115.
4. PACKER, L., WILLIAMS, M. A., AND CRIDDLE, R. S. (1973) Biochim. Biophys. Acta 205, 125.
5. MELNICK, R. L., AND PACKER, L. (1971) Biochim. Biophys. Acta 253, 503.
6. HACKENBROCK, C. R. (1971) J. Cell Bio. 51, 123.
7. PACKER, L., UTSUMI, K., AND MUSTAFA, M. G. (1966) Arch. Biochem. Biophys. 177, 381.
8. PACKER, L., AND MEHARD, C. W. (1974) Biochim. Biophys. Acta, in press.
9. FLEISCHER, S., AND FLEISCHER, B. (1967) in Oxidation in Phosphorylation, a volume of Methods in Enzymology (Estabrook, R. W., and Pullman, M. E., eds.) 10, 406-433, Academic Press.
10. JOST, P., GRIFFITH, O. H., CAPALDI, R. A., AND VANDERKOOI, G. Biochem. Biophys. Acta 73652.
11. STANCLIFF, R. C., WILLIAMS, M. A., UTSUMI, K., AND PACKER, L. (1969) Arch. Biochem. Biophys. 131, 629.
12. WILLIAMS, M. A., STANCLIFF, R. C., PACKER, L., AND KEITH, A. D. (1972) Biochim. Biophys. Acta 267, 444.
13. BRANTON, D., AND JAMES, R. (1971) Biochim. Biophys. Acta 233, 504.
14. TORRES-PEREIRA, J., MEHLHORN, R., KEITH, A. D., AND PACKER, L. (1974) Arch. Biochim. Biophys. 160, 90.
15. TINBERG, H. M., PACKER, L., AND KEITH, A. D. (1972) Biochim. Biophys. Acta 283, 193.
16. WANG, A. Y.-I., AND PACKER, L. (1973) Biochim. Biophys. Acta 305, 488.
17. OJAKIAN, G. K., AND SATIR, P., Proceedings of the National Academy of Science, in press.
18. MEHLHORN, R., PACKER, L., ELSAETER, A., AND BRANTON, D., in preparation.

FREEZE-ETCH IMAGES OF SARCOPLASMIC RETICULUM AND SUB-MITOCHONDRIAL MEMBRANES

D. W. Deamer and N. Yamanaka

Freeze-etch microscopy has provided important new information about membrane structure. (For reviews, see 1, 2, 3). In this technique, biological material is frozen at liquid nitrogen temperatures, then fractured under vacuum. The fracture surface is shadowed and replicated with platinum-carbon, and the replica is examined by transmission electron microscopy. Since the fracture plane often passes along interior hydrophobic regions of membranes, in a sense "splitting" them, the freeze-etch technique permits visualization of interior membrane architecture.

When various membranous systems are viewed by freeze-etching, the most ubiquitous features are particulate structures apparently embedded within the plane of the membrane. In a few instances, the particles have been correlated with specific proteins or enzymes. For instance, Deamer and Baskin (4) suggested that the particles found on sarcoplasmic reticulum represented the calcium transport ATPase of this membrane. Several recent studies now support this suggestion, and these will be discussed later. Hubbel and co-workers (5, 6) found that rhodopsin-lipid systems were highly particulate when viewed by freeze-etching. The particles were best explained as rhodopsin molecules embedded in lipid bilayers, and it was proposed that the particulate appearance of rod-outer segments may also represent rhodopsin-lipid complexes in the membrane. Marchesi (2, 7) and co-workers have suggested that the freeze-etch particles of erythrocyte membranes are produced by the major glycoprotein which seems to extend through the membrane. Goodenough and Stoekenius (8) found that the particles associated with gap junctions in mouse liver cells contained a polypeptide with a subunit of 20,000 daltons.

Although the results cited above provide strong indirect evidence that the freeze-etch particles arise from proteins embedded in the lipid matrix of membranes, there is still little understanding of how the protein and lipid are organized at the molecular level, and the manner in which this organization may produce a particulate freeze-etch image. Some insight into these questions may be gained by disturbing the protein moiety of membrane in various manners, then observing the freeze-etch image. For instance, Engstrom (9) digested erythrocyte ghost membranes with pronase, a proteolytic enzyme, and found that the freeze-etch particles disappeared. This result was among the first to suggest that the particles could be correlated with membrane proteins. In other work, Pinto da Silva (10) found that the erythrocyte ghost particles rapidly aggregated under some conditions. Thus the particles appear to be embedded in a fairly fluid lipid layer and have translational motion.

In the present study, we have compared several parameters of two different membrane species undergoing proteolysis. Sarcoplasmic reticulum was chosen as one of the membranes, since there is good evidence that the particle is the ATPase enzyme associated with calcium transport. This made it possible to correlate alterations of ATPase activity and protein composition with changes in the freeze-etch image. The second system chosen was sub-mitochondrial membranes. This is a far more complex membrane than sarcoplasmic reticulum, but much data on the enzyme composition and organization of the membrane has been provided by previous biochemical studies. We therefore hoped that controlled digestion of the membrane by a proteolytic enzyme, correlated with biochemical and freeze-etch studies, might provide information about which of the enzymatic proteins contributed to the particulate freeze-etch images obtained for this membrane.

METHODS

Sarcoplasmic reticulum was isolated in microsomal form from lobster abdominal muscle (11) and rabbit thigh muscle (12) by published procedures. Rat liver mitochondria were prepared by the method of Stancliff, *et al*. (13) and sub-mitochondrial membranes by the method of Gregg (14) using a 15-min. sonication and a 60-min. centrifugation of 100 Kg to form the final pellet.

Mitochondrial membrane protein was digested by suspending 30 mg of protein in 2.0 ml of 0.25 M sucrose, 25 mM Tris-Cl (pH 7.4), 5 mM EDTA and 15 mg of Nagarse enzyme. (Nagarse is a crystalline bacterial alkaline protease, produced by Nagase, Ltd., Osaka, Japan, and was a kind gift of Drs. Y. Takahashi and I. Nishigaki). After 30-min. digestion at 25°C, the samples were centrifuged for 90 min. at 100 Kg through a sub-phase of 0.5 M sucrose, 10 mM Tris-Cl, pH 7.4, in order to remove Nagarse. The pellets were rinsed once with the original sucrose solution and used for enzyme measurements, gel electrophoresis or freeze-etch preparations as described below. In controls, Nagarse was absent. Sarcoplasmic reticulum (20 mgs. protein) was digested in 4 ml of 10mM Tris-Cl (pH 7.4) with 10 mg Nagarse, then centrifuged as described above.

ATPase activity was measured by the pH change which occurs during hydrolysis (15) in media containing 5.0 ml 0.2 M sucrose, 20 mM KCl, 3 mM $MgCl_2$, 1 mM Tris-Cl, 5 mM ATP. The calcium-dependent ATPase of sarcoplasmic reticulum was measured in the same manner, but in media containing 0.1 mM $CaCl_2$, 5 mM $MgCl_2$, 1 mM Tricine, and 5 mM ATP. In both instances the pH of measurement was 7.5 $\pm$ 0.1. Succinic oxidase activity was measured with a Clarke oxygen electrode in a reaction medium containing 5 ml of 0.1 M phosphate buffer, pH 7.8, 2.8 µM cytochrome c, 38 mM succinate and 0.4 mg protein per ml. Protein was measured by the Lowry method (16).

Gel electrophoresis in polyacrylamide gels containing sodium dodecyl sulfate was carried out by the

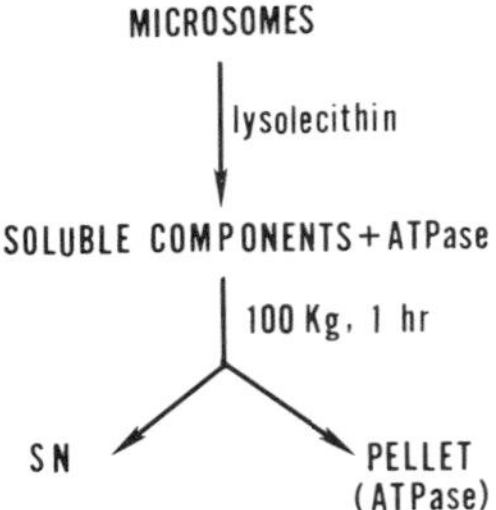

Fig. 1. Purification of ATPase from lobster muscle microsomes. Solid lysolecithin (1-2 mgs per mg protein) was added to 5 ml of microsome suspension (5 mg protein per ml). After 30 min incubation at 30° the solution was centrifuged. The resulting pellet contained much of the ATPase activity, with the specific activity increased two-fold.

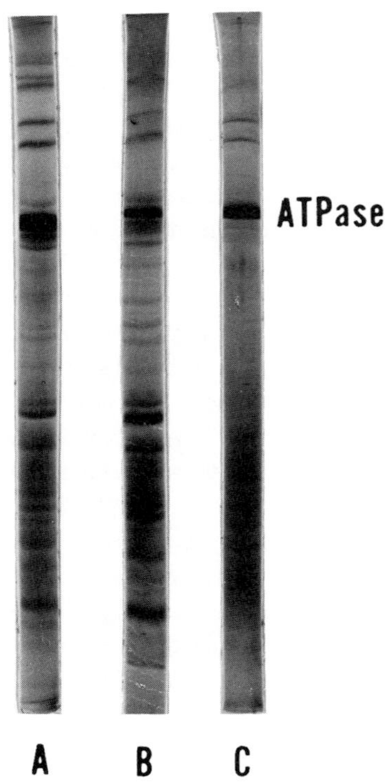

Fig. 2. (Legend on opposite page)

method of Melnick et al. (17) with 50 μg protein on each gel. For freeze-etching, pellets of membrane were mixed with enough glycerol to produce a final concentration of 50% by volume and equilibrated for 10 min. at 0°C. Freeze-etching was carried out by published procedures (18) with an etch time of 20 seconds.

RESULTS

Sarcoplasmic reticulum

The earlier suggestion (4) that the freeze-etch particles of sarcoplasmic reticulum represent a calcium transport ATPase has since been strengthened by several studies. MacLennan and co-workers (19, 20) isolated the ATPase of rabbit muscle microsomes and showed that freeze-etch preparations of the isolated material contained 80 A particles. Deamer (11) more recently devised an isolation method which permitted freeze-etching to be carried out at each step of the procedure. These results are summarized in Figs. 1-3. Figure 1 shows the steps of the relatively simple isolation method, which depends on addition of lysolecithin to the microsomes. The lysolecithin apparently disrupts membrane structure, releasing much of the original lipid and loosely bound proteins. However, the ATPase maintains its hydrophobic associations and will form a centrifugal pellet at 100,000 g. The specific activity of the pellet ATPase is approximately twice that of the microsomes, which would be expected since it is now generally accepted that the ATPase represents 50-70 percent of the microsomal protein (21). Figure 2 shows gel electrophoresis

---

Fig. 2. Band patterns of microsomal proteins on sodium dodecyl sulfate-polyacrylamide gels. Electrophoresis was carried out as described in the text. A, Total microsomal proteins. B, Supernatant from the preparation described in Fig. 1. C, Pellet from the preparation described in Fig. 1.

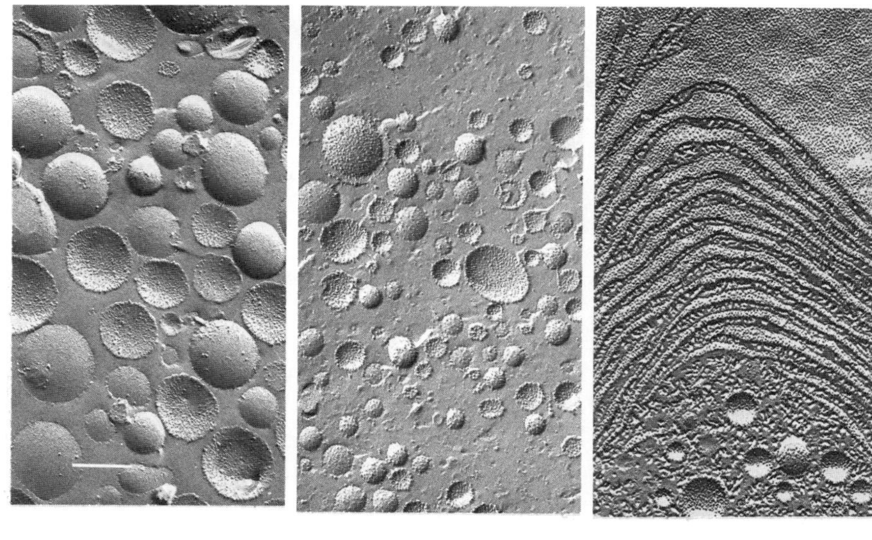

Fig. 3. Freeze-etch micrographs of lobster muscle microsomes. A, Original microsomes. Concave fracture faces are highly particulate, and convex faces are smooth. B, Lysolecithin-ATPase complex pellet described in Fig. 1, vesicular form. C, A lamellar form of the ATPase results if the vesicular form is partially dried. A few vesicles are still apparent. Bar equals 0.2 μm.

patterns of the original microsomes and purified ATPase. It is clear that the major band of the microsomal protein is identical to the band formed by the ATPase.

Figure 3A shows the original microsomes. As described in previous studies (4) the concave surfaces are highly particulate, while convex surfaces are smooth. Figure 3B and 3C show vesicular and lamellar forms of the isolated ATPase. The ATPase preparations are totally particulate, and both fracture faces contain equal numbers of particles. The particles are indistinguishable from those on the original microsomes, and we may conclude that the microsomal particles are correlated with the calcium-transport ATPase activity of sarcoplasmic reticulum.

Other proteins of sarcoplasmic reticulum

Although the 105,000 dalton band in the gels has been identified as the ATPase of sarcoplasmic reticulum, other bands have been assigned only tentative functions. MacLennan and Wong (22) isolated a 44,000 dalton protein from rabbit muscle microsomes which was termed "calsequestrin" because of its high affinity for calcium. It was proposed that this protein was involved in calcium uptake by sarcoplasmic reticulum, perhaps functioning as a calcium binding site on the inner surface of the vesicles. However, in lobster abdominal muscle microsomes, which have higher rates and total calcium binding capacity, this specific protein is not present in the gel patterns. Figure 4A compares gel patterns from rabbit and lobster muscle microsomes. The ATPase band in both systems is remarkably similar in its apparent molecular weight, but the lobster muscle microsomes have no significant band resembling calsequestrin in molecular weight. We are not able to exclude the possibility that some of the other proteins may have a calcium-binding function.

The high molecular weight bands seen in Figure 4A are always present in both lobster and rabbit muscle microsomes. It occurred to us that some of the bands may result from myosin which is trapped or adheres to the microsomes during the initial homogenization and later isolation procedures. Therefore we compared gels of lobster and rabbit myosin with the microsomal proteins (Figure 4B, C). There is no clear corresponance between the microsomal and myosin bands, although one of the bands in each instance may correspond to that of myosin subunits. The high molecular weight bands appear to be an integral part of the membrane structure, perhaps analogous to spectrin-like proteins which play a structural role in erythrocyte membranes (3).

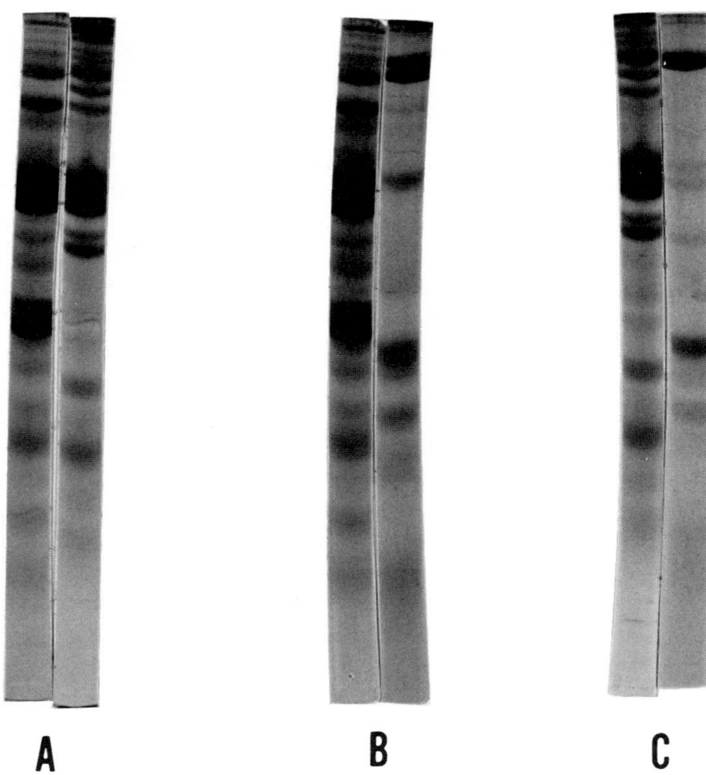

Fig. 4. Comparison of lobster and rabbit muscle microsome proteins by gel electrophoresis. A, The left gel shows the band pattern from rabbit microsomes, and the right gel shows lobster microsome patterns. The major ATPase band is very similar in both preparations, but the lobster microsomes lack the M 55 and calsequestrin bands described previously (22, 23). B, Rabbit microsome pattern compared with rabbit myosin. C, Lobster microsome pattern compared with lobster myosin. See text for details.

Effect of proteolysis on sarcoplasmic reticulum and sub-mitochondrial membranes

The effect of Nagarse treatment on membranes of sarcoplasmic reticulum is shown in Figure 5. Half the

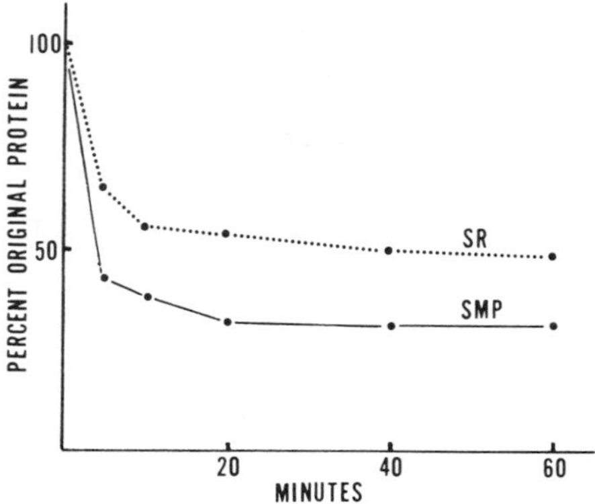

Fig. 5. Proteolytic digestion of sub-mitochondrial membranes (SMP) and lobster sarcoplasmic reticulum (SR) by Nagarse enzyme.

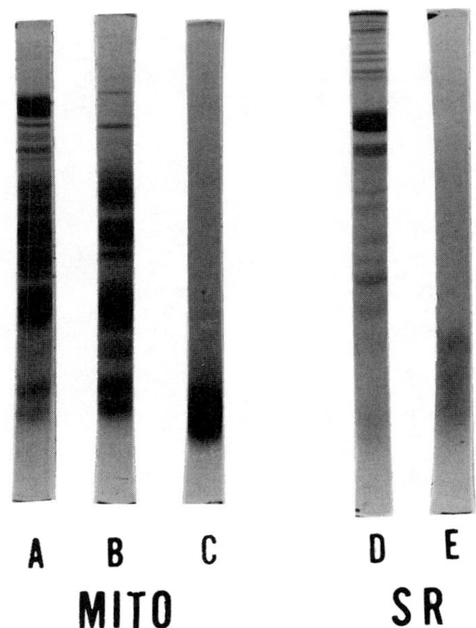

Fig. 6. Band patterns of mitochondrial membranes (A,B, C) and sarcoplasmic reticulum from lobster muscle (D,E). A, Total mitochondrial protein. B, Sub-mitochondrial membranes. C, Preparation B after one hour digestion. D, Sarcoplasmic reticulum membranes. E, Preparation D after one hour digestion with Nagarse.

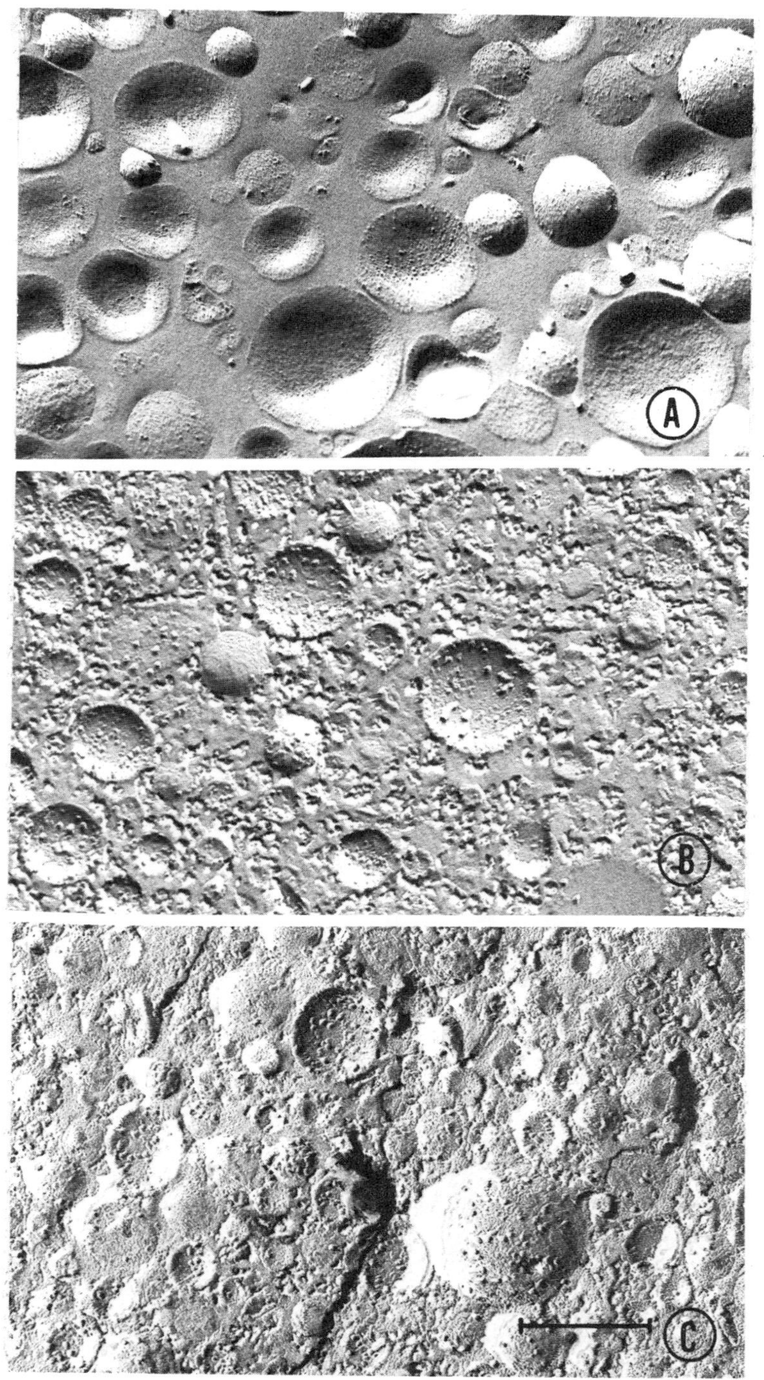
A
B
C

original protein was solubilized and did not appear in the pellet, and ATPase activity was rapidly destroyed. This was reflected in loss of the normal band pattern from disc gels, and appearance of a diffuse, lightly staining band in the region of 15,000 daltons (Figure 6). Since no polypeptides remained which were large enough to produce visible particles at the resolution of freeze-etch microscopy, we expected that freeze-etch images of the membranes would appear relatively smooth. In fact, this was the case, as shown in Figure 7A. A few particulate structures could be found, but most of the vesicles had featureless fracture faces.

Sub-mitochondrial membranes were also markedly affected by the Nagarse enzyme, and only 30% of the original protein remained after digestion (Figure 5). ATPase activity was entirely lost, although a measureable amount (17%) of the original succinic oxidase activity remained. The gel patterns of the digested membranes again showed a complete disappearance of the normal band pattern (Figure 6). A diffuse, darkly staining band, presumably low molecular weight polypeptide fragments, appeared in the region of 10,000 daltons. It is important to note that omission of 2-mercaptoethanol reduction in some experiments had little effect on the band patterns of the original or digested membranes. Negative stain preparations of the membrane before and after digestion are shown in Figure 8. Prior to digestion, the membrane fragments had the typical appearance of sub-mitochondrial inner membrane fragments, with attached ATPase head groups. After digestion, the membranes appeared entirely smooth.

Although the negative stain results could be understood in terms of the enzymatic data and electro-

Fig. 7 (opposite). Freeze-etch micrographs of sarcoplasmic reticulum and sub-mitochondrial membranes. A, Sarcoplasmic reticulum after one hour digestion with Nagarse. The membranes are relatively smooth. (Compare with Fig. 3A.) B, Sub-mitochondrial membranes. C, Preparation B after one hour digestion. Bar equals 0.2 μm.

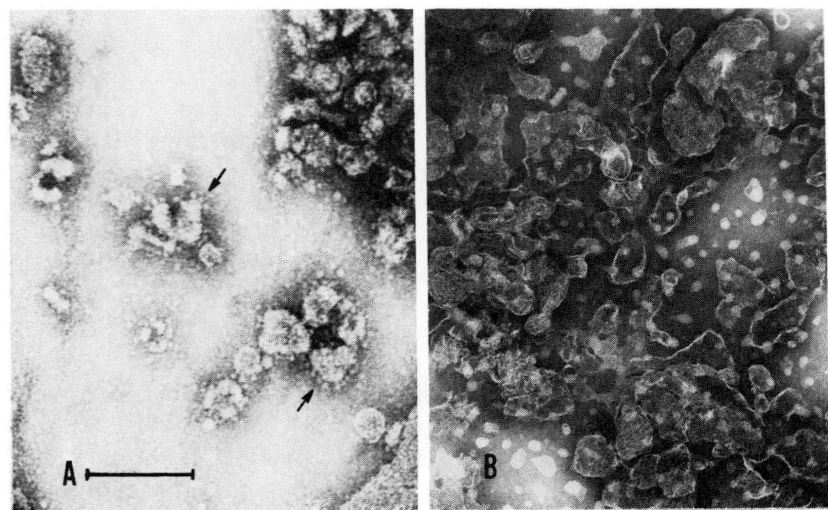

Fig. 8. Negative stains of sub-mitochondrial membranes. A, Before digestion, most of the membrane fragments have ATPase head groups (arrows). B, After one hour digestion, the membranes are entirely smooth. Bar equals 0.2 μm.

phoretic band pattern, the freeze-etch appearance of the digested membranes was unexpected. Instead of smooth membranes, as in the case of sarcoplasmic reticulum, particles could be found on most fracture faces (Figure 7B, C). A quantitative count of the particle distribution on the combined area of fracture faces (both concave and convex) showed approximately a 20% decrease in particle density. Thus, 80% of the original number of freeze-etch particles are apparent in Nagarse treated membranes, despite the fact that only a fraction (30%) of the original protein remained. Furthermore, the remaining fraction consisted of low molecular weight polypeptide fragments, and single polypeptide chains of this size would be too small to produce the particles seen in the freeze-etch preparations. It may be necessary to consider alternative explanations for mitochondrial freeze-etch particles, other than the simplest hypothesis that they represent high molecular weight enzymatic polypeptides embedded in a lipid matrix.

CONCLUSIONS

This investigation has established the following:

1) The membrane proteins of both sarcoplasmic reticulum and sub-mitochondrial fragments are readily attacked by Nagarse, a proteolytic enzyme of the subtilisin family.

2) This is reflected by loss of membrane protein and by disappearance of normal band patterns seen by gel electrophoresis. Nearly all bands disappear from digested sarcoplasmic reticulum and mitochondrial membranes, and are replaced by diffuse bands in the gels in the range of 10,000-15,000 daltons.

3) Freeze-etch fracture faces of the digested sarcoplasmic reticulum membranes appear smooth. However, the appearance of sub-mitochondrial membranes is little affected by digestion, and particles may be found on most fracture faces.

Earlier work with trypsin digestion (11) showed that the ATPase activity of this membrane could be attacked by trypsin, with a corresponding loss of the 105,000 dalton band from disc gels. However, other polypeptide bands were not affected by trypsin, and it was suggested that these were somehow resistant to trypsin action, perhaps by being trapped within the vesicular volume. The fact that Nagarse readily attacks all membrane proteins of sarcoplasmic reticulum makes this interpretation less certain, although it is possible that Nagarse may be able to penetrate the vesicle interior. In any event, the strong proteolytic activity of Nagarse provides a useful tool, as demonstrated by the present study.

The effect of proteolysis on freeze-etch images of sarcoplasmic reticulum was expected, and corresponds to earlier work of Engstrom (9) on erythrocyte ghost membranes treated with pronase. In the latter study, it was found that nearly all the freeze-etch particles had disappeared by the time 70% of the protein had been digested. The present results show that Nagarse is able to digest nearly all the membrane protein of sarcoplasmic reticulum into low molecular weight fragments, and the freeze-etch appearance was predictably that of relatively smooth membranous vesicles.

The result with sub-mitochondrial membranes under similar conditions was surprising. Nagarse was able to attack all the protein, and the remaining polypeptides were in the molecular weight range of 10,000 daltons. However, the polypeptides were capable of producing a fairly typical freeze-etch appearance. This result is reminiscent of the work of Packer *et al.*(24) who compared freeze-etch images of wild and petite mutant yeast mitochondria. Since the petite mutant is known to have lost many of its original mitochondrial enzymes, including ATPase activity and electron transport proteins, it was reasoned that this may be accompanied by significant alterations of the freeze-etch image. However, these investigators found no significant difference in the distribution and size of particles on the various fracture faces of the yeast mitochondria. This result might have been predicted from the present experiments, in which membrane proteins were even more dramatically disturbed with little effect on the freeze-etch image.

These findings suggest to us that at least portions of mitochondrial membrane proteins are specialized for interaction with membrane lipid. When proteolysis takes place in the presence of Nagarse, the lipophilic portions are protected and appear as low molecular weight fragments remaining in the lipid phase. Since the low molecular weight polypeptides singly are too small to produce visible freeze-etch particles, it may be that the mitochondrial particles result from lipid complexes formed around the embedded polypeptides. This may be analogous to the boundary lipid found by Jost *et al.* (25) in cytochrome oxidase preparations. A corollary of this model is that membranes such as sarcoplasmic reticulum and erythrocyte ghosts have freeze-etch particles which arise from lipoproteins containing high molecular weight polypeptides. Proteolytic digestion of the polypeptides would thus result in disappearance of the freeze-etch particles, as noted by Engstrom (9) and the present study.

SUMMARY

Membranes of sarcoplasmic reticulum and mitochondria were digested with Nagarse, a proteolytic enzyme. Changes in membrane protein composition were followed by polyacrylamide gel electrophoresis, and membrane structural changes were visualized by freeze-etch electron microscopy. The enzyme rapidly attacked all protein components of both membrane species. Hydrolysis products which remained attached to the membranes ranged from 10-20,000 in molecular weight. Freeze-etch particles of sarcoplasmic reticulum have been correlated with calcium transport ATPase, a 105,000 molecular weight protein, and in the present study sarcoplasmic reticulum appeared highly particulate in freeze-etch micrographs. Digested membranes had no ATPase activity, and the freeze-etch micrographs showed relatively smooth fracture faces. However, mitochondrial membranes were particulate before and after digestion. This result suggests that freeze-etch particles of mitochondrial membranes may be produced by relatively low molecular weight polypeptides. Since small polypeptides singly are not large enough to be resolved by the freeze-etch method, membrane lipid associated with the polypeptides may contribute to the apparent size of the particles.

REFERENCES

1. BRANTON, D., AND DEAMER, D. W. (1972) _Protoplasmatologia_ II, E, 1.
2. MARCHESI, V. T. JACKSON, R. L., SEGREST, J. P., AND KAHANE, I. (1973) _Fed. Proc._ _32_, 1833.
3. BRETSCHER, M. (1973) _Science_ _181_, 622.
4. DEAMER, D. W., AND BASKIN, R. J. (1969) _J. Cell Biol._ _42_, 296.

5. CHEN, Y. S., AND HUBBELL, W. L. Exptl. Eye Res., in press.
6. HONG, K., AND HUBBEL, W. L. (1972) Proc. Nat. Acad. Sci. U.S. 69, 2617.
7. MARCHESI, V. T., TILLACK, T. W., JACKSON, R. L., SEGREST, J. P., AND SCOTT, R. E. (1972) Proc. Nat. Acad. Sci. U.S. 69, 1445.
8. GOODENOUGH, D., AND STOEKENIUS, W. (1972) J. Cell Biol. 54, 646.
9. ENGSTROM, L. H. (1970) Ph.D. Dissertation, University of California, Berkeley.
10. PINTO DA SILVA, P. (1972) J. Cell Biol. 53, 777.
11. DEAMER, D. W. (1973) J. Biol. Chem. 248, 5477.
12. MARTONOSI, A. (1968) J. Biol. Chem. 243, 71.
13. STANCLIFF, R. C., WILLIAMS, M. A., UTSUMI, K., AND PACKER, L. (1969) Arch. Biochem. Biophys. 131, 629.
14. GREGG, C. T. (1967) in Methods in Enzymology (Estabrook, R. W., and Pullman, M. E., eds.) p. 181, Academic Press, New York.
15. NISHIMURA, J., ITO, T., AND CHANCE, B. (1962) Biochim. Biophys. Acta 59, 177.
16. LOWRY, O. H., ROSEBROUGH, N. J., FARR, A. L., AND RANDALL, R. J. (1951) J. Biol. Chem. 193, 265.
17. MELNICK, R. L., TINBERG, H. M., MAGUIRE, J., AND PACKER, L. (1973) Biochim. Biophys. Acta 311, 230.
18. FISHER, K., AND BRANTON, D. (1974) Methods in Enzymology, in press.
19. MAC LENNAN, D. H. (1970) J. Biol. Chem. 245, 4508.
20. MAC LENNAN, D. H., SEEMAN, P., ILES, G. H., AND YIP, C. C. (1971) J. Biol. Chem. 246, 2702.
21. MEISNER, G., AND FLEISCHER, S. (1971) Biochim. Biophys. Acta 241, 356.
22. MAC LENNAN, D. H., AND WONG, P. T. S. (1971) Proc. Nat. Acad. Sci. U.S. 68, 1231.
23. MEISNER, G., CONNER, G. E., AND FLEISCHER, S. (1973) Biochim. Biophys. Acta 298, 246.
24. PACKER, L., WILLIAMS, M., AND CRIDDLE, R. (1973)

Biochim. Biophys. Acta, **292**, 92.
25. JOST, P., GRIFFITH, O. H., CAPALDI, R. A., AND VANDERKOOI, G. (1973) *Biochim. Biophys. Acta* **311**, 141.

PHYSICOCHEMICAL ASPECTS OF ION TRANSPORT BY IONOPHORE CARRIERS

D. Balasubramanian and B. C. Misra

The family of macrotetrolides - the nactins (Figure 1) and the macrocyclic peptides/depsipeptides (Figure 2) - show the remarkable ability of complexing with alkali and alkaline earth cations with great selectivity. For example, valinomycin forms a K^+ complex with a stability constant of 2×10^6, while complexing with Li^+ or Na^+ occurs only with difficulty. Again, the nactins and the enniatins also prefer the K^+ ion over others. Considerable efforts have been spent in the understanding of the physicochemical basis of such complexing and ion-selectivity shown by these <u>ionophore</u> molecules. This phenomenon of alkali ion complexing by the macrocylic antibiotics

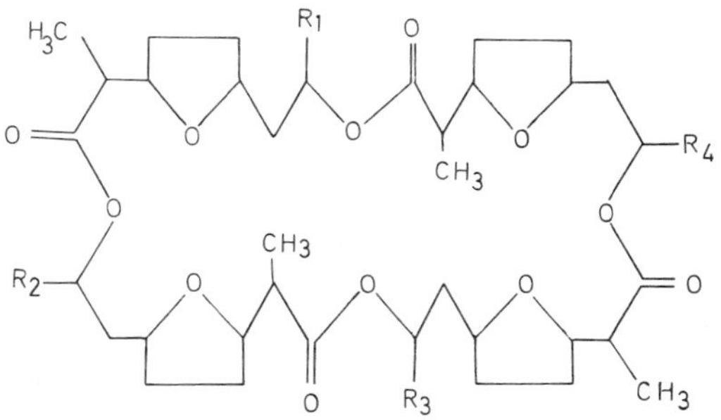

$R_1 = R_2 = R_3 = R_4 = CH_3$ NONACTIN

$R_1 = R_2 = R_3 = CH_3$ $R_4 = C_2H_5$ MONACTIN

$R_1 = R_3 = CH_3$ $R_2 = R_4 = C_2H_5$ DINACTIN

$R_1 = CH_3$ $R_2 = R_3 = R_4 = C_2H_5$ TRINACTIN

Fig. 1.

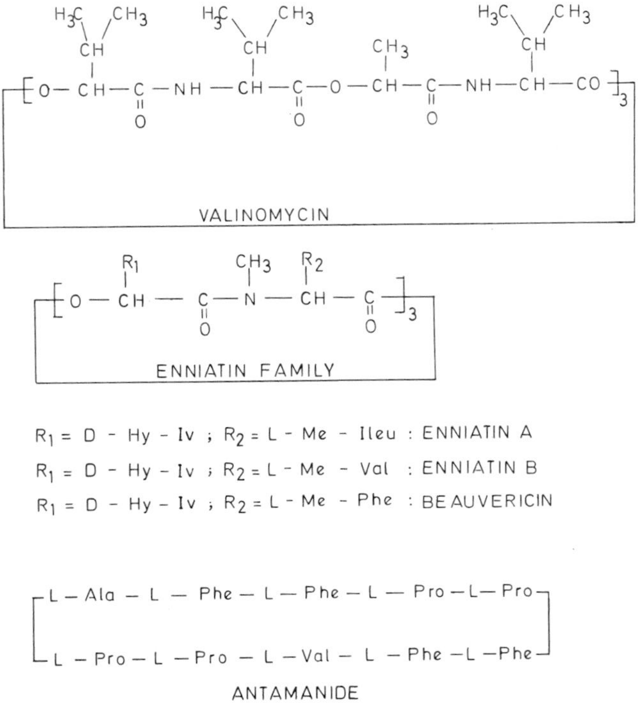

$[O-CH(CH(CH_3)_2)-CO-NH-CH(CH(CH_3)_2)-CO-O-CH(CH_3)-CO-NH-CH(CH(CH_3)_2)-CO]_3$ (cyclic)

VALINOMYCIN

$[O-CH(R_1)-CO-N(CH_3)-CH(R_2)-CO]_3$ (cyclic)

ENNIATIN FAMILY

R_1 = D - Hy - Iv ; R_2 = L - Me - Ileu : ENNIATIN A

R_1 = D - Hy - Iv ; R_2 = L - Me - Val : ENNIATIN B

R_1 = D - Hy - Iv ; R_2 = L - Me - Phe : BEAUVERICIN

L – Ala – L – Phe – L – Phe – L – Pro – L – Pro
L – Pro – L – Pro – L – Val – L – Phe – L – Phe (cyclic)

ANTAMANIDE

Fig. 2.

has been suggested to be of importance in carrier-mediated ion transport, ion permeability, oxidative phosphorylation and so on (1). In this paper, we present and analyze certain aspects pertaining to the physical chemistry of this complex formation exhibited by these carriers. Attention will be focussed on (i) the nature of the ion-ligand interaction, (ii) relative affinities of the ligands of interest and of water towards a given cation, and (iii) differential affinities for the ligands in carriers by members of the alkali group.

Table I is a compilation of the reported stability constants of the alkali metal cation complexes of some macrocycles. It is obvious from the Table that the complexing ability is different for different

TABLE I
STABILITY CONSTANT VALUES (K) OF SOME IONOPHORES

Ionophore	T,°C	Li^+	Na^+	K^+	Rb^+	Cs^+	Anion	Ref.
Nonactin	30		2×10^3	1.3×10^4			SCN	(2)
	17		5×10^3	5×10^3		1.1×10^3	ClO_4	(3)
Monactin	30		3.4×10^3	1.9×10^4			SCN	(2)
Dinactin	30		8.7×10^2	5×10^3			SCN	(2)
Enniatin A	25			1.2×10^3			I	(4)
Enniatin B	25	1.9×10^1	2.6×10^2	8.3×10^2	5.5×10^2	2×10^2	-	(5,6)
			1.4×10^3	3.7×10^3			CNS	(7)
Tri(N-des Me) Enniatin B			2.5×10^3	2.6×10^3			CNS	(7)
Beauvericin		10^2	3×10^2	3.1×10^3	3.5×10^3	3.5×10^3		(8)
Valinomycin	25		0	2×10^6	2.6×10^6	6.5×10^5	Cl	(9)
	25	5	4.7	8×10^4	1.8×10^5	2.6×10^4		(5,6)
Homodetic analog of valinomycin	25	(7-fold increase over valinomycin value)					Br	(10)
Antamanide			2.5×10^3	2.5×10^2				(11)
Dicyclohexyl 18-crown-6 A	25		1.2×10^4	1×10^6		4×10^4		(12)
Dicyclohexyl 15-crown-5	25		5.1×10^3	3.8×10^3		6×10^2		(12)
Dicyclohexyl 14-crown-4	25		1.5×10^2	2×10^1				(12)
Macrohetero- m=1 n=0	25	3.2×10^2	2.5×10^5	8.9×10^3	3.6×10^2	10^2		(13)
m=1 n=1	25	10^2	8×10^3	2.5×10^5	2.2×10^4	10^2		(13)

cations, and the value of the complex stability constant, K, for a given ionophore moiety is affected by (i) the anion, (ii) minor variations in the structure of the sidechains, (iii) the direction or "sense" of the same ionophore (e.g., enniatin and false enniatin), (iv) replacement of a given ligand group by another (e.g., ester by amide), and (v) the ring size of the macrocycle (e.g., macrobicyclic heteroligand, crown compounds). Table II presents some selected data on analogs of valinomycin and enniatin, where some of the above points are exemplified in greater detail.

It may be noticed from Table II that (i) a change in the side chain groups is sufficient to alter the complex stability constant values (items a-e and o-s

TABLE II
SELECTED VALUES OF K FOR K^+:IONOPHORE ANALOG COMPLEXES

Item	Compound	K_{K^+} at 25°C
a.	valinomycin	2.0×10^6
	analogs	
b.	1 D-val → 1-D-ala	2.0×10^6
c.	3 D-val → 3 D-ala	1.6×10^2
d.	3 L-lac → 3 L-Hyval	3.7×10^5
e.	"false retro" valino.	2.9×10^6
f.	1 D-val → 1 L-val	1.1×10^4
g.	3 D-val → 3 L-val	5×10^1
h.	1-D-HyIv → 1 L-HyIv	1×10^2
i.	1-L-lac → 1 L-ala	3×10^5
j.	2 L-lac → 2 L-ala(*)	2×10^5
k.	homodetic analog (**)	7 × value of a
l.	1 L-val → 1 L-HyIv	2.5×10^2
m.	48 member ring analog	5×10^1
n.	24 member ring analog	5×10^1
o.	Enniatin B	3.7×10^3
	analogs	
p.	enniatin A	9.8×10^3
q.	enniatin C	5.5×10^3
r.	beauvericin	3.1×10^3
s.	false enniatin A	1.7×10^3
t.	tri (N-des Me) enniatin B	2.6×10^3(***)
u.	L-Meval → D-Meval	5×10^2
v.	3-Meval → 3-HyIv	1×10^3
w.	"tetra" enniatin (12 membered ring analog)	5×10^1
x.	24 membered ring analog	5×10^3

(*) shows increase in K_{Na^+}; (**) item k from ref. (10), Rest from ref. (14); (***) for item t., K_{Na^+} about same as K_{K^+} (ref. (7)).

in Table II); (ii) a change in the optical configuration even in a few residues drastically reduces K (f-h, u); (iii) an exchange of the peptide and ester groups, particularly when an amide is replaced by an ester moiety, reduces complexing ability (i,j,l,v). However, note that the reduction in K is greater when peptide to ester replacement occurs, and also that the synthetic homodetic analog of valinomycin (k) increases the complex stability. Incidentally, it is also noteworthy that the analogs (i) and (j) have increased affinities for sodium ions compared to the parent, and the enniatin B analog (t) seems to have a largely reduced K^+, Na^+ selectivity; and (iv) a variation in ring size decreases the value of K for potassium ion (m,n,w). Any physicochemical explanation of the action of ionophores must be able to account for these variegated properties of these ion carriers and analogs.

LIGANDING GROUPS IN IONOPHORES

Inspection of the structures of the ionophores shows that the liganding groups of interest are the amide (peptide) group, the ester group, and the ether linkage, besides water itself in aqueous media. Insofar as much of the complexing by the ionophores proceeds from the aqueous phase, a knowledge of the differential affinities of these ligands and water to the cation is important in order to understand the mechanism of carrier mediated ion transport. We shall look at these ligands one by one.

Amide-Cation Interactions

We have considered the problem of amide groups coordinating to alkali metal ions in some detail, both theoretically (15) and by experiments (16). Molecular orbital calculations using Pople's CNDO/2 method were performed on the interaction of several amides with the lithium ion. Table III briefly summarizes the relevant results of such calculations.

These calculations suggest that the basis of the interaction is the ion-dipole type, with the interaction energies becoming stronger with increasing dipole

TABLE III
RESULTS OF CNDO/2 CALCULATIONS ON Li^+ INTERACTIONS
WITH AMIDES AND OTHER RELATED LIGANDS

Ligand	$r_{Li...O}$ (A)	Energy per Li..O bond kJ/mole	Δr		Force constant (mdyne/A)	Charge transfer	Li-O-C angle
			C=O	C-N			
1:1 adducts							
formamide	2.2.2	300	+0.02	-0.02		-0.24	180
N-methylacetamide	2.2	365	+0.02	-0.03		-0.28	180
dimethylformamide	2.2	345	+0.02	-0.03		-0.27	180
1:2 adducts							
formamide	2.4	464	+0.03	-0.02	1.2	-0.31	60
formaldehyde	2.4	417	+0.01		1.1	-0.30	60
dimethyl ether	2.4	390			1.1	-0.25	along lone pair
water	2.4	194			1.0	-0.17	"

Note: Similar results are obtained with 1:4 adducts of Li^+:formamide.

moments of the ligands. Among several possibilities that one may envisage for such interactions, the calculations suggest the following mode:

$$\text{>C(=O)—N} + Li^+ \longrightarrow \text{>C(—O}\cdots\text{Li)}{=}N^+\text{<}$$

Balasubramanian and Shaikh (16) have studied the interaction between alkali metal salts and several monomeric amides by measuring the heats of interaction, the infrared spectral changes upon coordination, and nuclear magnetic resonance. In all cases, the heats of interaction of the metal ion with the amide are more negative than with water, and the heats of interaction with amides decrease in the sequence Li > Na > K, and vary with the counterion in the sequence ClO_4 > Br > Cl > OAc. A similar dependence of the complexing constants of ionophores upon the anion has been noted in Table I. Investigations of the variation of the heats of mixing of amides with aqueous solutions of salts of varying molarity shows that a progressive and competitive replacement of water by amide can occur in the solvation sphere of the cation.

Infrared spectral investigations on the amide: salt system indicate that the cation binds at the carbonyl oxygen site, and in so doing, alters the molecular geometry and spectral properties of the ligand. The stretching vibration corresponding to the newly formed Li...O bond, predicted by theory, is seen to occur around 400 cm^{-1} in the lithium:amide adducts. Nuclear magnetic resonance studies support these observations, and reveal the change in the bond orders in the amide tending towards an increased rigidification of the central CN bond in the amide.

A few points emerge out of these studies that are of relevance to ionophore systems. The heats of interaction of lithium salts increase as we go from primary to secondary to tertiary amides, and upon increasing the methyl groups in the amide ligand (the heats of interaction of aqueous LiCl, under comparable conditions, with formamide, N-methyl formamide, N-methyl acetamide, Dimethyl formamide, Dimethyl acetamide, and Dimethyl propionamide are: -450, -1400, -2400, -3100, -3300, -3600, and -3800 cals/mole amide, respectively). This increase upon increasing the methylation and substitution may owe its origin to increase in the dipole moments of the ligand, as well as to changes in the hydrophobic interactions that are likely to occur in substituted amides in water. It is interesting in any case that the ionophores carry branched or substituted amino acid residues such as Ileu, Val, N-Methyl Val, which might offer greater complex stability. The decrease in the complexing ability seen in, e.g., tri N-des methyl enniatin B and in the 3 D-ala analogs of valinomycin, compared to the respective parents in Table II, may reflect to some extent the interaction differences among amide ligands. It is also interesting that our calorimetric studies on amides with salts show that the amide group is able to strip water off and solvate itself around the metal ion under appropriate concentration conditions. NMR studies on nonactin-metal complexes (3), and on valinomycin-ion complexes (17) show that the alkali cation is dehydrated and is fully enclosed or solvated by the ligands of the macrocycle in the complex. Considering the amount of water present in a membrane, such dehydration and ligand substitution process should be facile in carrier mediated ion transport. Another point of some significance is the Li...O distance calculated by the CNDO/2 calculations. This predicted distance of 2.2 A for the amide:Li^+ adduct agrees well with the distance seen in the Li^+ complex of the antibiotic antamanide, i.e., 2.11 A (18). Similar calculations have been made by Kostetsky _et al._ (19), and by Perricaudet and Pullman (20), for both amide and ester ligands, with similar results. Based on our calculations and the observation of the Li...O stretching band in model systems, we have interpreted the far-infrared bands seen in metal-ionophore complexes by Ivanov _et al._ (21) as due to the metal-oxygen stretching frequencies. We thus conclude that the mode of binding of the metal ion to ionophores is the same as that with the model amides.

Ester Group as a Ligand

Until recently, relevant data on the nature and strength of the interaction between alkali metal ions and the ester group as a ligand have been meagre. In order to understand the complexing phenomenon by ionophores, a comparative study of the abilities of the ester, ether, and amide groups will be of value. Krasne and Eisenman (22) have been forced, in the absence of such data, to regard the ester and amide groups as equipotent ligands. As will be shown below, this assumption does not appear valid.

Quantum chemical calculations (20,19) have compared the ability of the amide and ester groups as ligands, and have predicted the cation sequestering ability of the amide group to be better than that of the ester. This is also what should be expected qualitatively on the basis of the dipole moments of the two ligands. Experimental studies on alkali cation-ester interactions, similar to those with amide ligands, conducted in our laboratory (23), support the predictions of theory. That the site of coordination of the cation in the ester ligand is also the carbonyl oxygen is established by infrared spectral studies on the shifts of the ester carbonyl absorption upon metal attachment, and by the presence of metal-oxygen stretching bands. The heats of interaction of alkali metal salts with the model ester, methyl acetate, are decidedly lower than those with model amides. The order of interaction of alkali salts with ester follows essentially the same as the sequence with amides.

These model studies suggest that the ligand strength of the amide group for alkali metal cations is greater than that of the ester group. While Perricaudet and Pullman (20) suggest the ligand strengths to decrease in the order amide, water, and ester, the heats of mixing of aqueous salt solutions with methyl acetate are observed to be exothermic, suggesting that the ester group may be roughly comparable or somewhat better than water. An assessment of the relative affinity of the other ligand of interest, i.e., the ether linkage, must await further experimental data, though one expects on the basis of CNDO/2 calculations (Table III) the ether group to be a weaker ligand than the keto group. In light of this, it is intriguing to note that the K^+ complexing constant of the macrocyclic polyether dicyclohexyl 18-crown-6 is larger than that of enniatins. Clearly the stability constant values reflect factors such as cavity size, snugness of fit, and conformational aspects of the macrocycle besides the ion-ligand interaction energy. Yet it is interesting to observe that the crystal structure analysis of the K^+-nonactin complex (24) reveals the ether..K distance (2.82, 2.83 A) to be larger than the keto...K distance (2.75, 2.79 A). Compared to this, the O...K distances in the K^+-enniatin B complex (25) are reported to be 2.6 - 2.8 A; in enniatin B the ligands are three amide and three ester groups.

ION SELECTIVITIES

Model compounds, by their very nature, might not be able to offer direct insights into all factors that govern the complexing behavior of ionophores, particularly their ion selectivities. Despite this, attempts have been made, notably by Eisenman and coworkers (22,26), and by Simon and Morf (27), to explain the specificities on the basis of the "asymmetries of interactions" of amides/esters, and on the basis of steric factors, respectively.

Krasne and Eisenman (22) have plotted the difference between ion solvation energies in amide and in water ($\Delta H^{i}_{am} - \Delta H^{i}_{w}$) versus the reciprocal cation radius ($1/r^{+}$), and have demonstrated that the magnitude of this differential is maximum in the case of potassium, and varies in the order K > Rb > Cs > Na > Li. This is the same sequence seen in the sequestering of these ions by nonactin. This differential increases with increasing methylation of the amide, and is attributed to an increase in the dipole moment upon increasing methylation. Further corroboration of this ion selectivity or "asymmetry" has been presented by their theoretical results on the electrostatic interaction of the ion with the carbonyl group as the ligand, and with water. It is worthwhile to note that Eisenman _et al._ postulate this "asymmetric" interaction sequence of the ions with C=O groups as the principal factor in the ion selectivity of ionophores, and that steric aspects are not emphasized.

The rationale for this greater emphasis on "asymmetry" of interaction comes from the earlier paper of Eisenman _et al._ (26), and the argument is as follows: The free energy of complexing of a given ion I^+ with a carrier S is the total of contributions from (i) the respective free energies of hydration ($\Delta F_{hyd,I}$, $\Delta F_{hyd,S}$); (ii) the free energy of ion sequestering which in itself arises from electrostatic attractions (I^++S) and repulsions (I^+I^+,SS) termed ΔF_{el}; and conformational differences between S and SI^+ (ΔF_{conf}); and (iii) the Born charging process of SI^+ and from the free energy of cavity formation in water around

SI^+. The relative selectivity ratio for a given carrier for two ions I^+ and J^+ is then simplified to be of the form:

$$RT \ln \frac{K_{SI^+}}{K_{SJ^+}} = [\Delta F_{hyd,I^+} - \Delta F_{hyd,J^+}] - [\Delta F_{el,SI^+} - \Delta F_{el,SJ^+}]$$

provided one assumes that the complexes SI^+ and SJ^+ are "isosteric," i.e., of the same stereochemical geometry, so that conformational terms, Born terms, and cavity terms for both SI^+ and SJ^+ are the same. The ΔF_{el} are measures of the _in vacuo_ attractive energies between the ion and carrier, once isostericity is granted. The ΔF_{el} values can be evaluated either using a coulomb model or an ion-dipolar interaction model. The difference ($\Delta F_{hyd} - \Delta F_{el}$) for a given ion I^+ is the differential free energy between when water, and the carrier molecule is the "solvent" around the ion, respectively. Since experimental data are available more conveniently as the enthalpies (rather than free energies) of ion solvation, the ion selectivity is equated approximately as the differential enthalpies of solvation: $[\Delta H_{hyd,I^+} - \Delta H_{el,SI^+}]$. It is in such plots of these differential enthalpies versus reciprocal cation radius that "asymmetries" are seen. ΔH_{hyd} values are available from literature. ΔH_{el} are estimated from (i) data on amides as model compounds, and (ii) electrostatic calculations using the carbonyl group as the dipolar ligand.

On the basis of the data and arguments presented earlier, a few comments are in order. Krasne and Eisenman have been forced to treat amides and esters as equipotential, and have performed model calculations using the bare carbonyl group with its partial charges on the C and O atoms as the ligand. However, it is important to recognize that the ester group is a weaker ligand than the amide, and it is not necessarily true that both ligands show the same degree and sequence of "asymmetry" and selectivity. Krasne and Eisenman have also not considered the ether linkage explicitly, while this group is the primary ligand in the crown polyethers, and in part in the nactins. More importantly, this approach lends a unique place to the K^+ ion in the selectivity sequence; using the model amide and carbonyl groups as ligands, the asymmetry of interaction is maximum with K^+. And yet, it is known that compounds such as antamanide, actinomycin, and monensin sequester the sodium ion preferentially (see Table I, or ref. 7). The selectivity and binding constants of isomers of the same carrier are different towards a given ion. For example, false enniatin B is claimed to lose its K^+ selectivity, while enniatin B prefers K^+ over other ions; this is most likely due to conformational reasons. And there are cyclic peptides such as alamethicin that show no selectivity for potassium. Finally, there may be some questions raised about whether complexes of the same ionophore S with two different cations I^+ and J^+ are indeed isosteric. Grell, Funck, and Eggers (5) have presented circular dichroism spectra of the complexes of several alkali and alkaline earth cations with valinomycin, and with enniatin B. The major differences seen between the circular dichroic profiles of the K^+, Rb^+, and Cs^+ complexes on the one hand and the Na^+ complex on the other have led them to suggest that the ligand conformation is considerably different depending on the cation bound by the carriers.

A somewhat different approach that emphasizes the steric considerations and coordination characteristics has been taken by Diebler *et al.* (29), and elaborated in greater detail by Simon and Morf (27). Diebler *et al.* have proposed general rules for the design of carrier molecules that emphasize factors such as solvation sphere replacement, cavity size, optimal fit as opposed to minimal size of the cavity based on solvation and ligand-ligand repulsion energies, flexibility of conformation in the ionophore to allow for progressive desolvation/complexation, and the distribution of hydrophobic and polar groups in the overall molecular conformation. Simon and Morf have amplified these factors in detail and argue that "since a multidentate ligand system (macrocycle) - contrary to monodentate solvents - is not flexible enough to offer to all cations the optimal coordination (and the most negative free energy of interaction), but is restricted to a fixed number n of coordinating atoms ('strait jacketing' of cations), the selectivity orders change drastically." These authors have attempted to calculate the interaction free energy between the ionophore and the cation on the basis of several terms, which notably include interaction of the complex with the bulk ligand, conformational distortions that depend on, as well as are independent of, the ion size, cavity formation, and the number and geometry of coordination. In principle, this approach ought to be able to account for the ion selectivity. Unfortunately, however, it is very difficult in practice to estimate contributions from conformational distortions that arise out of the generation of the "equilibrium cavity," and the ion dependent "optimal fit" ligand distortions. As a result, an explanation of the variations in selectivity and binding constants that arise out of minor changes in sidechains, anions, and so on is yet to come.

There is one other intriguing feature about the complexing abilities of ionophores. Notice from Table I that the K values and also the selectivity ratios K_{K+}/K_{Na+} vary widely in the ionophores valinomycin, enniatins, antamanide, and the synthetic macrocyclic polyether. What causes this variation in the values of K and the selectivities between valinomycin, enniatin, and the polyether, all of which are hexacoordinated to the central cation? Inference on this point comes from the excellent paper of Ovchinnikov, Ivanov, and Shkrob (14), who have summarized the exhaustive and admirable work done by the Russian group over the past decade or so. They note that in general the enniatins have lower complexing constants and lower selectivity when compared to valinomycin, and suggest that the reason for this may be two-fold: (i) even though the cavity sizes in the two ionophores are comparable, the orientation of the liganding C=O groups in the two towards the cation are different. In the K^+:valinomycin complex, the ester carbonyls point directly towards the cation, while in the enniatin B complex the "ion-dipole" bonds are almost normal to the C=O bonds, thereby weakening the bond strength; and (ii) while valinomycin has a rigid hydrogen-bonded conformation, enniatin B possesses considerable conformational mobility. Hence in the formation of the sodium complex, the higher hydration energy may be compensated by more effective ion-dipole interactions arising out of a reorientation of the carbonyl groups towards greater colinearity.

Lastly, while the phenomenon of ion selection and permeation has been explained on the basis of the selectivities in the complexing by the ionophores, recent experiments by de Gier and coworkers (30) suggest that complex formation may not be the only factor that determines the extent and rate of such facilitated transport. These experiments reveal that the valinomycin induced exchange of $^{86}Rb^+$ over the bilayers of liposomes is vastly changed (enhanced) in the presence of unsaturated lipids (linoleate) in the paraffin core of the lipid bilayers. Changes in the Rb permeability were also noticed in liposome bilayers in the presence of phosphatidyl glycerol (an influence of size and charge of the polar headgroup), and in the presence of cholesterol (attenuation of carrier transport in presence of the sterol). These suggest that lipid-ionophore interactions may be of significance in conferring ion-selectivity and in the permeation rates, and that the mobility of the lipid chain might be expected to alter the effectiveness of carrier induced cation specificity. It seems important from these experiments that one may not be quite justified in treating the problem of specificity by considering the isolated carrier molecule alone, and more effort ought to be spent in the area of carrier-lipid interactions.

REFERENCES

1. For a recent set of reviews, see (1972) Molecular Mechanisms of Antibiotic Action on Protein Biosynthesis and Membranes (Munoz, E., Garcia-Ferrandiz, F., and Vazquez, D., eds.), Elsevier Scientific Publishing Co., Amsterdam.
2. ZUST, CH., AND SIMON, W., (1972) Helv. Chim. Acta, in preparation; ZUST, CH., Dissertation ETHZ, Zurich.
3. PRESTEGARD, J. H., AND CHAN, S. I. (1969) Biochemistry 8, 3921; (1970) J. Am. Chem. Soc. 92, 4440.
4. WIPF, H.-K., PIODA, L. A. R., STEFANAC, Z., AND SIMON, W. (1968) Helv. Chim. Acta 51, 377.
5. GRELL, E., FUNCK. TH., AND EGGERS, F. (1972) in Molecular Mechanisms of Antibiotic Action on Protein Biosynthesis and Membranes (Munoz, E., Garcia-Ferrandiz, F., and Vazquez, D., eds.), p. 646-685, Elsevier, Amsterdam.
6. EGGERS, F., FUNCK, TH., AND GRELL, E., in preparation.
7. JAIN, M. (1972) The Biomolecular Lipid Membrane: A Systems Study, Van Nostrand Reinhold Co., N.Y.
8. OVCHINNIKOV, YU. A., IVANOV, V. T., AND MIKHALEVA, I. I. (1971) Tetrahedron Letters 2, 159.
9. SHEMYAKIN, M. M., OVCHINNIKOV, YU. A., IVANOV, V. T., ANTONOV, V. K., VINOGRADOVA, E. I., SHKROB, A. M., MALENKOV, G. G., EVSTRATOV, A. V., LAINE, I. A., MELNIK, E. I., AND RYABOVA, I. D. (1969) J. Membrane Biol. 1, 402.
10. GISIN, B. F., AND MERRIFIELD, R. B. (1972) J. Am. Chem. Soc. 94, 6165.
11. IVANOV, V. T., MIROSHNIKOV, A. I., ABDULLAEV, N. D., SENYAVINA, L. B., ARCHIPOVA, S. F., UVAROVA, N. N., KHALILULINA, K. KH., BYSTROV, V. F., AND OVCHINNIKOV, YU. A. (1971) Biochem. Biophys. Res. Comm. 42, 654.
12. FRENSDORFF, H. K. (1971) J. Am. Chem. Soc. 93, 600.
13. LEHN, J. M., AND SAUVAGE, J. P. (1971) Chem. Commun. 440.

14. OVCHINNIKOV, YU. A., IVANOV, V. T., AND SHKROB, A. M., in ref. 1, p. 459-52.
15. BALASUBRAMANIAN, D., GOEL, A., AND RAO, C. N. R. (1972) Chem. Phys. Letters 17, 482.
16. BALASUBRAMANIAN, D., AND SHAIKH, R. (1973) Biopolymers 12, 1639.
17. HAYNES, D. H., PRESSMAN, B. C., AND KOWALSKY, A. (1971) Biochemistry 10, 852.
18. KARLE, I., KARLE, J., WIELAND, TH., BURGERMEISTER, W., FAULSTICH, H., AND WITKOP, B. (1973) Proc. Natl. Acad. Sci. U. S. 70, 1836.
19. KOSTETSKY, P. V., IVANOV, V. T., OVCHINNIKOV, YU. A., AND SHCHEMBELORT, G. (1973) FEBS Letters 30, 205.
20. PERRICAUDET, M., AND PULLMAN, A. (1973) FEBS Letters 34, 222.
21. IVANOV, V. T., ET AL. (1973) FEBS Letters 30, 199.
22. KRASNE, S., AND EISENMAN, G. (1973) in Membranes: A Series of Advances (Eisenman, G., ed.) Vol. 2, p. 277, Marcel Dekker, N.Y.
23. BALASUBRAMANIAN, D., AND MISRA, B. C., FEBS Letters, submitted.
24. KILBOURN, B. T., DUNITZ, J. D., PIODA, L. A. R., AND SIMON, W. (1967) J. Mol. Biol. 30, 559.
25. DOBLER, H., DUNITZ, J. D., AND KRAJEWSKI, J. (1969) J. Mol. Biol. 42, 603.
26. EISENMAN, G., SZABO, G., MCLAUGHLIN, S. G. A., AND CIANI, in ref. 1, p. 545.
27. SIMON, W., AND MORF, W. E., in ref. 22, p. 329.
28. DIEBLER, H., EIGEN, M., ILGENFRITZ, G., MAASS, G., AND WINKLER, R. (1969) Pure and Applied Chem. 20, 93.
29. DE GIER, J., DEMEL, R. A., HAEST, C., VAN ZUPTEN, H., VAN DER NEUT-KOK, E., MANDERSLOOT, G., DE KRUYFF, B., NORMAN, A. W., AND VAN DEENEN, L. L. M., in ref. 1, p. 709.

CYTOTOXIC PROTEINS FROM COBRA VENOM AS PROBES FOR THE STUDY OF MEMBRANE STRUCTURE

Beatriz M. Braganca

As is evident from other chapters in this book, a great deal of effort involving a variety of technical approaches is presently directed at unravelling the details of molecular structure of biomembranes.

Cobra venom is a complex mixture of biologically active proteins, many of which have been shown to have specific effects on cell membranes. In this context cytotoxins and phospholipases isolated from cobra venom which are free of contaminants and have specific action on membrane components can be very promising as probes for the study of protein-lipid interactions in biomembranes. This approach is free of the possible artifacts associated with chemical treatments.

In this chapter I shall discuss observations relating to intact cells and the membrane bound enzyme $Na^{+} \cdot K^{+}$-ATPase of cells obtained using a pure cytotoxic protein and phospholipase A_2 both isolated from cobra (Naja naja) venom.

The cytotoxin (P_6) preferentially cytotoxic to certain cells was isolated from cobra venom by conventional protein fractionation procedure (1). Figure 1 shows the acrylamide gel electrophoretic patterns of

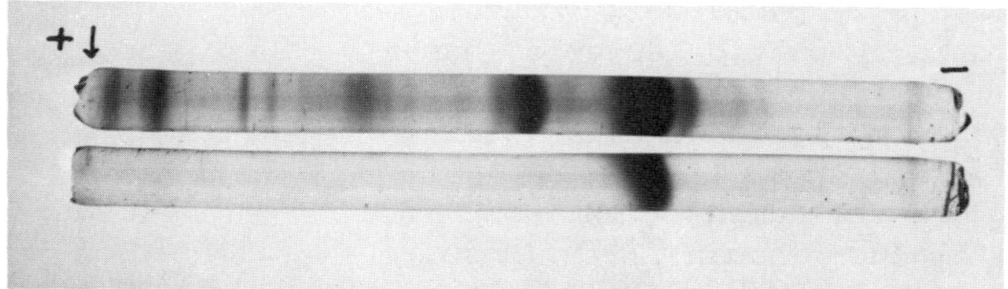

Fig. 1. Acrylamide gel electrophoresis of crude cobra venom and P_6.

crude venom and the cytotoxin P_6 at pH 2.4 which has been used in most of the experiments described here. The relative cytotoxicity of P_6 to various discrete and intact cell preparations was assayed by the dye exclusion method. Lissamine green, the dye employed here, as well as many other dyes, is known to penetrate damaged cells and is excluded by viable and intact cells (2). The results given in Table I are expressed in ug P_6/ml required to make 50% of the cells permeable to the dye. The effects on erythrocytes were determined by direct count in the haemocytometer. The results of few human tumor samples are based on analysis of aspirated fluid containing tumor cells from cancer patients. It is seen that P_6 shows a wide spectrum of activity when examined on a

TABLE I

CYTOTOXICITY OF P_6 TO VARIOUS CELL TYPES

Cell type	Cytotoxicity (ug Protein/ml)
Erythrocytes (rat)	>6000.0
" (human)	>4500.0
" (hamster)	>2500.0
" (chicken)	480.0
" (bull)	400.0
Leucocytes (human)	50.0
Lymphocytes (rat)	4.3
Bone marrow cells (rat)	3.4
Y.S.cells (rat)	0.4
L 1210 ascites cells (mouse)	25.0
Dalton's lymphoma cell (mouse)	5.0
Chronic lymphatic leukemia (human) (J.J. Hosp.)	>50.0
Chronic myeloid leukemia (human) (J.J. Hosp.)	5.0
CA. breast (human) (T.M. Hosp.)	10.0
CA. bladder (human) (T.M. Hosp.)	5.0
Ovary ascites (human) (T.M. Hosp.)	2.5

variety of cells. The Yoshida sarcoma cells are most susceptible to destruction by this cytotoxin. Observations on the cytotoxic behavior of P_6 on WBC have also shown preferential destruction of granulocytes. The lymphocytes were relatively more resistant to the lytic effect of P_6. As seen in Fig. 2 addition of graded amounts of the cytotoxin greatly increased the proportion of lymphocytes. This property of P_6 may

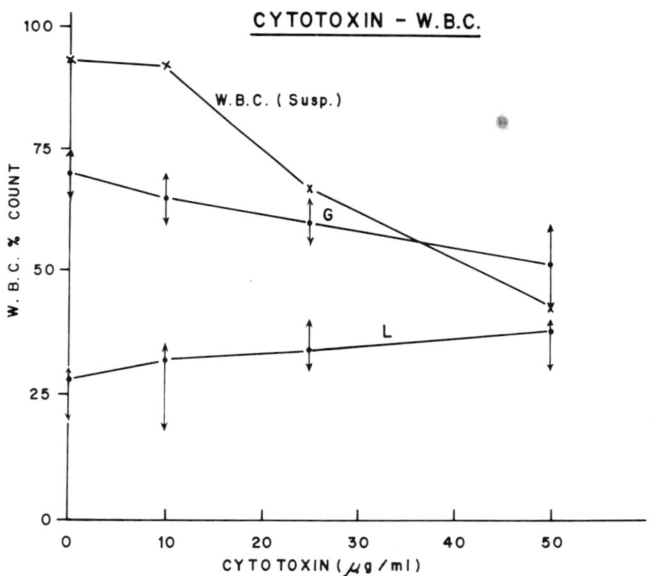

Fig. 2. Differential effects of P_6 on human WBC.

be useful for the development of procedures for the preparation of pure lymphocytes required in many studies. It was of interest that P_6 was also effective in vivo. Experiments described in Fig. 3 demonstrate that 1/7 the LD_{50} dose of P_6 could prevent the growth of tumor in rats inocculated with Yoshida sarcoma cells. Since many of the proteins isolated from cobra venom are cytotoxic to various cells it was of interest to compare the properties of P_6 with

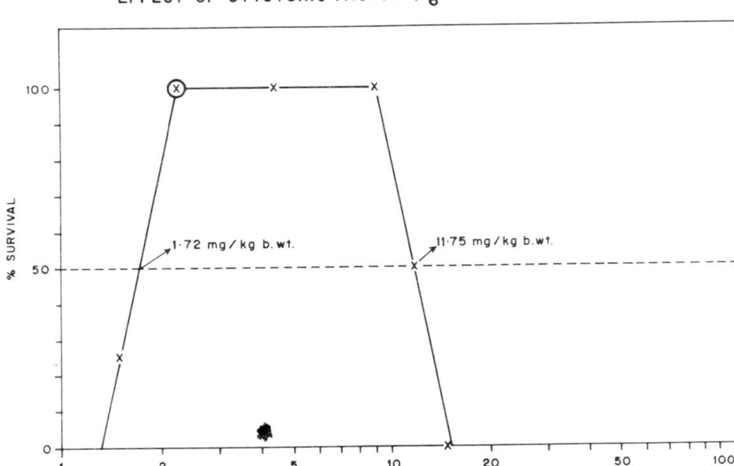

Fig. 3. Effect of P_6 on growth of Yoshida sarcoma in rats.

the major fraction of phospholipase A_2 isolated from cobra venom and two neurotoxins which were lethal to mice when injected individually. It is evident from Table II that the four proteins show different biological effects on cells and that they are not contaminated with one another to any significant extent. They all have similar molecular weights as determined

TABLE II
DIFFERENCES IN BIOLOGICAL PROPERTIES OF COBRA VENOM FRACTIONS

Materials	LD_{50} Prot. mg/kg body weight	Hemolytic activity ug Prot./unit	Cytotoxicity to Yoshida sarcoma Dilution 1 mg/ml	Molecular weight
Crude venom	0.71	1.12	-	-
Phospholipase A_2 (P_1)	>19.20	0.06	1/2	10,000
Neurotoxin (P_4)	0.12	23.00	1/2	10,000
Neurotoxin (P_5)	0.45	18.00	1/32	-
Cytotoxin (P_6)	44.50	>32.00	1/2000	10,500

by Sephadex elution pattern method (3). Among some of its other properties it may be mentioned that P_6 is relatively non-toxic to rats. In doses which could prevent the growth of Yoshida sarcoma in rats the erythrocyte and WBC counts and hemoglobin remained within the normal range. In attempts to determine the nature of its combination with cell constituents, the cytotoxin was complexed with the fluorescent dye, Dansyl (4). Fluorescent microscopy of cells to which dansylated P_6 was added showed that in the first instance it binds with constituents in the plasma membrane of cells which are susceptible to its destructive effects. No binding was observed in cells which were more resistant to its cytotoxic action. For a further insight into its mode of action the effect of the cytotoxin was examined on a membrane bound enzyme (5). $Na^+ \cdot K^+$-ATPase preparations were made from cells which were resistant as well as susceptible to the cytotoxic action of P_6. The results summarized in Table III demonstrate that P_6 strongly inhibited the $Na^+ \cdot K^+$-ATPase activity of Yoshida sarcoma cells. Results on the effects of P_6 on erythrocytes from human and dog blood demonstrate that dog

TABLE III

INHIBITION OF $Na^+.K^+$-ATPase AND CYTOTOXICITY BY P_6 TO DIFFERENT CELLS

Cytotoxicity: Yoshida sarcoma cells - P_6 required to release cytoplasmic contents (260 mu) from 50% cells in the system.
Tissue culture cell lines - P_6 required to inhibit growth by 100%.

Cell type	No.	Preparations	$Na^+.K^+$-ATPase activity uMoles P_i/H/ mg Protein	80-100% inhibition Ouabain (uM)	50% inhibition by P_6 ug/ml	Cytotoxicity to intact cells P_6 ug/ml
Yoshida sarcoma cells	1	Y.S. cell extract (105,000×g pellet)	10.0	50.0	9.0	1.5
RBC	2	Human RBC ghost (25,000×g pellet)	3.75	40.0	42.5	40.0
"	3	Dog RBC ghost (25,000×g pellet)	2.15	37.5	17.0	10.0
Tissue culture cell lines	4	KB cell extract (47,000×g pellet)	1.6	50.0	0.75	5.0
"	5	Hela cell extract (47,000×g pellet)	3.0	50.0	1.6	10.0
"	6	L-132 cell extract (47,000×g pellet)	1.12	50.0	4.5	15.0

erythrocytes are more susceptible to lysis by the cytotoxin and also that $Na^{+} \cdot K^{+}$-ATPase of this species of erythrocytes is inhibited to a greater extent. Experiments using 3 different tissue culture cell lines show clear correlation between inhibition of growth and the inhibitory effects of P_6 on $Na^{+} \cdot K^{+}$-ATPase of these cell lines. Observations on all cells examined have consistently demonstrated a clear correlation between susceptibility to lysis and inhibition of $Na^{+} \cdot K^{+}$-ATPase activity by the cytotoxin P_6. In attempts to determine the type of constituents in the plasma membrane which may be involved in the reaction with P_6 it was found that phosphatidyl ethanolamine and phosphatidyl serine were able to reverse the cytotoxic action of P_6 on Yoshida sarcoma cells but lecithin was without effect (Table IV).

TABLE IV
EFFECT OF PHOSPHOLIPIDS AND THEIR DERIVATIVES ON THE LYSIS OF YOSHIDA SARCOMA CELLS BY P_6

Phospholipid or derivative	Net absorbancy at 260 mu		Percentage activation/ inhibition
	P_6	P_6+phospholipid	
PC	0.525	0.530	+ 1
PE	0.550	0.278	- 50
PS	0.550	0.116	- 79
LPE	0.550	0.560	+ 2
LPS	0.550	0.605	+ 10
OPE	0.550	0.480	- 13
OPS	0.550	0.310	- 44
E	0.555	0.595	+ 8
S	0.555	0.635	+ 14

This possibly resulted through competition with sites in the membrane linked with similar phospholipids and suggests that the mechanism of P_6 action may involve combination with membrane lipoproteins containing these acidic phospholipids. It may be noted here that P_6 is very basic with an isoelectric point above pH 8.4. It has not been possible to detect degradation of phosphatidyl ethanolamine, phosphatidyl serine, phosphatidyl choline or phosphatidyl inositol by P_6 using pure phospholipids. Many phospholipid degrading enzyme preparations (Phospholipase A_2, C and D) have been shown to inhibit the $Na^+ \cdot K^+$-ATPase activity of different cell systems and phosphatidyl serine is known to reverse the inactivation produced by phospholipases on this system (6). Although it has not been possible to detect phospholipase A_2 activity in P_6 using various pure phospholipids as substrates, the inhibitory effect of P_6 on $Na^+ \cdot K^+$-ATPase of Yoshida sarcoma cells was also reversed by phosphatidyl serine and phosphatidyl ethanolamine but lecithin had no effect. Phosphatidyl serine was most effective in reactivating the enzyme inhibited by P_6. Thus it is apparent that factors which protected the enzyme $Na^+ \cdot K^+$-ATPase also protected the cell membranes against the lytic action of P_6. The $Na^+ \cdot K^+$-ATPase is the enzyme concerned with the transport of Na^+ and K^+ ions across the cell membrane and it is conceivable that the disturbance in the balance of these ions may be responsible for the lytic effects of P_6. Phospholipids are mostly confined to cellular membranes and are known to provide structural support which maintains the spatial organization of many complex membrane bound enzyme systems. Observations which have shown great differences in the degree of inhibition by P_6 on $Na^+ \cdot K^+$-ATPase activity of different cells and cell organelles reflect the great heterogeneity existing in the arrangement of protein-phospholipid environment present in the close neighborhood of this enzyme in different cells.

Other investigations have demonstrated the presence of multiple forms of phospholipase A_2 in cobra venom (7) and some of these have been obtained in a pure state. A study of their properties shows that they differ in their specificity for different phospholipid substrates. It is also of interest that none could lyse intact human erythrocytes. In common with P_6 they have similar molecular weights ranging from 10,000 to 11,000 and are all stable to high temperatures. A review of the literature indicates that phospholipase A_2 purified from pancreas (8) can liberate lysolecithin from red cell ghosts but has no lytic action on intact cells. However, enzyme preparations isolated from cobra venom (Naja naja) have been reported to bring about degradation of phospholipids from intact human erythrocytes (9). The availability of phospholipase A_2 and cytotoxin P_6 both from same venom (Naja naja) which were uncontaminated with each other made it possible to study the effects of these two proteins individually and in combination on intact human erythrocytes. For these studies the major fraction of the phospholipase A_2 from cobra venom was employed (10). The lysis of cells was determined from haemoglobin liberated as measured by O.D. at 540 mu. The results summarized (11) in Table V show that up to 40 ug of phospholipase A_2 produced no lysis nor did it show any significant spot for lecithin or lysolecithin in TLC; 2 ug of P_6 by itself also had no significant effect on the intact cells. However, mixtures of only 1 ug phospholipase A_2 and 2 ug P_6 led to considerable lysis and distinct spots for lecithin and lysolecithin were seen in TLC plates.

TABLE V

COMBINED EFFECTS OF P_6 AND PHOSPHOLIPASE A_2 (P_1) ON INTACT HUMAN RBC UNDER ISOTONIC CONDITIONS

Tube no.	Reaction mixture		O.D. at 540 mu	TLC results	
	P_1 ug	P_6 ug	15% RBC	Lecithin	Lysolecithin
1	nil	nil	nil	-	-
2	20	nil	0.06	-	-
3	40	nil	0.06	-	±
4	20	2	0.60	+++	+++
5	40	2	1.14	++++	++++
6	nil	2	0.15	-	-
7	1	2	0.52	+++	+++

In the light of these observations showing that pure phospholipase A_2 from cobra venom neither degrades phospholipids nor lyses intact human erythrocytes conclusions regarding the location of phospholipids in the erythrocyte membrane (whether on the inner or outer surface) which are based on results obtained with less pure preparations of venom enzymes need a reappraisal. Degradation of phospholipids and lysis produced by mixture of P_6 and phospholipase A_2 could arise through the inhibition of $Na^+ \cdot K^+$-ATPase by P_6, which brings about an imbalance of Na^+K^+ ions, and in turn leads to penetration of water making the cells swell. Under these circumstances phospholipids may become accessible to the phospholipase action. Thus it appears that some linkages in the outer surface of human erythrocyte membrane have to be loosened before the lecithin becomes available to degradation by the phospholipase.

REFERENCES

1. BRAGANCA, B. M., PATEL, N. T., AND BADRINATH, P. G. (1967) Biochim. Biophys. Acta 136, 508.
2. HOLMBERG, G. (1961) Exptl. Cell Res. 22, 406.
3. ANDREWS, P. (1964) Biochim. J. 91, 222.
4. PATEL, T. N., BRAGANCA, B. M., AND BELLARE, R. A. (1969) Exptl. Cell Res. 57, 289.
5. ZAHEER, A., NORONHA, S. H., AND BRAGANCA, B. M. (in preparation).
6. ZWAAL, R. F. A., ROELOFSEN, B., AND MICHAEL, COLLEY (1973) Biochim. Biophys. Acta 300, 159.
7. BRAGANCA, B. M., AND SAMBRAY, Y. M. (1967) Nature 216, 1210.
8. ROELOFSEN, B., ZWAAL, R. F. A., COMFURIUS, P., WOODWARD, C. B., AND VAN DEENEN, L. L. M. (1971) Biochim. Biophys. Acta 241, 925.
9. GUL, S., AND SMITH, A. D., (1972) Biochim. Biophys. Acta 288, 237.
10. BRAGANCA, B. M., SAMBRAY, Y. M., AND GHADIALLY, R. C. (1969) Toxicon 7, 151.
11. KAKIRDE, M. B., NORONHA, S. H., AND BRAGANCA, B. M. (in preparation).

PART III

MEMBRANE BIOENERGETICS

Bioenergetics, particularly membrane bioenergetics, represents one of the most sophisticated specializations of the living world, as biological membranes make possible the controlled channeling of energy into different modalities. Consider, for example, inner mitochondrial and chloroplast membranes, where energy is transduced into transport functions, synthesis of ATP, development of ionic and electric potentials across the membrane, and formation of heat. (Transport includes mono- and divalent cations, weak acid anions, which lend specificity to carbon cycle reactions, and adenine nucleotide exchange. Potential energy is developed in the form of ionic concentrations and electric potentials across the membrane.) These energy transductions are interconvertible except, of course, the formation of heat. Resolution of this unique feature of a highly developed biological energy-transducing membrane requires knowledge of structural and functional organization with which many of the chapters in this section deal. A fuller understanding of the basis of energy-transduction in membranes remains one of the most intriguing and unsolved current problems in biology.

MITOCHONDRIAL SULPHATE AND SULPHITE TRANSPORT

M. Crompton, F. Palmieri, M. Capano, and
E. Quagliariello

Interest in the permeability of mitochondria to sulphite and sulphate stems from the fact that a considerable part of the metabolism of these two anions in liver occurs in the mitochondria. Sulphate is the major end-product of the degradation of the sulphur-containing amino acids in animals; it is formed from sulphite by the mitochondrial enzyme, sulphite oxidase. The site of sulphate formation is almost certainly outside the inner mitochondrial membrane, since two groups of workers have provided good evidence that sulphite oxidase is located in the intermembrane space of liver mitochondria (1,2). However, the formation of sulphite may occur intramitochondrially, e.g., from thiosulphate catalyzed by rhodanese (3), and from cysteine sulphinate by the action of aspartate aminotransferase (4); in this last reaction, sulphite is released by the spontaneous hydrolysis of the immediate product of the transamination, sulphinyl pyruvate. It is clear that the oxidation of the sulphite formed in these reactions necessitates its permeation outwards across the inner membrane to the site of sulphite oxidase.

The capacity of sulphate to permeate mitochondria from both liver and kidney has been apparent for quite some time. In 1962, Winters _et al._ (5) detected an accumulation of radioactive sulphate by isolated kidney mitochondria against a concentration gradient. The simplest interpretation of these results, i.e., that the accumulation was due to an equilibration of radioactive sulphate between the external sulphate and the endogenous sulphate of the mitochondrial preparation, was ruled out by the fact that the endogenous content was insufficient to account for the concentration gradients observed. The possibility that externally-added sulphate was able to exchange with intramitochondrial anions other than sulphate was not considered at this time. Two years later, Rasmussen et al. (6) demonstrated an uptake of sulphate by rat liver mitochondria and, moreover, that the total sulphate uptake was diminished by both phosphate and arsenate. This suggests, at least with hindsight, that perhaps the influxes of sulphate, phosphate, and arsenate occur in exchange for the same intramitochondrial anions.

In the years following, the ways in which phosphate, arsenate, dicarboxylates, and tricarboxylates are transported across the inner mitochondrial membrane were elucidated (7-9). The carriers catalyzing these processes, their substrates, and some inhibitors used in the present study are summarized in Table I. In spite of this explosion of knowledge, little is known of the mechanisms by which sulphite and sulphate permeate.

A relatively simple approach to the study of sulphate and sulphite permeation is to look at the rate of mitochondrial swelling under conditions which employ these two anions. Rapid mitochondrial swelling in an isoosmotic solution of an ammonium salt indicates that this anion permeates in exchange for hydroxyl ions or together with protons, e.g., ammonium phosphate (7). The key point is that the permeation of the cation (in this case, as NH_3) creates a pH gradient across the inner membrane, alkaline inside. However, if the cation of the isoosmotic salt permeates electrogenically, then salt influx, and hence mitochondrial swelling, only follows if the anion is able to permeate electrogenically as well. This is the case with the nitrate anion; mitochondria swell in isoosmotic potassium nitrate in the presence of valinomycin, which allows electrogenic permeation of K^+ (10), but do not swell in ammonium nitrate (11).

TABLE I
THE TRANSPORTING SYSTEMS FOR PHOSPHATE AND TRICARBOXYLIC ACID CYCLE INTERMEDIATES IN RAT LIVER MITOCHONDRIA

Carrier	Exchange catalyzed	Inhibitor
Phosphate	Between phosphate or arsenate and hydroxyl (7)	N-ethylmaleimide, mersalyl* (12, 16,17)
Dicarboxylate	Between phosphate, arsenate, malonate, succinate, and malate (8)	Mersalyl* (12), butylmalonate (18)
Oxoglutarate	Between oxoglutarate, malonate, succinate, and malate (9,18)	
Tricarboxylate	Between citrate and some other tricarboxylates, malate (8), and phosphoenolpyruvate (19)	

*25 nmole/mg mitochondrial protein; at concentrations greater than 50 nmole/mg protein, mersalyl also inhibits the oxoglutarate and tricarboxylate carriers (13).

Figure 1 shows the swelling of rat liver mitochondria when suspended in isoosmotic solutions of sulphite and sulphate salts. The data indicate that sulphite, but not sulphate, is able to permeate by a mechanism equivalent to an exchange for hydroxyl ions.

Chappell and Haarhoff (8) showed that mitochondria do not swell in isoosmotic solutions of ammonium malate, succinate, or malonate until a low concentration of phosphate, insufficient in itself to cause swelling, is added. In this case, ammonium phosphate permeates, and the dicarboxylate anions enter on the dicarboxylate carrier in exchange for phosphate. The net influx is the ammonium salt of the dicarboxylic

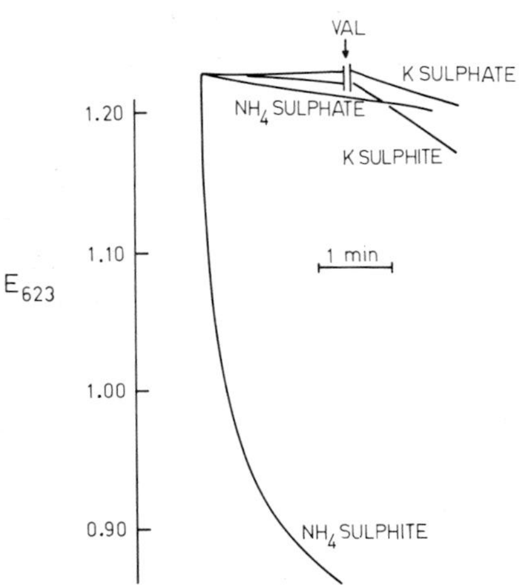

Fig. 1. Mitochondrial swelling in sulphite and sulphate salts. The incubations contained the ammonium or potassium salt of sulphurous or sulphuric acid, conc. 120 mM, and 20 mM tris HCl, 0.5 mM EDTA, 1 mM KCN, 2.4 mg mitochondrial protein and, where indicated, 1 μg valinomycin (val.). Final vol., 2.5 ml; pH 7.4.

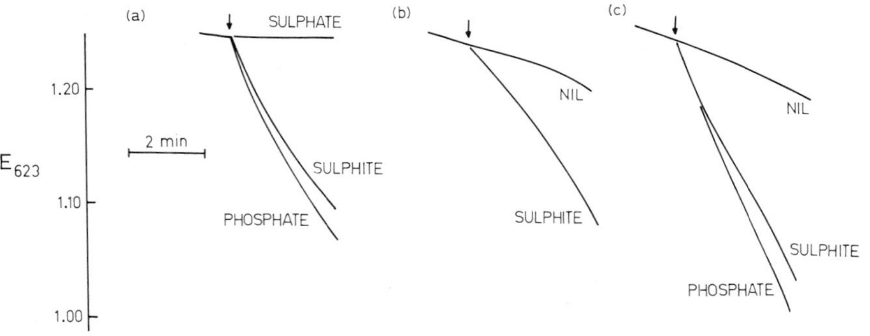

Fig. 2. Mitochondrial swelling in ammonium malate, sulphate, and phosphate. The incubations contained either 100 mM ammonium malate (a), phosphate (b), or sulphate (c), 20 mM tris HCl, 0.5 mM EDTA, 1 mM KCN, 2.5 mg mitochondrial protein and, in (b) 2 mM N-ethylmaleimide. Vol., 2.5 ml; pH 7.4. The further additions of 50 uL, 250 mM ammonium phosphate, sulphate and sulphite were made at the arrows; final conc., 5 mM.

acid. Figure 2a shows that sulphite, but not sulphate, is almost as effective in eliciting swelling in ammonium malate as phosphate. Similar experiments have confirmed that sulphite is equally able to produce mitochondrial swelling in isoosmotic ammonium malonate and succinate. Sulphite also stimulates swelling in isoosmotic ammonium phosphate in the presence of N-ethylmaleimide (Fig. 2b); under these conditions, the phosphate carrier is inhibited and the only known way in which phosphate can enter is by the dicarboxylate carrier. The simplest way to interpret these data is that the influxes of phosphate, malate, succinate and malonate occur in exchange for an efflux of sulphite and are catalyzed by the dicarboxylate carrier. An additional point is that, although ammonium sulphate itself does not permeate, swelling in ammonium sulphate may be induced by the addition of phosphate or sulphite (Fig. 2c), implying that sulphate influx can occur in exchange for both phosphate and sulphite.

These possibilities were further investigated by measuring the exchanges in a more direct manner. Mitochondria were loaded with various labelled metabolites and the ability of other, unlabelled, externally-added anions to elicit an efflux of the intramitochondrial, labelled anions was investigated. It is reasonable to assume that any efflux occurring does so by exchange with the external anion. In the case where the externally-added anion is the same as the internally loaded anion and in the absence of inhibitors (control exchange), the exchange is the maximum obtainable, and the other exchanges have been represented as a percentage of the control exchange. The results are contained in Table II.

It can be seen that externally-added sulphite and sulphate exchange to approximately the maximum degree with internal malonate, succinate, and malate. These exchanges are inhibited by mersalyl at a concentration of 45 nmole/mg mitochondrial protein; this amount of mersalyl inhibits the dicarboxylate carrier (12), but does not inhibit either the oxoglutarate (13) or tricarboxylate carriers (14). From this it may be

TABLE II
THE EXCHANGES BETWEEN INTRAMITOCHONDRIAL ANIONS AND EXTRAMITOCHONDRIAL SULPHITE AND SULPHATE

External anions and inhibitors	% exchange with internal anions						
	malonate	succinate	malate	phosphate	sulphate	citrate	oxoglutarate
Sulphate	98	94	94	91	99	2	2
Sulphate, mers	1	1	2	3	7		
Sulphate, NEM	96	96	95	83	96		
Sulphate, NEM, BM	-5	2	-5	8	-1		
Sulphite	88	93	102	93	100	4	8
Sulphite, mers	0	1	0	9	8		
Sulphite, NEM	83	90	91	98	94		
Sulphite, NEM, BM	-2	6	1	9	-1		

Mitochondria (2-2.5 mg protein), loaded (see ref. 20) with the anion indicated (internal anion), were incubated in a medium containing 100 mM KCl, 20 mM tris-HCl, 1 mM EGTA, and 1 mM KCN at 8°; final vol., 1 ml. Some incubations contained 0.1 mM mersalyl (mers), 2 mM N-ethylmaleimide (NEM), and 20 mM butylmalonate (BM) as indicated. After 2 min incubation, the exchanges were initiated by adding 2 mM unlabelled anion (external anion) and terminated 1 min later by sedimentation of the mitochondria by centrifugation for 1 min. The radioactivity content of the pellet was determined. The values are expressed as a percent of the relevant control exchanges in which the unlabelled external anion added was the same as the internal anion loaded and no inhibitor was added. Negative values indicate that the radioactivity content of the pellet was greater in the presence of external anion than in its absence.

concluded that only the dicarboxylate carrier is active in exchanging the dicarboxylate anions for sulphite and sulphate. The exchanges are also inhibited by butyl-malonate.

In addition, external sulphite and sulphate exchange with phosphate, the other known substrate of the dicarboxylate carrier, and sulphate, and these processes are also sensitive to mersalyl and butyl-malonate.

In contrast, little exchange occurs with internal citrate or oxoglutarate, which provides additional evidence against the involvement of the tricarboxylate carrier and oxoglutarate carrier in sulphate or sulphite transport.

It should be noted that the amount of sulphate formed by sulphite during those exchange experiments involving external sulphite is very small, and measurements indicate that the externally-formed sulphate would account for less than 15% of the total exchange observed.

The above experimental data appear to justify the postulate that both sulphite and sulphate may be transported by the dicarboxylate carrier of rat liver mitochondria. This, perhaps, explains the observations of Rasmussen *et al.* (6) alluded to earlier, since both phosphate and arsenate are substrates of the dicarboxylate carrier, and would be expected, therefore, to decrease sulphate uptake. One interesting feature of the dicarboxylate carrier is the apparent existence of different binding sites for phosphate and the dicarboxylate substrates (15), and knowledge of the sites occupied by sulphite and sulphate should further our understanding of the properties of the substrate-binding sites.

The additional mechanism permeation to sulphite seems to be equivalent to an exchange for hydroxyl ions. The true nature of this process remains to be elucidated. However, we have shown that the influx of ammonium sulphite is quite insensitive to inhibition by both N-ethylmaleimide and mersalyl, both of which inhibit ammonium phosphate influx (12,16,17). It appears therefore that the only other carrier known to catalyze an equivalent process, i.e., the phosphate carrier, is not involved in sulphite permeation.

REFERENCES

1. WATTIAUX-DeCONNINCK, S., AND WATTIAUX, R. (1971) *Eur. J. Biochem.* *19*, 552-556.
2. COHEN, H., BETCHER-LANGE, S., KESSLER, D. L., AND RAJAGOPALAN, K. V. (1972) *J. Biol. Chem.* *247*, 7759-7766.
3. KOJ, A., FRENDO, J., AND JANIK. Z. (1967) *Biochem. J.* 103, 791-795.
4. SINGER T. P., AND KEARNEY, E. B. (1956) *Arch. Biochem. Biophys.* *61*, 397-409.
5. WINTERS, R. W., DELLUVA, A. M., DEYRUP, I. J., AND DAVIES, R. E. (1962) *J. Gen. Physiol.* *45*, 757-775.
6. RASMUSSEN, H., SALLIS, J., FANG, M., DELUCA, H. F., AND YOUNG, R. (1964) *Endocrinology* *74*, 388-394.
7. CHAPPELL, J. B., AND CROFTS, A. R. (1966) *in* Regulation of Metabolic Processes in Mitochondria

(Tager, J. M., Papa, S., Quagliariello, E., and Slater, E. C., eds.) p. 293-316, Elsevier, Amsterdam.

8. CHAPPELL, J. B., AND HAARHOFF, K. M. (1967) in Biochemistry of Mitochondria (Slater, E. C., Kaniuga, Z., and Wojtczak, I., eds.) p. 75-91, Academic Press, London and New York.

9. MEIJER, A. J., AND TAGER, J. M. (1966) Biochem. J. 100, 79P.

10. HENDERSON, P. J. F., McGIVAN, J. D., AND CHAPPELL, J. B. (1969) Biochem. J. 111, 521-535.

11. CROMPTON, M., McGIVAN, J. D., AND CHAPPELL, J. B. (1973) Biochem. J. 132, 27-34.

12. MEIJER, A. J., GROOT, G. S. P., AND TAGER, J. M. (1970) FEBS Letters 8, 41-44.

13. QUAGLIARIELLO, E., AND PALMIERI, F. (1971) Biochimica Applicata 18, 191-219.

14. PALMIERI, F., STIPANI, I., QUAGLIARIELLO, E., KLINGENBERG, M. (1972) Eur. J. Biochem. 26, 587-594.

15. PALMIERI, F., PREZIOSO, G., QUAGLIARIELLO, E., AND KLINGENBERG, M. (1971) Eur. J. Biochem. 22, 66-74.

16. FONYO, A., BESSMAN, S. (1968) Biochem. Med. 2, 145-163.

17. TYLER, D. D. (1968) Biochem. J. 107, 121-123.

18. ROBINSON, B. H., AND CHAPPELL, J. B. (1967) Biochem. Biophys. Res. Commun. 28, 249-255.

19. ROBINSON, B. H. (1969) FEBS Letters 4, 251-254.

20. PALMIERI, F., QUAGLIARIELLO, E., AND KLINGENBERG, M. (1972) Eur. J. Biochem. 29, 408-416.

THE INTERACTION OF Ca^{2+} WITH THE MITOCHONDRIAL MEMBRANE AND WITH A SOLUBLE MITOCHONDRIAL GLYCOPROTEIN

Ernesto Carafoli

The evidence for the existence of a carrier that mediates the translocation of Ca^{2+} across the inner mitochondrial membrane has been reviewed recently (1-3) and is presented in Table I. Clearly, the energy-

TABLE I
EVIDENCE FOR A Ca^{2+} CARRIER IN THE MITOCHONDRIAL MEMBRANE

1	The transport shows saturation kinetics.
2	The transport system shows specificity. Ca^{2+} and Sr^{2+} are transported with equivalent efficiency. Mn^{2+} and Ba^{2+} are transported less efficiently. Be^{2+} and Mg^{2+} are not transported.
3	The transport of Ca^{2+} is competitively inhibited by Sr^{2+}.
4	The transport of Ca^{2+} is inhibited by La^{3+} and ruthenium red at low concentrations.
5	No energy-linked transport Ca^{2+} is present in some mitochondrial types.
6	Sites having a very high affinity for Ca^{2+}, and sensitive to La^{3+} and ruthenium red, have been identified in the inner mitochondrial membrane.

linked translocation process satisfies all of the classical parameters of carrier-mediated transport processes. The high-affinity Ca^{2+} binding sites, first described by Reynafarje and Lehninger (4), deserve particular mention, since they share with the energy-linked transport process many of the most important properties (Table II). The suggestion that

TABLE II
THE HIGH AFFINITY Ca^{2+} BINDING SITES OF THE INNER MITOCHONDRIAL MEMBRANE

Maximal capacity	About 1 μmole of Ca^{2+} per mg of protein (rat liver mitochondria).
Affinity for Ca^{2+}	$K_d < 1$ μM
Specificity	Ca^{2+} and Sr^{2+} are bound very efficiently. Mn^{2+} less efficiently. Other Me^{2+} are not bound.
Intermitochondrial distribution	Present only in the inner membrane.
Species distribution	Present in all mammalian mitochondria. Absent from yeast and blowfly flight muscle mitochondria.
Inhibitors	Sensitive to μM concentrations of ruthenium red and lanthanides. Also sensitive to uncouplers of oxidative phosphorylation.

they participate in the transfer of Ca^{2+} in intact mitochondria is thus rather reasonable. The most obvious suggestion in this respect is that they represent the Ca^{2+} carrier, be it a mobile carrier which shuttles between the two faces of the inner mitochondrial membrane, or a fixed hydrophylic channel which spans across the entire width of the membrane. It is, however, also possible that the system for the transfer of Ca^{2+} has more than one component; in addition to the carrier, there could be a "recognition" site, which confers Ca^{2+} specificity to selected areas of the membrane surface, and thus facilitates the interaction of Ca^{2+} with the carrier. The high-affinity sites could in this case thus represent the recognition site, and convey Ca^{2+} to the carrier which will translocate it.

One important common property of the transport system and of the high affinity sites is the inhibition by ruthenium red, a widely used histochemical stain which has been introduced to mitochondria by Moore (5), and which has rapidly become a very important tool for research in the area of Ca^{2+} transport (6). Whereas the absolute specificity of ruthenium red for mucopolysaccharides and glycoproteins has recently been questioned (7), its effects certainly indicate that carbohydrate-containing compounds are involved in the translocation of Ca^{2+}, and in its binding to the high-affinity sites. It is interesting in this respect that a protein fraction isolated by Gomez-Puyou et al. (8) by ammonium-sulphate fractionation of osmotic extracts of liver mitochondria, as well as several partially purified liver mitochondrial fractions obtained by various treatments in our laboratory (9), contained measurable amounts of carbohydrates.

The demonstration by Sottocasa et al. (10) of the existence of glycoproteins in liver mitochondria offered precise indications on the nature of the carbohydrate-containing compound of the Ca^{2+} transfer system, and has provided a powerful incentive for our efforts to isolate and characterize glycoproteins from mitochondria. The isolation procedure normally used is described in a series of publications from our Laboratory (9,11-13), and will not be reviewed here in detail. Essentially, mitochondria are exposed to hypotonic solutions under conditions which limit greatly the leakage of enzymes from the matrix (10), and a very acidic glycoprotein is isolated from the suitably concentrated osmotic extract by preparative polyacrylamide gel electrophoresis. Additional amounts of glycoprotein can be isolated from the mitochondrial residues, after extraction of the glycoprotein with osmotic shocks, by a vigorous sonication. Still more glycoprotein can be liberated from the mitochondrial membranes after collection of the protein solubilized

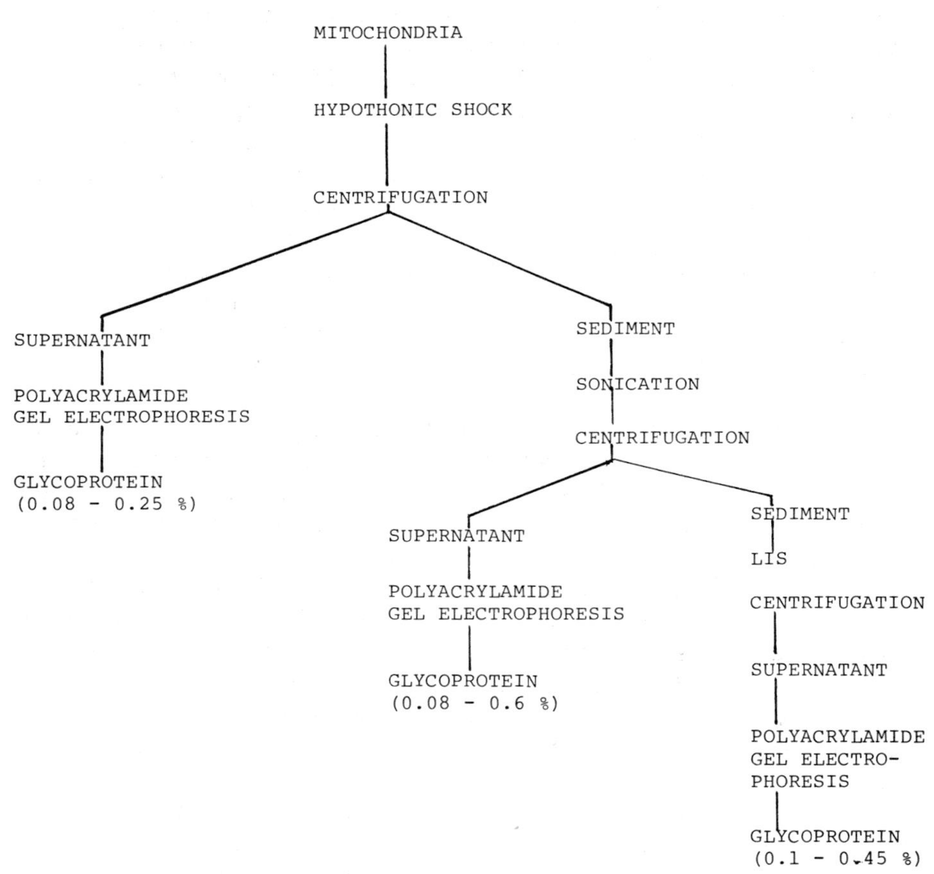

Fig. 1. Preparation of glycoproteins from mitochondria. Explanations in the text.

by sonication. This requires treatment with lithium di-iodosalycilate (LIS), a chaotropic agent specific for membrane glycoproteins (14). The extraction and purification scheme is summarized in Fig. 1, and shows that the 3 treatments described liberate from mitochondria roughly the same amounts of glycoprotein, up to a total amount which in most experiments corresponds to about 1% of the total mitochondrial proteins. The problem of the intramitochondrial location of the glycoprotein has been discussed by Sottocasa et al. (10,15), and it now appears clear that of the four mitochondrial compartments, i.e., the inner membrane, the outer membrane, the intermembrane space, and the matrix, the latter can be ruled out as a possible source of glycoprotein. Both the inner and outer membrane, and the intermembrane space, however, contain measurable amounts of it. According to Melnick et al. (16) the glycoproteins from the inner and outer membranes are different. The chemical properties of the glycoprotein are summarized in Table III. Particularly noteworthy is its pronounced acidic

TABLE III
A SUMMARY OF THE PROPERTIES OF THE Ca^{2+}-BINDING GLYCOPROTEIN

Monomer molecular weight	About 35,000
Sugars	About 10%
Phospholipids	Up to 1/3 in weight
Endogenous Ca^{2+} and Mg^{2+}	3-5 moles per mole of protein
Amino acids	1/3 glutamic and aspartic acids
Binding sites for Ca^{2+}	
1) high affinity	2-3 per mole
2) low affinity	20-30 per mole
Inhibition of Ca^{2+} binding	By La^{3+} and ruthenium red
Species distribution	No glycoprotein capable of binding Ca^{2+} with high affinity in yeast and blowfly flight muscle mitochondria

character, which seems to be a common property to most Ca^{2+} binding proteins. Indeed, most of them bind large amounts of Ca^{2+} at sites having a low affinity for it, and very small amounts at sites which have a much stronger affinity. It seems reasonable to suggest that the carboxylic groups are involved in the low affinity binding, whereas the binding with high affinity possibly requires different and specific groups in each protein. The properties outlined in Table III refer to the protein separated from the inner membrane by osmotic shocks. No significant differences have been found in the LIS protein, which justifies the conclusion that the two fractions are the same. One possible difference is the content in phospholipids, which is currently under study in our laboratory, and which could conceivably determine the different tightness of association of the two proteins with the membrane.

The isolated glycoprotein binds Ca^{2+} very actively, apparently at two classes of sites widely differing in affinity. The high affinity class has a low capacity for Ca^{2+} (2-3 moles per mole of protein), binds it with a K_d of less than 1 μM, and is abolished by La^{3+} and ruthenium red. The reaction of Ca^{2+} with the low affinity sites, however, is also markedly depressed by the two inhibitors.

The last property mentioned in Table III, namely, the absence of glycoproteins with high affinity for Ca^{2+} from yeast and blowfly mitochondria, speaks in favor of the involvement of the glycoprotein in the transport process in intact mitochondria. So does also the effect of inhibitors. The evidence is indirect, however, and in any case the problem still remains that of deciding whether the glycoprotein acts as a superficial recognition site for Ca^{2+}, or as a carrier for it.

The pronounced polar character of the glycoprotein would seem an argument against its location (and, possibly, free motion) within the hydrophobic domain of the mitochondrial membrane. It must be recalled, however, that Gitler and Montal (17) have recently succeeded in solubilizing polar proteins in apolar solvents by neutralizing the negative charges of the molecule with Ca^{2+}, and the positive charges with acidic phospholipids. By using the same method, Montal and Korenbrot (18) have recently incorporated rodopsin into artificial lipid bilayers and have shown that it remains functional after incorporation. In principle, therefore, the glycoprotein could well act as a mobile carrier (or an immobile channel) in the hydrophobic region of the mitochondrial membrane, and we have indeed obtained indications that, under certain circumstances, the LIS protein can be solubilized in η-decane.

We have recently carried out experiments (19) on artificial lipid bilayers, and have obtained indications compatible with the suggestion that the glycoprotein may act as a Ca^{2+} translocator. Figure 2

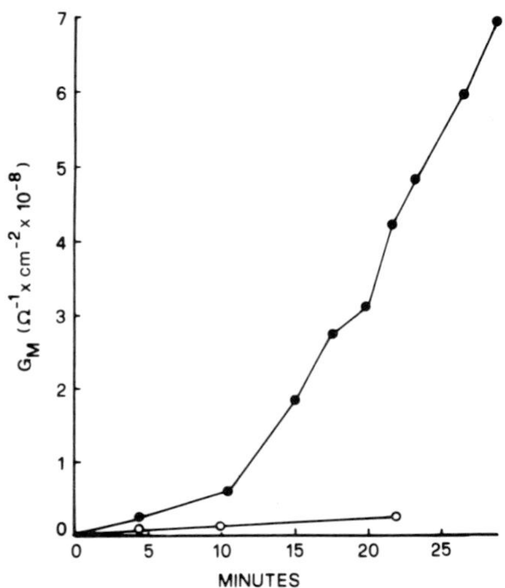

Fig. 2. Change in conductance of a lecithin bilayer upon addition to the aqueous medium of the mitochondrial Ca^{2+}-binding glycoprotein. Effect of ruthenium red. The preparation of the bilayer and the measurement of electrical conductance are described in (19). 10^{-8} M glycoprotein was added at zero time to both sides of the bilayer membrane. The aqueous medium contained 4 mM $CaCl_2$. Ruthenium red was 2 μM.

shows that the glycoprotein greatly enhances the electrical conductance of lecithin bilayers when added to the medium containing, as the only cation, 4 mM Ca^{2+}. The figure also shows that ruthenium red abolishes the effect of the glycoprotein, indicating that the increase of the electrical conductance is related to the ability of the glycoprotein to complex Ca^{2+}. The specificity of Ca^{2+} is, however, not absolute. Figure 3 shows that monovalent cations also increase, although considerably less than Ca^{2+}, the electrical conductance of the bilayer in the presence of the glycoprotein. This rather surprising finding may be taken to indicate that monovalent cations may also be transported by the glycoprotein, at least in the artificial

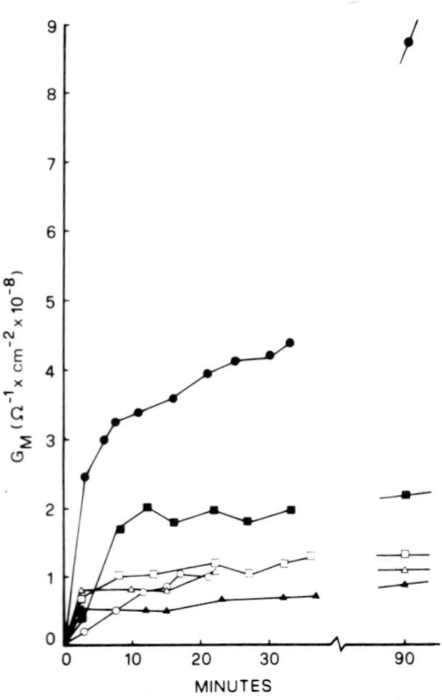

Fig. 3. Changes in membrane conductance of lecithin bilayers induced by the addition of 10^{-8} M Ca^{2+}-binding glycoprotein (zero time) in the presence of 4 mM Ca^{2+} or equal concentrations of various monovalent cations.

membrane system. Alternative explanations are, however, possible. Ca^{2+} (and, less efficiently, monovalent cations) could for example permit the specific interaction of the glycoprotein with the bilayer. Once associated with the bilayer, the glycoprotein could induce the increase in electrical conductance by facilitating the transport of different species, including protons.

REFERENCES

1. LEHNINGER, A. L., AND CARAFOLI, E. (1969) in Biochemistry of the Phagocytic Process (Schulz, J., ed.) p. 9, North Holland, Amsterdam.
2. CARAFOLI, E., AND LEHNINGER, A. L. (1971) Biochem. J. 122, 681.
3. CARAFOLI, E. (1973) Biochimie 55, 755.
4. REYNAFARJE, B., AND LEHNINGER, A. L. (1969) J. Biol. Chem. 244, 589.
5. MOORE, C. (1971) Biochem. Biophys. Res. Commun. 42, 298.
6. VASINGTON, F. D., GAZZOTTI, P., TIOZZO, R., AND CARAFOLI, E. (1972) Biochim Biophys. Acta 256, 43.
7. LUFT, H. J. (1971) Anat. Rec. 171, 347.
8. GOMEZ-PUYOU, A., TUENA DE GOMEZ-PUYOU, M., BECKER, G., AND LEHNINGER, A. L. (1972) Biochem. Biophys. Res. Commun. 47, 814.
9. CARAFOLI, E., GAZZOTTI, P., VASINGTON, F. D., SOTTOCASA, G. L., SANDRI, G., PANFILI, E., AND DE BERNARD, B. (1972) in Biochemistry and Biophysics of Mitochondrial Membranes (Azzone, G. F., Carafoli, E., Lehninger, A. L., Quagliariello, E., and Siliprandi, N., eds.) p. 623, Academic Press, N.Y.
10. SOTTOCASA, G. L., SANDRI, G., PANFILI, E., AND DE BERNARD, B. (1971) FEBS Letters 17, 100.
11. SOTTOCASA, G. L., SANDRI, G., PANFILI, E., DE BERNARD, B., GAZZOTTI, P., VASINGTON, F. D., AND CARAFOLI, E. (1972) Biochem. Biophys. Res. Commun. 47, 808.
12. CARAFOLI, E., GAZZOTTI, P., SALTINI, C., ROSSI, C. S., SOTTOCASA, G. L., SANDRI, G., PANFILI, E.,

AND DE BERNARD, B. (1973) *in* Mechanisms in Bioenergetics (Quagliariello, E., Ernster, L., Azzone, G. F., and Siliprandi, N., eds.) p. 293, Academic Press, N.Y.

13. CARAFOLI, E., AND SOTTOCASA, G. L., 9th Intern. Congress of Biochem., Stockholm, 1973, Symposium in honor of B. Chance, in press.
14. MARCHESI, V. T., AND ANDREWS, E. P. (1971) *Science* *174*, 1247.
15. SOTTOCASA, G. L., SANDRI, G., PANFILI, E., GAZZOTTI, P., AND CARAFOLI, E., 9th Intern. Congress of Bichem., Stockholm 1973, Abstract 4 f 12.
16. MELNICK, R. L., TINBERG, H. M., MAGUIRE, J., AND PACKER, L. (1973) *Biochim. Biophys. Acta* *311*, 230.
17. GITLER, C., AND MONTAL, M. (1972) *FEBS Letters* *28*, 329.
18. MONTAL, M., AND KORENBROT, J. I. (1974) *Nature*, in press.
19. PRESTIPINO, G. F., CECCARELLI, D., CONTI, F., AND CARAFOLI, E., submitted.

A MOLECULAR BASIS FOR INTERACTION OF BIGUANIDES WITH THE MITOCHONDRIAL MEMBRANE

Guenter Schaefer

INTRODUCTION

The blood sugar lowering activity of derivatives of guanidines, diguanidines, and biguanides *in vivo* is useful in the treatment of diabetes mellitus. On the metabolic level, two different phenomena which have been studied extensively *in vitro* can be distinguished: (1) the inhibition of mitochondrial oxidative phosphorylation (1-3) and (2) the inhibition of glucose production in gluconeogenetic tissues (4,5). Both may be correlated via the cellular ATP/ADP ratio, a very simple concept. A more common basis for both inhibitory activities may be seen in the interaction of these drugs with biological membranes. Such an assumption is justified because most of the processes sensitive to guanidines or biguanides are either directly membrane bound or involve membrane linked reactions.

Inhibitory actions of guanidine derivatives have been observed on the following processes: respiration and oxidative phosphorylation (1,2,3,6,7,19); specificity to site I and/or II in oxidative phosphorylation, specific interaction with the energized state (2,8); turnover of TCA cycle, ^{14}C-incorporation into liver lipids (9,10); gluconeogenesis in liver and kidney (4,5,11,12); intestinal glucose uptake (13,13a); ^{32}P/ATP exchange in isolated mitochondria (14); energy transfer in chloroplasts (15); phosphorylation in chromatophores (site specific) (16); cation-exchange at the mitochondrial membrane Mg^{++}, Ca^{++}, (K^{+}?) (17, 18).

The site specificity of certain guanidines against phosphorylation sites I and II in mammalian mitochondria as well as in the energy coupling system of chromatophores is one of the most interesting features of these inhibitors. It has been interpreted by several investigations to support the chemical theory of energy coupling and oppose the chemiosmotic hypothesis, as the discussion will show, although other explanations are possible.

Since it has been repeatedly observed that the onset of the guanidine induced inhibitions has a considerable lag-phase and requires the existence of a so-called "energized state" of mitochondrial or chromatophore membranes, it has been concluded that guanidines either bind to a high energy intermediate (8,16,19) or prevent its utilization in the sequence of energy conservation reactions.

A number of different possibilities for biguanide interaction with energy conservation may be considered.

1) The inhibitor reversibly or irreversibly binds to small molecules participating in the reaction (e.g., postulated high energy intermediates).
2) The inhibitor competes with smaller molecules for binding sites at macromolecular structures.
3) The inhibitor dissipates membrane potentials by active permeation.
4) The inhibitor generates or alters a membrane potential by binding to the surface.
5) The inhibitor causes changes of the dielectric properties of a membrane.
6) The inhibitor causes ultrastructural changes by perturbation of molecular organization of a membrane.

Taking into account that the inhibitory concentrations are much higher than expected on the basis of a stoichiometric interaction with any functional intermediate, possibility 1) seems rather unlikely unless the affinity constant at the binding sites is extremely low. As a rough estimate the concentration of high energy intermediates should be in the range of the cytochrome concentrations in mitochondrial membranes.

All other possibilities stated above may result from 6), which assumes that merely by the presence of biguanides changes occur within the membrane itself, creating the micro-environment for the energy conservation reactions. This may lead to a simultaneous modification of several membrane linked processes thus yielding a general basis for a variety of observed metabolic effects.

An exact discrimination between the different postulated mechanisms requires detailed knowledge of binding and distribution of the inhibitor molecules. Such information was missing until specific studies were carried out in our laboratory (20,21). In the following I will therefore concentrate on two topics:

1) The properties of biguanide binding to various types of membranes.
2) Physical membrane properties in the presence of bound biguanides.

BINDING OF BIGUANIDES TO NATURAL AND SYNTHETIC MEMBRANES

With natural membranes such as mitochondria, submitochondrial particles or microsomes, biguanide binding can be studied by means of either isotope uptake using ^{14}C-labeled compounds or by measurement of ANS fluorescence following addition of biguanides to the suspended particles. If a large-scale uptake of biguanides to the inside of particles is assumed either by passive or active permeation processes (22), the results will differ. Whereas ^{14}C-binding measures the total amount of biguanide associated with the particles, ANS should respond only when biguanides bind to the membrane-solution interface. Bound ANS has been shown to be located at the aqueous/lipid interface in the region of the phosphoester bonds within the headgroups of polar phospholipids (23).

Interestingly enough, both methods agree that binding can be saturated and that affinity constants are obtained. For this and several other reasons, we conclude that biguanides are almost completely membrane bound even when accumulated inside the particles.

Figure 1 shows titrations of mitochondria with different biguanides using the ANS-fluorescence method.

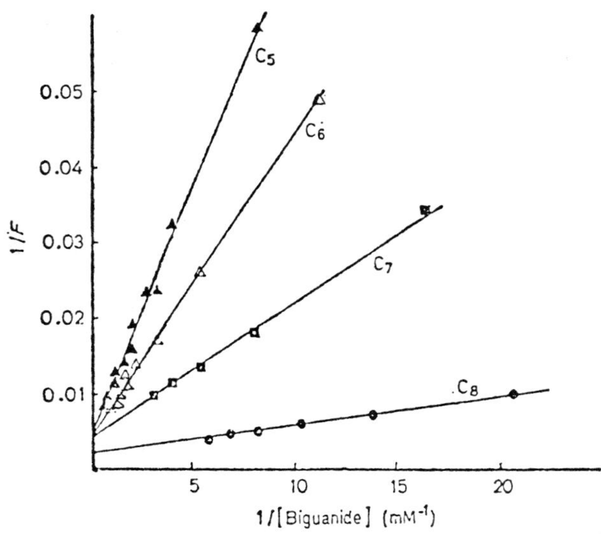

Binding of different biguanides to rat liver mitochondria as indicated by anilinonaphthalene sulfonate fluorescence. Conditions were sucrose-Tris medium pH 7.3; 2 μm rotenone, 30 μM anilinonaphthalene sulfonate; total volume 2.08 ml; mitochondrial protein 3.2 mg. C_5—C_8 signifies length of hydrocarbon side chain; C_6 is phenethyl biguanide

Fig. 1

TABLE I

BINDING AFFINITY OF BIGUANIDES TO MITOCHONDRIAL MEMBRANES

Biguanide derivative	$C_{1.1}$	C_4	C_5	C_6*	C_7	C_8
K_{ass} (ANS) mM^{-1}	0.012	0.312	0.505	0.681	1.49	3.18
K_{ass} (14-C) mM^{-1}	--	0.312	--	0.63	1.46	4.38
P_o (n-octanol/ phosph. buffer)	0.037	0.063	0.096	0.148	0.247	0.605

*C_6 signifies phenethylbiguanide (DBI).

The subscripts give the length of the hydrocarbon side-chain of the compounds. The excellent linearity of the double reciprocal plot yields affinity constants which are given in Table I together with those obtained from isotope distribution studies in mitochondria. Figure 2 clearly demonstrates that binding of biguanides is essentially hydrophobic as seen from the excellent correlation between affinity-constants and the partition coefficients of these compounds. The extremely small ordinate-intercept indicates that there is

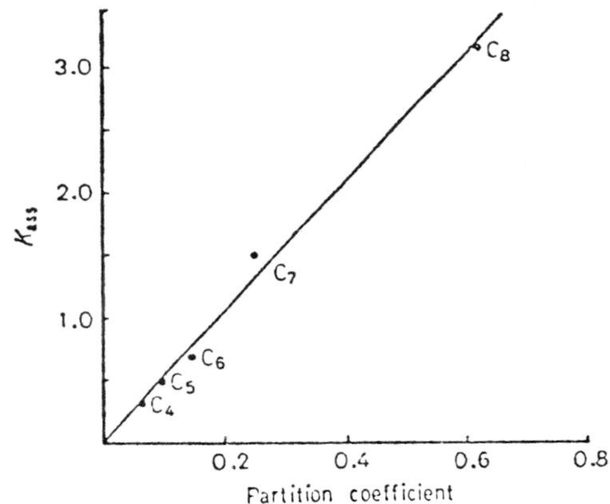

Correlation of affinity constants for biguanide binding and partition coefficients. Affinity constants were determined from reciprocal plots of biguanide titrations of anilinonaphthalene sulfonate fluorescence. ($K_{ass} = 5.17\ P_0 + 0.018$; $r = 0.995$)

Fig. 2

almost no electrostatic contribution to binding. Electrostatic forces, however, may modify the binding characteristics, for example during energization of the membrane as we have shown elsewhere (20,21). It could be demonstrated that the affinity constants were changed when the particles became energized although the total number of binding sites was unchanged. This has to be expected when binding occurs essentially to hydrophobic sites.

The method of titration of ANS-fluorescence with biguanides has proved to be of special value for comparison of different types of lipid membranes and for the identification of the natural binding sites.

ANS binds equally well to mitochondria, submitochondrial particles, microsomes, synthetic lipid micelles or bimolecular membranes. The titration of fluorescence of ANS with biguanides lead to comparable binding constants with all of these membranes, suggesting that the mechanism of binding is identical. Table II lists affinity constants obtained with different biguanides and different membranes. The data

TABLE II
AFFINITY CONSTANTS OF BIGUANIDES TOWARDS DIFFERENT MEMBRANES

Compound	C_6*	C_7	C_8	Type of Membrane
K_{ass}** (ANS)	0.603	1.49	3.38	Mitoch. (liver)
K_{ass} (14 C)	0.63-0.68	1.46	4.38	Mitoch. (liver)
K_{ass} (ANS)	0.521	--	--	Sub.mitoch. particles
K_{ass} (ANS)	0.51-0.63	--	4.96	Liposomes
K_{ass} (ANS)	0.4	1.64	4.39	Microsomes

*The index gives the number of carbon atoms in the side chain; C_6 = phenethylbiguanide (DBI).

**The affinity constants are given as mM^{-1}.

are in very good agreement, especially when one takes into account that membranes of completely different origins and compositions were compared. More attention has been given to phenethyl-biguanide since the data for this compound regarding metabolic models for correlation to binding properties are available. In comparing phosphatidyl-ethanol-amine, phosphatidyl choline, cardiolipin or sphingomyelin, the data vary only within the error of the method.

The close relationship of binding to natural membranes or synthetic phospholipid structures strongly suggests that membrane phospholipids are the natural binding sites. Additional evidence could be accumulated by successive lipid extraction from mitochondrial membranes resulting in a gradual decrease of biguanide binding. Figure 3 shows the double reciprocal plot of fluorescence enhancement over biguanide concentration for two different biguanides, using mitochondria which were extracted from phospholipids by different degrees. The extraction procedure was essentially that reported by Fleischer (24). It is important to note that by lipid extraction the ordinate intercept

BINDING OF BIGUANIDES TO PHOSPHOLIPID DEPLETED AND LECITHIN RESTORED RAT LIVER MITOCHONDRIA

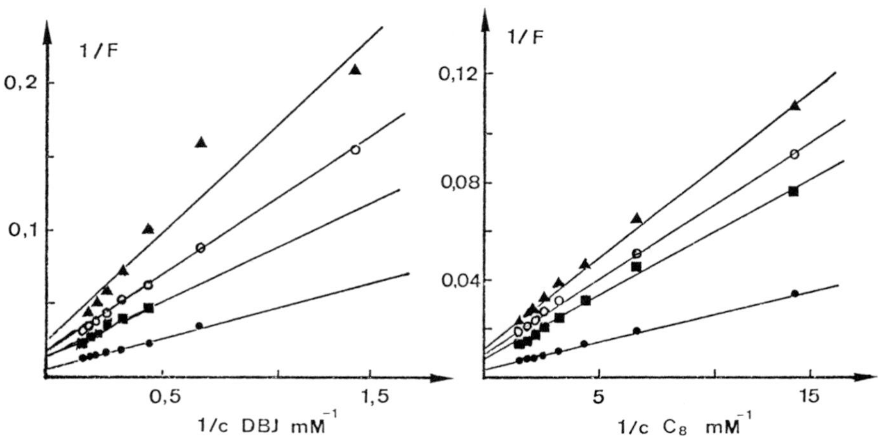

Phospholipid content:

- ●–● 100 % - norm. mitochondria
- ▲–▲ 16.4 %
- o–o 24.8 %
- ■–■ 34.4 %

Binding was monitored by ANS fluorescence.
(ref.: Schäfer, G., and Bojanowski, D., 1972)

Fig. 3

of the linearized adsorption-isotherms varies with the degree of extraction, thus indicating a change in the number of binding sites. In contrast, the abscissa-intercept remains fairly constant. This is expected when there is no change of affinity.

For these experiments the relative number of binding sites is compared to the phospholipid content of the membrane preparation (Table III). The actual

TABLE III
CORRELATION OF PHOSPHOLIPID CONTENT AND CAPACITY FOR BIGUANIDE BINDING TO LIVER MITICHONDRIA

Phospholipid content		Number of binding sites			
μg/mg protein	% of control	Phenethyl-biguanide $n_{relat.}$*	nMol/mg**	n-octyl-biguanide $n_{relat.}$	nMol/mg
12.6	100	1.0	82	1.0	130
4.2	33.4	0.40	33	0.47	62
3.1	24.6	0.35	29	0.38	50
1.8	14.3	0.25	20	0.32	41

*Data taken from ANS fluorescence titrations.

**Data based on 14-C-biguanide binding to normal rat liver mitochondria (= 100%).

number of binding sites was calculated on the basis of the isotope distribution data obtained with non-extracted particles. A regression analysis gave an absolutely linear correlation between phosphate contents of the membrane and the capacity for biguanide binding. The regression line, however, does not go to 0. There remains a considerable ordinate intercept of about 10-20% of the total binding capacity. This clearly points out that binding also occurs to other than phospholipid sites. These sites may be assumed to be of lipid nature as well, since it has been shown that only negligible amounts of biguanides bind to proteins (25). It should be stressed that biguanide binding to lipid extracted mitochondrial membranes could be gradually restored by reintegration of lipids into mitochondrial membranes.

From the experiments described so far it appears that phospholipids in fact represent the main binding sites for biguanides as inhibitors of oxidative phosphorylation. Additional evidence is given by another analogy of binding to synthetic membranes and mitochondrial membranes. When biguanides are added to phospholipid-containing membrane preparations a release of protons occurs. This can be monitored by absorbancy changes of bromcresolpurple (BCP). It is a fast reaction, in which the kinetics are identical in mitochondrial and model lipid-membranes. The reaction can be described by a second order mechanism. Under appropriate conditions, however, a pseudo-first order reaction results as shown in Fig. 4, yielding the oscilloscope trace of a stopped flow experiment. As depicted from the figure, the reaction is biphasic in this particular experiment where lecithin dispersions

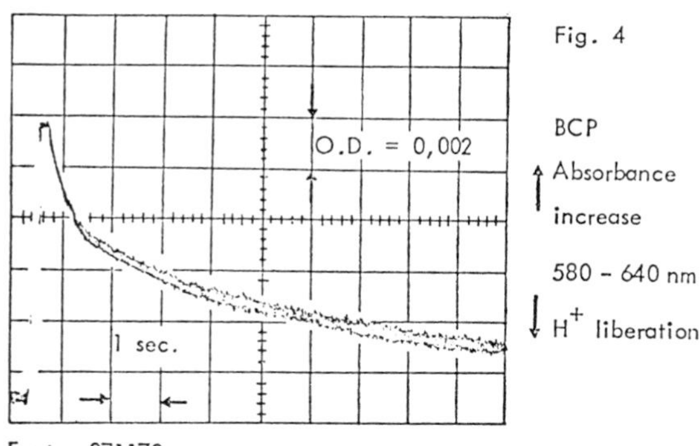

Stopped-flow measurement of liberation of hydrogen ions from phosphoöipid vesicles. Equal volumes of vesicle suspension and of n-octylbiguanide solution were mixed at 25°C; d=1cm; flow stops as indicated by the arrow.

Fig. 4

have been used. The kinetic resolution under pseudo-first order conditions is shown in Fig. 5 with the velocity constant k_1 corresponding to the constant measured with mitochondria. The slower phase with the constant k_2 only occurs with synthetic lipid vesicles. This part of the reaction may be due to the fact that multilayer particles are present which are gradually penetrated by the hydrophobic guanidines. In this case k_2 would be proportional to the diffusion constant. Another explanation for the dye response would be a redistribution between membrane and aqueous phase subsequent to binding of biguanides. This will be discussed elsewhere in more detail (25a).

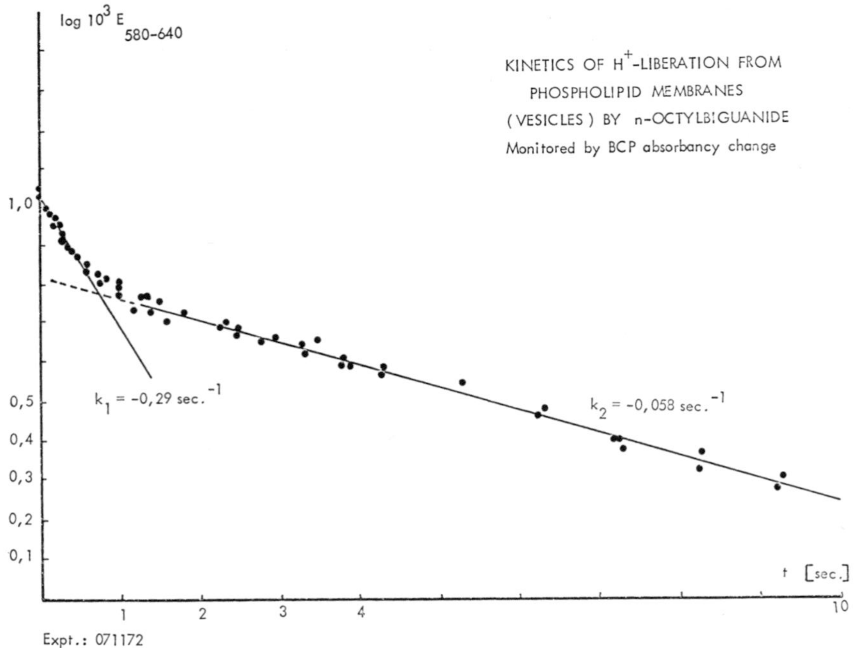

Fig. 5.

By the first set of experiments evidence has been accumulated that biguanides bind to the phospholipid moiety of biological and synthetic membranes. These results are perfectly in line with morphological properties of mitochondria from different sources and with their cytochrome content, for example, which we can take as a measure of the inner membrane surface. Cytochrome content and capacity for biguanide binding appear to be closely correlated. Let me turn now towards the functional alterations of membranes after binding of biguanides.

MITOCHONDRIAL MEMBRANE PROPERTIES AND ION FLUXES IN THE PRESENCE OF BOUND BIGUANIDES

Experiments will be described which show that binding of biguanides to biological membranes such as mitochondria or submitochondrial particles leads to pronounced changes in interactions of membranes with smaller molecules. These findings will later be compared with the properties of phospholipid bilayers under the influence of biguanides.

Whereas inhibition of respiration in phosphorylating particles by guanidines and biguanides has been demonstrated many times, little is known about the interaction with cation transport. Figure 6 shows

INFLUENCE OF BIGUANIDES ON MITOCHONDRIAL VOLUME OSCILLATIONS

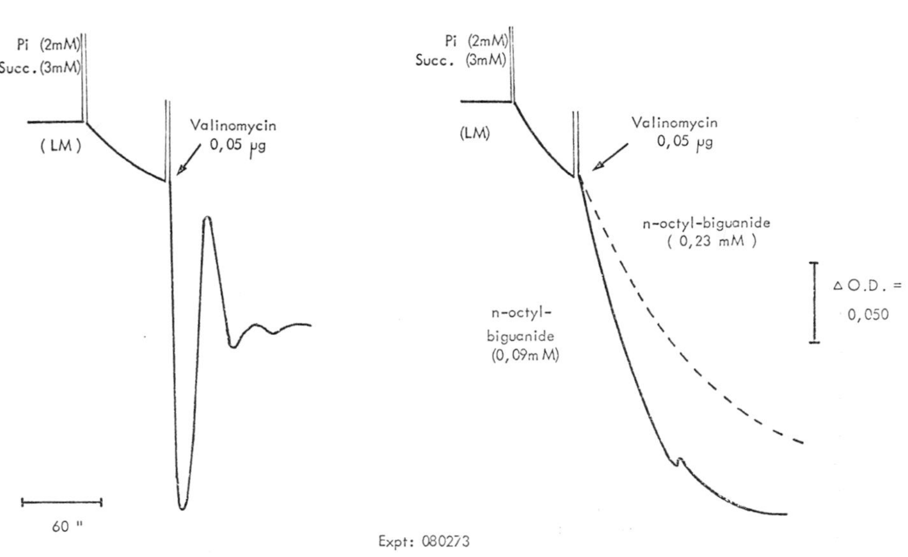

Fig. 6

swelling-shrinkage cycles of rat liver mitochondria during energy-linked uptake of phosphate induced by the addition of potassium/valinomycin. Potassium concentration in both cases is 15 mM. The right-hand trace shows that the presence of n-octyl-biguanide not only stops the oscillation, but drastically reduces the initial rate of ion flux. This is the first time that direct interaction of guanidine has been demonstrated with a system dependent on the flux of potassium ions.

Competition with potassium could also be shown during titration of biguanide binding to mitochondrial membranes by means of the ANS fluorescence procedure. As depicted in Fig. 7 the effect of potassium is more pronounced when valinomycin is present, giving the potassium ions access to more hydrophobic regions. The result of this experiment cannot be interpreted simply in terms of competition for identical and specific binding sites; one should keep in mind that

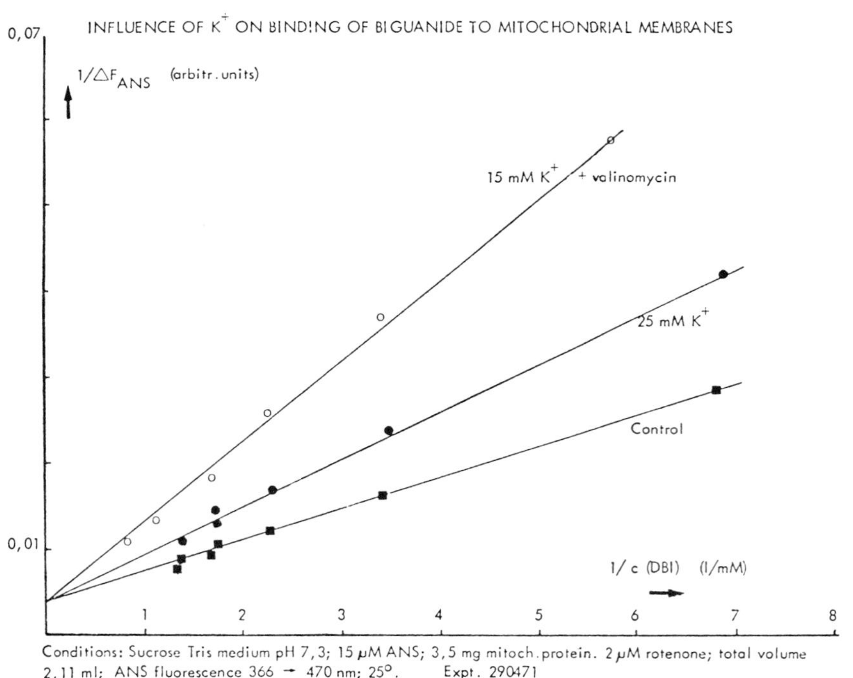

Fig. 7

screening effects play an important role in ion sorption studies using ANS as fluorescence probe.

Another effect of biguanide binding on the intrinsic properties of mitochondrial membranes is shown in Fig. 8. The fluorescence of antimycin in free solution

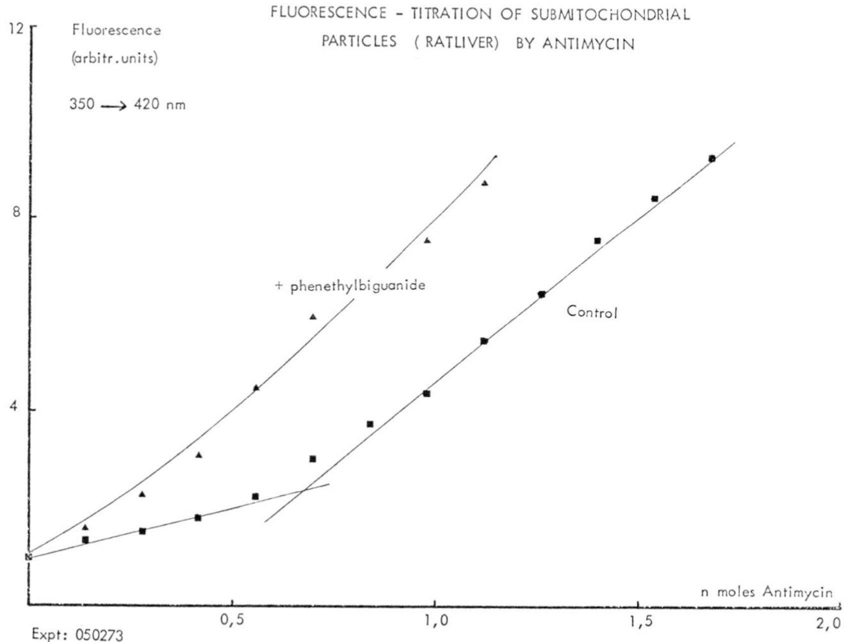

Fig. 8

or in its albumin-bound form is quenched by binding to submitochondrial particles. This quenching is due to energy transfer from antimycin to the heme of cytochrome b since the emission band overlaps with the Soret band of cytochrome b. Only antimycin added in excess of that necessary for saturating the specific binding sites will cause fluorescence with normal intensity. The experiments clearly demonstrate that after preincubation of the particles with phenethylbiguanide much less antimycin can be bound and the characteristic break point in the titration curve is practically absent. In this particular experiment serum albumin was present which makes the method more sensitive for detection of free antimycin.

These differences of antimycin binding to the site II region of oxidative phosphorylation are indicative of perturbation of normal molecular interactions in that area of the membrane where cytochrome b is located. This is of particular interest since the inhibitory effect of phenethyl-biguanide on mitochondrial electron transfer could be located in the same region.

An understanding of the molecular mechanism of action can only be expected when we know more about the physical properties of membranes in the presence of biguanides. An approach to this question requires experiments with simpler models. We have selected lipid bilayers for this purpose and have measured the

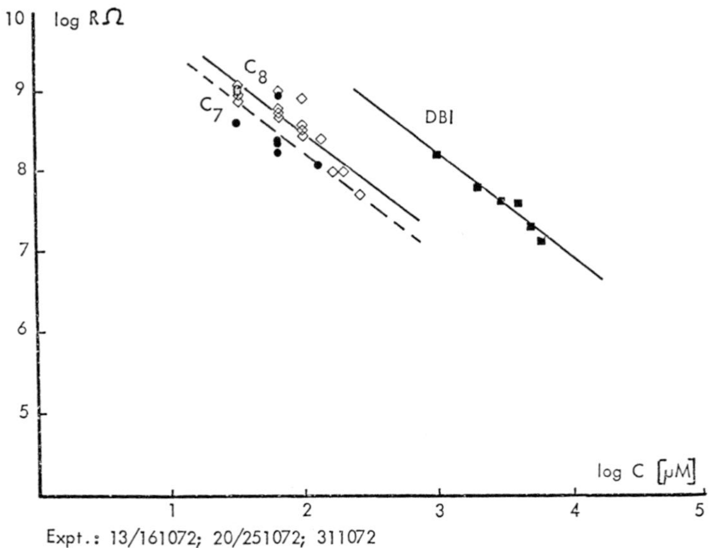

Fig. 9

electrical properties of these membranes under various conditions.

EFFECTS OF BIGUANIDES ON LIPID BILAYERS

With lecithin membranes bathed in a 0.1 M solution of potassium chloride, medium to high (mM) concentrations of biguanide cause a decrease in ohmic resistance (Fig. 9). The material used was purified egg lecithin. The additions were made to both sides of the membrane. Reproducibility was fairly good as seen from the scattering of the points. It was excellent with membranes made from absolutely pure synthetic dioleyl-phosphatidyl-choline. Nevertheless, quantitative evaluation of these titrations has shown that a binding phenomenon may provide the molecular basis of the conductivity enhancement. A double reciprocal plot of the relative change of ohmic resistance over the concentration of biguanide is given in Fig. 10 for n-octyl-biguanide as an example. It should be mentioned that at very high concentrations of biguanides

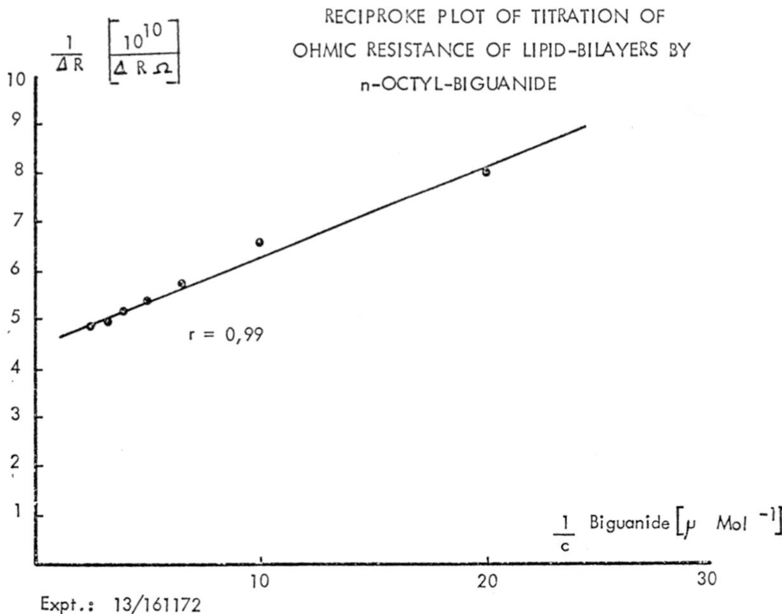

Fig. 10

the membranes become unstable and tend to break. From Fig. 9 it may also be noted that the relative activity of the compounds again follows their affinity for hydrophobic binding to lipids.

At high biguanide concentrations a considerable change of electrical capacity could also be observed (26). Capacity was measured by applying squarewave-pulses to the membrane in parallel to a 900Ω resistor. The basic capacity of the untreated membranes was in the range of 0.4 μF/cm^2. Figure 11 shows the increase of the time constants achieved by addition of increasingly larger amounts of biguanides. The relative activity of the compounds again is very well correlated to their lipid-affinity and is most pronounced with n-octyl-biguanide. For this latter compound the

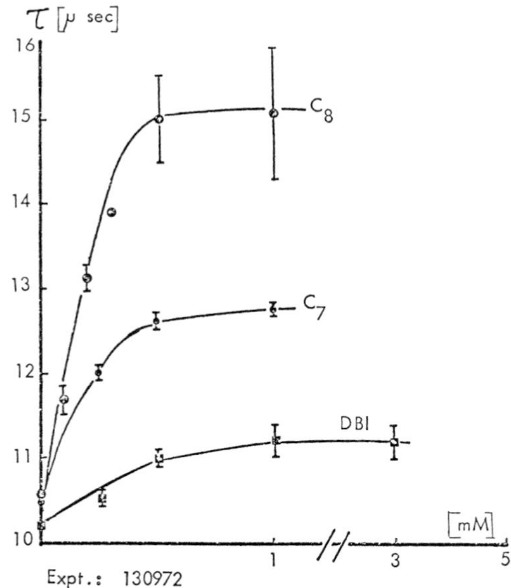

Fig. 11

absolute increase in electrical capacity is documented in Fig. 12.

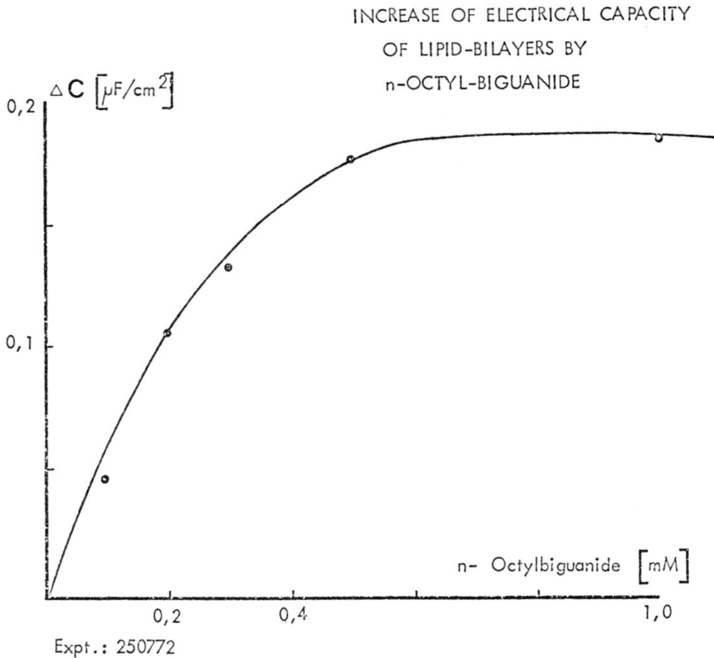

Fig. 12

An interpretation of this result has to consider two parameters which can hardly be distinguished without application of additional methods of investigation. Basically the increase of capacity may be due to a change in the dielectric properties of dipole structures including the amount of water present in the membrane. With n-octyl-biguanide the respective increase of the dielectric constant would be from 2.1 to 3.4.

On the other hand using a dielectric constant of 2.1 (as for medium chain aliphatic hydrocarbons) and assuming that it remains unchanged, a thinning of the membrane would result. Theoretically the membrane-core diameter would decrease from about 46 Å to 31 Å.

Such a dramatic thinning in the presence of a saturating biguanide concentration could certainly be responsible for the propagation of structural changes from lipids to proteins associated with the membrane. Preliminary X-ray studies do indeed suggest that considerable perturbation of the membrane structure occurs under the above conditions including changes in the spacing between the phosphate headgroups at the aqueous lipid interface. It should be mentioned at this stage that these effects can only be observed at rather high concentrations of the inhibitors.

MEMBRANE SURFACE-POTENTIALS MODIFIED BY BIGUANIDES

Biguanides are strong bases which in the protonated state form resonance-stabilized cations of a pK of about 10.5 to 12. As such they represent hydrophobic cations which can bind to lipid containing membranes and generate a surface potential at the aqueous lipid interface. This may arise from hydrophobic fixation of excess amounts of cations or simply from screening of negative charges already present in the membrane.

In order to prove this hypothesis, conductance measurements were made with very well-defined positively or negatively charged permeant species using neutral phospholipid membranes formed from synthetic dioleyl-phosphatidyl-choline.

Potassium-nonactin or potassium-valinomycin complexes were used as the positively charged permeant ions. Uncouplers of oxidative phosphorylation can be used as negatively charged carriers. These are lipid-soluble weak acids penetrating in the ionic form or carrying the charge as a dimer of the anionic with the protonated form as suggested by others (27-29). In the latter case the slope of double logarithmic conductivity plots over the concentration of the uncoupler should be n = 2. This has been found with the compound in the present experiments (an uncoupler of the NH-acidic-phenylhydrazone-type). With potassium-valinomycin a slope of n = 1 should be obtained, which was normally the case.

Figures 13 and 14 show current-voltage diagrams for both positively and negatively charged permeant species in the absence and presence of n-octyl-biguanide. It can easily be recognized that n-octyl-biguanide causes a decrease of electrical current with a positively charged permeant ion in contrast to an increase of electrical current when the permeant species is negatively charged. It should be mentioned that the current-voltage diagrams are linear and symmetric at least within a voltage range of 100 mV.

SHIFT OF THE CURRENT VOLTAGE DIAGRAMM OF A CATION-CONDUCTING LIPIDBILAYER BY n-OCTYLBIGUANIDE

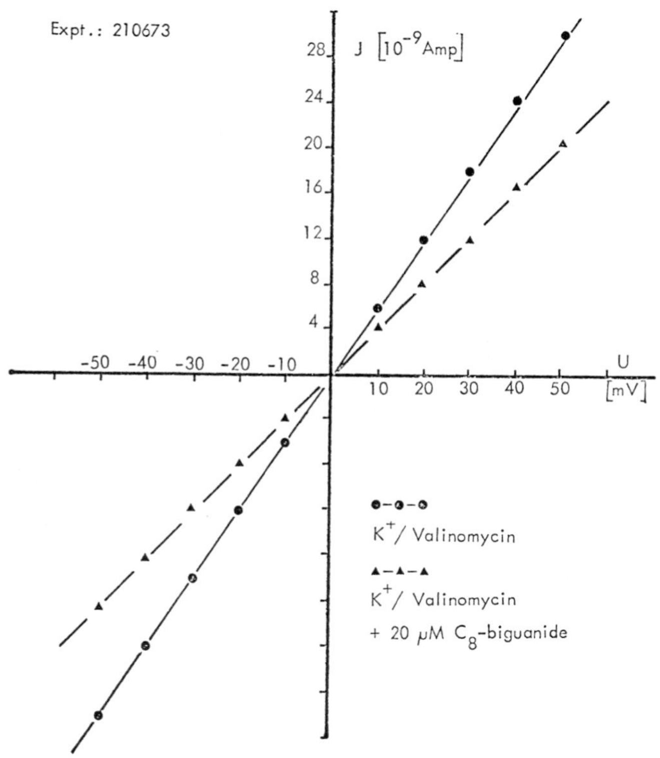

Fig. 13

SHIFT BY n-OCTYL-BIGUANIDE OF THE CURRENT VOLTAGE DIAGRAMM OF A LIPIDBILAYER CONDUCTING ANIONS.

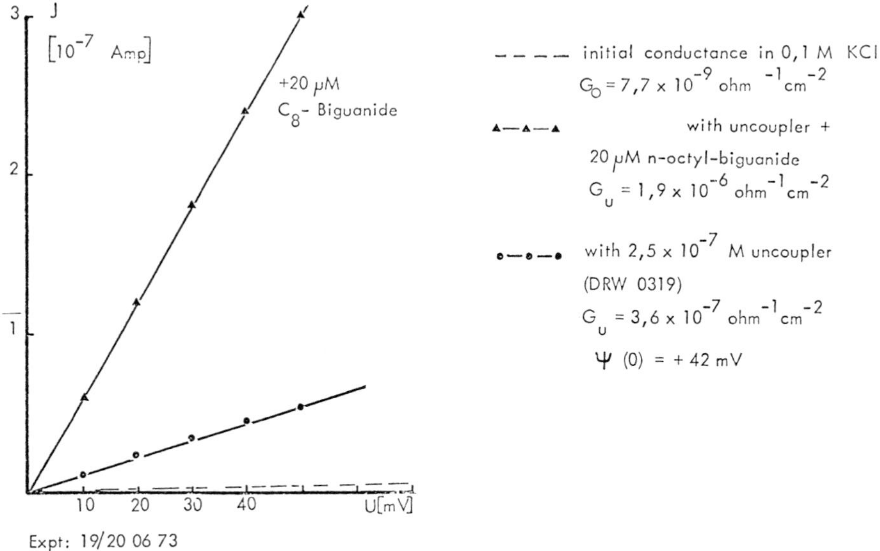

Fig. 14

Additions were always made symmetrically to both sides of the membrane. It is relevant for evaluation of these measurements that the increase of, for example, the conductivity with permeant uncoupler anions occurs at very low concentrations of biguanide. The added concentration by itself would only cause a negligible change in conductivity and can therefore be excluded as an additional charge carrying species in the experiment. In other words, the effect of uncoupler ions and of biguanides is not additive and the increased flux of charges through the membrane must result from an equivalent increase in anion flux only.

A complete conductivity titration with uncoupler is given in Fig. 15 as a double logarithmic plot of conductivity over uncoupler concentration. The points are the average of three to four single measurements. The lines drawn through the points are computer fitted.

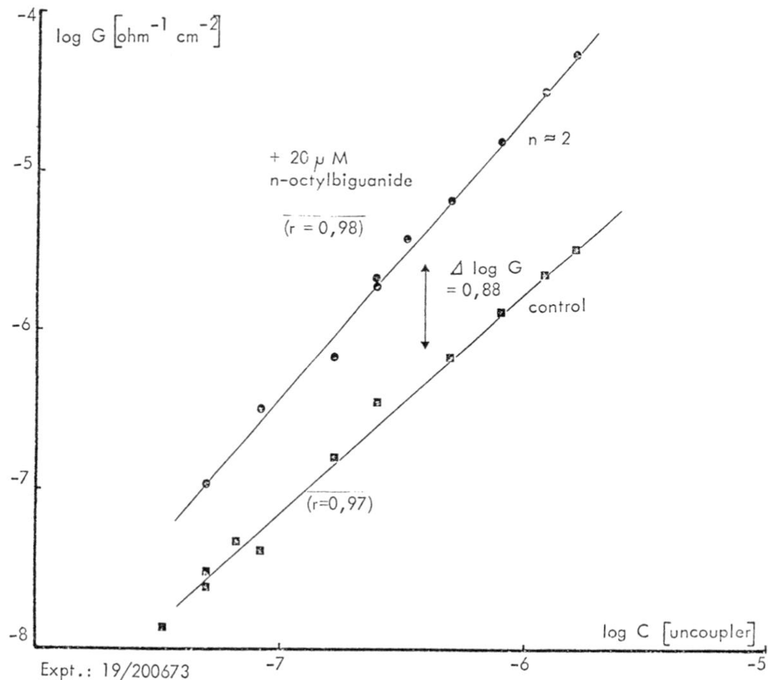

Fig. 15

In the presence of n-octyl-biguanide (upper line), membrane conductance is shifted to higher values over the whole range with an average of 0.88 log-units of conductance between the two cases. There is a slight difference in the slope of both curves with the slope of the lower curve (biguanides absent) being remarkably lower than the theoretical value of n = 2. This may be interpreted by assuming that at higher concentrations uncoupler anions adsorb to the membrane and generate a negative surface potential. This assumption is in line with the observation that in the presence of biguanides the slope is much closer to the theoretical value assuming that the guanidinium ions may produce a positive surface potential and screen the negative membrane charges.

Since the membrane under consideration is neutral at pH 7.2, the shift to higher conductivities with biguanide cations present suggests that the membrane has been transformed from a neutral to a positively charged bilayer. If sufficient guanidinium molecules are bound to the membrane the observation can be explained in terms of double-layer-theory and equations may be applied derived by Neumcke (30) and McLaughlin (31) for the conductance of a charged bilayer, which is equal to the conductance of a neutral membrane times the exponent of the potential at the surface of the membrane, $\psi(0)$:

$$(G_o^-)^{charged} = (G_o^-)^{neutral} \cdot \exp \frac{+F\psi(0)}{RT}$$

The sign of the exponent has to be negative when a positive permeant species is considered and has to be positive when a negative permeant species is used.

Treating the above experiments according to this rationale and solving for $\psi(0)$, a surface potential of +51 mV can be calculated under the applied conditions. Interestingly enough practically the same value of $\psi(0) = +51$ mV was obtained when the titration was performed with potassium/valinomycin as a positive permeant species. There is no doubt that this result lends strong support to the above concept. An experiment of the latter type is shown in Fig. 16 with phenethyl-biguanide as the membrane modifying agent, the upper line being the conductance curve of the neutral membrane and the lower line the conductance in the presence of biguanide. There is a remarkable decrease of conductance mediated by positively charged species which represents a surface potential of $\psi(0) = +84$ mV when treated according to the above equation.

A consideration of the charge density on the membrane should be possible on the basis of the Graham equation:

$$\sigma = \frac{1}{272} \cdot \left[\Sigma_i C_i \left(\exp \frac{-z_i F\psi(0)}{RT} - 1 \right) \right]^{1/2}$$

which reduces to the well known Gouy expression if only ions of one valence z are present.

$$\sinh \frac{zF\psi(0)}{2RT} = \frac{136\sigma}{\sqrt{C}}$$

where $\psi(0)$ is the surface potential, σ is the surface charge density in electronic charges per Å^2 and C is the concentration of the charged species in the bulk

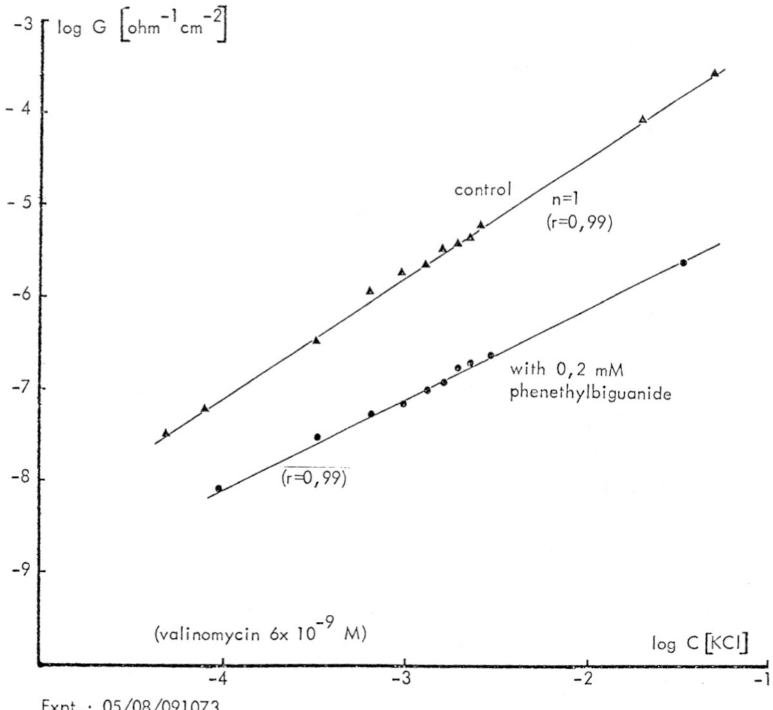

Fig. 16

solution in moles/l. R, T and F have their usual meanings.

Quantitative evaluation of our experiments has shown that in the concentration range tested σ values on the order of 5×10^{-4} to 3×10^{-4} were obtained.

Comparing different biguanides under conditions where they produce equal charge densities on an originally neutral membrane, the apparent ratio of the necessary concentrations almost exactly equals the ratio of the respective affinity constants measured with the ANS method. In view of the assumptions inherent in the derivation of the relevant equations and in view of the manifold sources of experimental error it is extremely gratifying that a reasonable and almost perfect agreement of these ratios was obtained.

CONCLUSIONS

Summarizing the experiments reported here, evidence has been presented that biguanides as inhibitors of oxidative phosphorylation are bound to the phospholipids of mitochondrial membranes mainly by hydrophobic interaction. The same may hold for guanidines or diguanidines which exert analogous effects differing only in degree.

When these inhibitors have been bound the subsequent alterations of membrane properties must be taken as the general basis for a variety of effects on membrane linked enzymatic functions. This is a reasonable assumption in view of the fact that the inhibitory concentrations of the respective compounds are orders of magnitude higher than those of the catalytic sites which may be involved in the process of energy conservation in mitochondrial respiration.

The relative effectiveness of the individual inhibitors closely follows their binding affinity not only with respect to oxidative phosphorylation but also regarding metabolic effects in general (5).

Although it has been shown that biguanides can influence the large scale configuration of mitochondria in different metabolic states (e.g., orthodox or condensed configuration) (20) modification of the molecular arrangements within the inner membrane will be out of the scope of electron microscopy. It became obvious, however, that cytochrome b interaction with antimycin is, for example, modified in the presence of biguanides. Moreover, processes involving ion fluxes as for instance energy dependent swelling-shrinkage-cycles definitely respond to the presence of biguanides.

What are the changes in membrane structure and properties introduced by the binding of these inhibitors? From binding studies and model experiments one may conclude that indeed the aqueous lipid interface of a membrane is the location of these drugs. When we locate the hydrocarbon chain of the inhibitors between the hydrocarbons of the phospholipid membrane core, the positively charged guanidinium group (very well resonance-stabilized) forms a spacy molecule capable of effectively screening negative charges of the phosphatidyl-choline dipole for example. Adsorption of the hydrophobic guanidium ions in any case will convert a neutral membrane into a positively charged membrane generating a net surface potential.

Due to the existence of a surface potential, ion fluxes will be affected in biological membranes in ways similar to those shown for lipid bilayers. The physiological significance of an interaction with ion movements may be different for particular species depending on whether they are translocated by a specific carrier or permeate membranes by passive diffusion. In mitochondria, carrier transport has been established for calcium. Potassium, however, may be a candidate especially sensitive to the surface potential. One should also remember that the respiratory phosphorylation-mechanism as well as photosynthesis, or energy conservation in chromatophores, involves the translocation of protons. The mechanism of biological proton conductance through the membrane certainly differs from the conductance of positive charges via protons in aqueous solutions, since we do not assume an aqueous continuum extending through a membrane at all. Protons presumably travel as bound to nucleophilic sites which may be screened or occupied by the biguanide or guanidinium structure. Further details in this respect have to be established.

The development of a membrane surface potential due to biguanide uptake also provides the correct explanation for the excellent utility of the ANS fluorescence method for studies of biguanide binding, because McLaughlin et al. (32) have clearly shown that ANS follows exactly the changes of surface potential as bromthymolblue probably does equally well.

The application of results obtained with phospholipid bilayers to biological membranes appears to be justified in the present study for two reasons. First, there is a complete analogy of binding properties of the inhibitors to both types of membrane. Second, evidence has been presented in the past two years that major portions of phospholipids in natural membranes may exist in a bilayer form (33). Furthermore one should keep in mind that the physical effects demonstrated by application of the double-layer theory on phospholipid membranes do not only apply to large scale two-dimensional spaces such as membranes, but also to very local events which include charge-transfer or ion-transfer reactions within smaller membrane-bound catalytic units.

The effect of surface charge in bilayer-like membrane areas may be considered of significant physiological importance in view of recent studies of Graham and collaborators (34). Their studies suggest that by changes of surface charge subsequent changes of the surface area per molecule of phospholipid may be induced, leading to a corresponding increase or decrease of configurational freedom of the hydrocarbon chains in the membrane. Structural modifications of this type may provide a molecular basis for the observed inhibitory actions in the region of cytochrome b and other respiratory pigments.

The "site-specificity" of biguanides as inhibitors of oxidative phosphorylation is hard to reconcile with the above theory. It may, however, accidentally occur by a different sensitivity of individual segments of the respiratory chain due to their sideness and specific location within the lipid protein assembly.

It should be mentioned that similar considerations regarding analgetic salicylates have been presented by McLaughlin (35). These studies perfectly explain the differences in sensitivity of specific nerve membranes against the negatively charged salicylates. In some situations one should expect that the guanidine derivatives we are dealing with will be inactive with biological membranes thought to be positively charged; further, they may be especially active with negatively charged membranes. Actually the membrane of intact mitochondria is thought to bear a negative surface charge. The observation that biguanides specifically require an "energized state" for onset of inhibition may be due to the fact that during energy conservation the outer surface of the inner membrane seems to expose negative charges which would be open for interaction with membrane bound guanidinium ions.

REFERENCES

1. SCHÄFER, G. (1963) Biochem. Z. 339, 46.
2. SCHÄFER, G. (1964) Biochim. Biophys. Acta 93, 279.
3. SCHÄFER, G. (1969) Biochim. Biophys. Acta 172, 334.
4. ALTSCHULD, R. A., AND KRÜGER, F. A. (1968) Ann. N. Y. Sci. 148, 612.
5. SCHÄFER, G. (1971) in R. Beckmann, "Biguanides", Handbook of Experimental Pharmacol. (Maske, H., ed.) p. 510, vol. 29, Springer-Verlag, Berlin, Heidelberg, New York.
6. PRESSMAN, B. C. (1963) J. Biol. Chem. 238, 401.
7. CHAPPELL, J. B. (1963) J. Biol. Chem. 238, 410.
8. PRESSMAN, B. C. (1963) Energy-Linked Functions of Mitochondria (Chance, B., ed.) p. 181, Academic Press, N.Y.
9. UNGAR, G., PSYCHOYOS, S., AND HALL, H. A. (1960) Metabolism 9, 36.
10. VON BRANDT, V. (1961) Drug Research 11, 739.
11. GORDON, E. E., AND DE HARTOG, M. (1973) Diabetes 22, 50.

12. SCHÄFER, G. (1973) Int. Congr. Biochem., Abstr. 873.
13. KRÜGER, F. A., ALTSCHULD, R. A., HOLLOBAUGH, S. L., AND JEWETT, B. (1970) Diabetes 19, 50.
13a. CZYZYK, A., TAWECKI, J., SADOWSKI, J., POVIKOWSKA, I., AND SZCZEPANIK, Z. (1968) Diabetes 17, 492.
14. FALCONE, A. B., MAO, R. L., AND SHRAGO, E. (1962) J. Biol. Chem. 237, 904.
15. GROSS, E., SHAVIT, N., SAN PIETRO (1968) Arch. Biochem. Biophys. 127, 224.
16. KEISTER, D. L., AND MINTON, N. J. (1970) in Electron Transport and Energy Conservation (Tager, J. M., Papa, S., Quagliariello, E., and Slater, E. C., eds.) p. 409, Adriatica Editrice, Bari.
17. PRESSMAN, B. C., AND PARK, J. K. (1963) Biochem. Biophys. Res. Commun. 11, 182.
18. DAVIDOFF, F. (1969) Diabetes 18, Suppl. 1, 331.
19. CHANCE, B., AND HOLLUNGER, G. (1963) J. Biol. Chem. 238, 432.
20. SCHÄFER, G., BOJANOWSKI, D., AND SCHLIMME, E. (1972) in The Biochemistry and Biophysics of Mitochondrial Membranes (Azzone, F., and Siliprandi, N., eds.) Academic Press, N.Y.
21. SCHÄFER, G., AND BOJANOWSKI, D. (1972) Eur. J. Biol. Chem. 27, 364.
22. DAVIDOFF, F. (1971) J. Biol. Chem. 246, 4017.
23. LESSLAUER, W., AND BLASIE, J. K. (1971) Biochim. Biophys. Acta 241, 547.
24. FLEISCHER, S., FLEISCHER, B., AND STOECKENIUS, W. (1967) J. Cell. Biol. 32, 193.
25. SHEPHERD, H. G., AND McDONALD, H. J. (1958) Clinical Chemistry 4, 496.
25a. SCHÄFER, G. (1974) Europ. J. Biochem., in press.
26. SCHÄFER, G., AND RIEGER, E. (1973) Abstracts 17th Ann. Meeting Biophys. Soc.; Biophys. J. 13, 13a.
27. FINKELSTEIN, A. (1970) Biochim. Biophys. Acta 205, 1.
28. NEUMCKE, B., AND BAMBERG, E. (1973) in "Membranes" (G. Eisenmann, ed.) Vol. 3; M. Dekker Inc., N.Y.

29. FORSTER, M., McLAUGHLIN, S., AND HARARY, H. (1973) Biophysical Society Abstracts in *Biophys. J.* *13*, 173a; and *J. Membrane Biol.*, in press.
30. NEUMCKE, B. (1970) *Biophysik* *6*, 231.
31. McLAUGHLIN, S., SZABO, G., EISENMANN, G., AND CIANI, S. M. (1970) *Proc. Nat. Acad. Sci.* *67*, 1268.
32. McLAUGHLIN, S., SZABO, G., AND EISENMANN, G. (1971) *J. Gen. Physiol.* *58*, 667.
33. SINGER, S. J., AND NICOLSON, G. L. (1972) *Science* *175*, 720.
34. GRAHAM, D. E., AND LEA, E. J. (1972) *Biochim. Biophys. Acta* *274*, 286.
35. McLAUGHLIN, S. (1973) *Nature* (New Biology) *243*. 234.

A MECHANISM OF THERMOGENESIS BY MODIFICATION OF SUCCINATE DEHYDROGENASE

R. Ramasarma and L. Susheela

INTRODUCTION

A mechanism is proposed for increased thermogenesis involving succinate dehydrogenase and ubiquinone under conditions of environmental stress such as cold exposure. Succinate dehydrogenase behaves as a half-site enzyme, being activated specifically by ubiquinol, which presumably binds to one of the two active sites - a property shared by the competitive inhibitors. Consequently, NADH, the quantitatively significant substrate, is diverted to shunt pathways in the mitochondria and the microsomes which bypass coupled phosphorylation and liberate heat. The shunt pathway of succinate oxidation is activated by ubiquinone in the oxidized form, and compounds known to be thermogenic presumably act by making available more of ubiquinone at the appropriate site or altering its redox-state in the membrane.

THERMOGENESIS IN COLD EXPOSURE

The ability of a homeotherm to survive the stress of environmental cold depends upon the capacity to produce extra heat to compensate for the loss due to increased temperature differential between the body and the surrounding environment. In chronic exposures, adaptation occurs in animals by increasing the heat production by "non-shivering" or chemical thermogenesis (1) involving the internal organs, the liver,

Abbreviations: Q - ubiquinone-9; QH_2 - ubiquinol-9; SDH - succinate dehydrogenase; PMS - phenazine methosulfate; NT - neotetrazolium chloride; DCI - 2,6,dichlorophenolindophenol; DNP - 2,4,dinitrophenol.

the intestines, and the brain (2,3). The process of adjustment starts with the diversion of food stuff towards combustion rather than synthetic reactions and growth. In a sense, the initial stages of exposure to cold can be considered to be equivalent to semi-starvation with negative caloric balance. Adaptation thereof seems to be initiated by increased food intake followed by adjustments in metabolism directed towards maintaining the thermal state of the body at the expense of less vital functions. The first obvious signs of these processes are the increased basal metabolic rate and O_2 uptake in tissues.

CALORIGENIC SHUNT PATHWAYS

Cold exposure favors gluconeogenesis and the three key enzymes, phosphoenolpyruvate carboxykinase, fructose-1,6-diphosphatase, and glucose-6-phosphatase are considerably increased in the liver. This is in line with the utilization of non-carbohydrate sources for calorigenic purposes (4). The flow of carbohydrate material through glycolysis seems to be enhanced by way of activation of the enzymes. A number of mitochondrial electron transport components and enzyme activities also increase during cold exposure usually in the range of 20-30% (5). The electron transport through the coupled pathway conserves energy in the form of ATP with an efficiency of about 50%. The balance of the energy is released as heat which is considered the wasteful by-product and therefore had not attracted attention. But this heat serves to keep the thermal state of the body and increased thermogenesis becomes most vital under cold exposure.

Continued high turnover of ATP or loosening of the respiratory control imposed by ATP/ADP charge are discounted as the mechanisms for the production of extra heat. It is postulated that the extra heat can be generated by augmentation of alternate pathways of electron transport or "calorigenic shunts" (6) which are not restricted by coupled phosphorylation. In order to account for increased utilization of caloric sources and consumption of O_2, diversion of electrons away from the phosphorylation sites, most probably at the dehydrogenase level, would be necessary. An attempt has been made in this study to implicate ubiquinone (Q) and succinate dehydrogenase (SDH) for such a role in thermogenesis.

LOCATION OF UBIQUINONE

Q is the only lipid component of the respiratory chain. It seems to be the mobile link in the lipophilic environment of the membrane-bound electron transport system. It has a unique location as the common hydrogen acceptor and is the point of convergence of a number of mitochondrial dehydrogenases including those of NADH and succinate. After Q, a shift occurs from 2-electron carriers to 1-electron carriers - the cytochrome system, as shown schematically below:

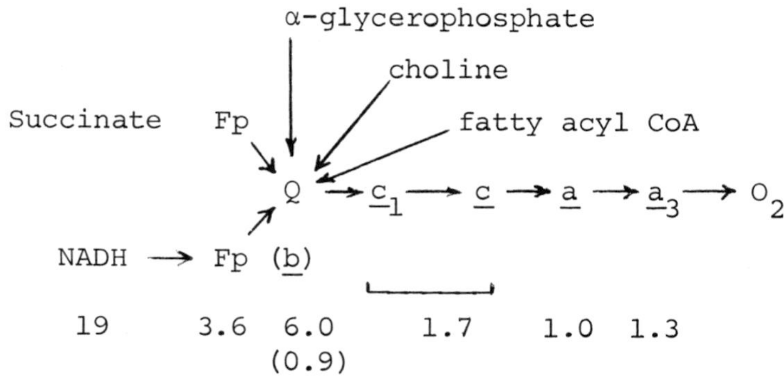

(The figures represent the molar proportions with respect to cytochrome a).

Q, however, occurs far in excess of the molar proportions of the cytochromes and it is possible that it may have additional structural or regulatory functions in cellular oxidations.

TABLE I
ACTIVATION OF SUCCINATE OXIDATION BY Q/QH_2

Additions (0.6 mM)	% Control	
	PMS reduction	NT reduction
Ubiquinone	90	250
Ubiquinol	190	88

TABLE II
DIFFERENTIAL RESPONSES SHOWN BY THE TWO ASSAYS OF SUCCINATE DEHYDROGENASE

Conditions	PMS assay	NT assay
Acceptor site	Flavoprotein	Presumably a Q-lipoprotein
Rate	Fast	Slow
<u>Additions in vitro</u>		
1. Dicarboxylates	Activated	No activation
2. Ubiquinol	Activated	No activation
3. Ubiquinone	No change	Activated
4. Antimycin A	Insensitive	Sensitive
5. Tyramine	No change	Activated
6. Tryptamine	No change	Activated
<u>Treatment in vivo</u>		
a. Thyroxine	Marginal stimulation	Stimulated 2-3 fold
b. Ubiquinone	No change	Increased
c. Cold exposure	Marginal stimulation (20-30%)	Increased 2-3 fold
d. Hypobaria (0.5 atm)	Increased	No change
e. Hypoxia (10% O_2)	Increased	No change

SUCCINATE DEHYDROGENASE

SDH is a mitochondrial inner membrane-bound flavoprotein. Its catalytic activity can be measured by the reduction of the dyes phenazine methosulfate (PMS) and neotetrazolium chloride (NT), which are considered to represent the main respiratory chain and a shunt pathway, respectively.

Interestingly, the reduction of NT appears to be dependent on Q and is activated by adding exogenous Q (7). The reduction of PMS is activated by ubiquinol (QH_2) (8,9). It is already known that the ratio of Q/QH_2 undergoes large changes during the transition from state 3 to state 4 (10) and corresponding changes dependent on the QH_2 concentration were found in SDH (11). The two activations are specific to the two forms of Q. Thus QH_2 has no effect on the reduction of NT and Q on the PMS assay (Table I).
The differential responses shown by the two assays are given in Table II.

MODULATION OF SUCCINATE DEHYDROGENASE - A HALF-SITE ENZYME

The activation of SDH (PMS assay) by QH_2 seems to be a part of a general phenomenon. Four diverse classes of compounds, dicarboxylates, nitrophenols, quinols, and pyrophosphates activate the enzyme when mitochondria are preincubated with them (9). The only common feature among these appears to be the presence of two ionizable oxygen atoms spatially separated by about a distance of 6-7 Å (Fig. 1). It appears that

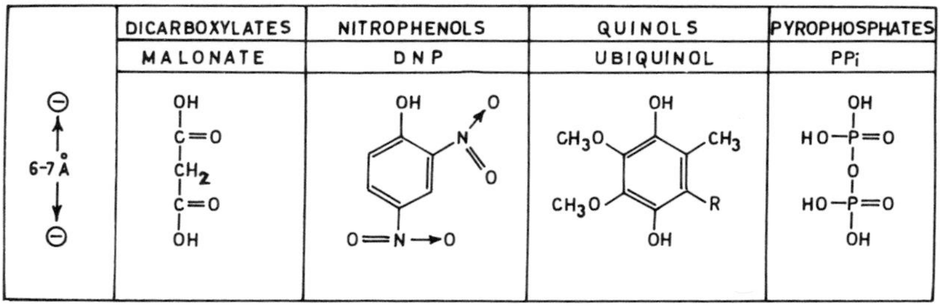

	DICARBOXYLATES	NITROPHENOLS	QUINOLS	PYROPHOSPHATES
	MALONATE	DNP	UBIQUINOL	PPi
⊖ ↕ 6-7 Å ⊖				

Fig. 1. Activators of succinate dehydrogenase having a distance of about 6 Å between the ionizable oxygen atoms.

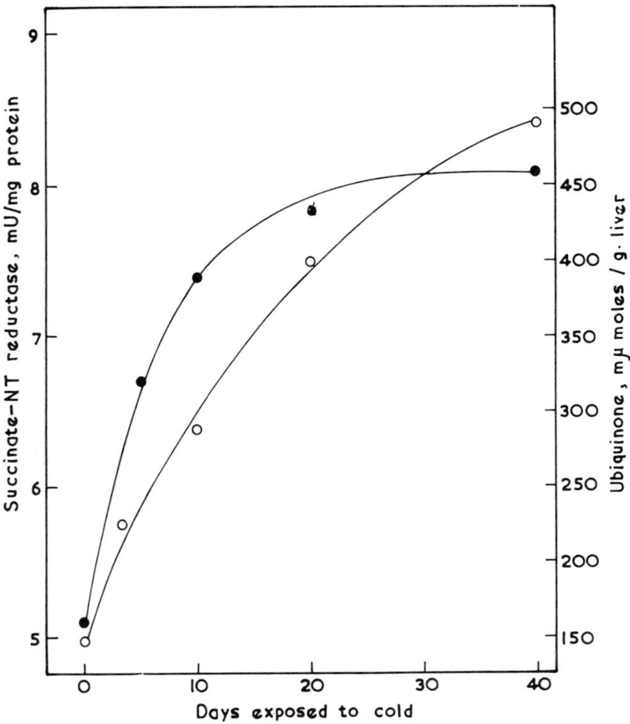

Fig. 2. Changes in mitochondrial ubiquinone concentration (o—o) and succinate-NT reductase activity (●—●) in liver of rats exposed to cold (from ref. 15).

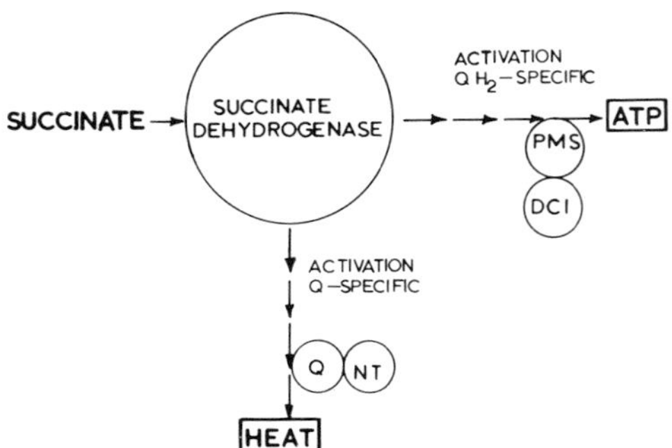

Fig. 3. A schematic representation of the modulation of succinate dehydrogenase by the redox-state of ubiquinone.

these facilitate the appropriate interaction of the effectors with specific sites on the enzyme protein which may then cause the conformational change resulting in increased molecular capacity of the enzyme known to occur (12). Some of these compounds are well known competitive inhibitors of the enzyme. But there is no contradiction as the activation is obtained at a very low concentration as compared to the inhibition.

Kinetic analysis showed that this type of activation does not fit either the "K" or "V" types described by Monod et al. (13) as both K_m and V_{max} are changed and Lineweaver-Burk plots were nearly parallel (Susheela, L., unpublished data). The properties fit well with the requirement of a half-site enzyme with a flip-flop mechanism proposed by Lazdunski (14) wherein only half the number of the active sites is catalytically active and the other half is bound by the substrate or competitive inhibitors resulting in activation. This type of enzyme modulation offers an excellent mode of storing the protein in an inactive form and activating it when need arises.

ACTIVATION OF SUCCINATE OXIDATION IN COLD EXPOSURE

Exposure of rats to low environmental temperature resulted in increase in activities of hepatic mitochondrial succinate oxidase and SDH (15) and also Q (16,17). Simultaneous with the increase in Q, succinate-NT reductase activity progressively increased with the period of cold exposure (Fig. 2). Succinate-NT reductase is known to be a Q-dependent enzyme system (7) and increased activity of this enzyme through increased Q may act as a bypass of electrons through a shunt pathway generating heat in contrast to the main-chain which produces ATP (Fig. 3). It should be emphasized that the redox-state of Q has a decisive role to play. Only the oxidized form activates the shunt pathway and the activation of the main pathway represented by the reduction of PMS is equally specific for QH_2 (see Table I). One advantage with the liver tissue is that it can be enriched with

Q by absorption from an exogenous supply and this is equally effective in increasing the activity of succinate-NT reductase (17). This offers a possibility of artificially increasing the hepatic Q should this prove to be thermogenic as anticipated.

SOME CONDITIONS OF THERMOGENESIS

The calorigenic action of administered thyroxine and catecholamines is well known. It was therefore of interest to test the effect on succinate-NT reductase of administration of thyroxine and two of the drugs which increase the catecholamines - imipramine (5-(3-dimethylaminopropyl)-10,11-dihydro-5H-dibenz (b,f) azepine), an antidepressant which selectively blocks the reuptake of norepinephrine even in tissues other than the nerve tissue (18,19) and RO-4-1284 (2-hydroxy-2-ethyl-3-isobutyl-9,10-dimethoxy,1,2,3,4, 6,7-hexahydrobenzoquinolizine), an amine releaser with quick onset and short duration of action (20). The results in Table III show that succinate-NT reductase was similarly activated in all the conditions tested - cold exposure and treatment with thyroxine, imipramine, and RO-4-1284. Except for imipramine, other treatments increased the activity of succinate-PMS reductase as well.

TABLE III
HEPATIC MITOCHONDRIAL SUCCINATE OXIDATION IN VIVO UNDER SOME CONDITIONS OF THERMOGENESIS

Treatment	% Stimulation	
	PMS assay	NT assay
Cold exposure (0-5°, 10 days)	55	50
Thyroxine (25 μg/rat, 2 hr)	41	55
Imipramine (15 mg/kg, 2 hr)	12	40
RO-4-1284 (10 mg/kg, 2 hr)	39	63

The drugs were administered intraperitoneally.

EFFECT OF BIOGENIC AMINES

It is known that catecholamines widely exert their functions through the activation of adenyl cyclase although they are also implicated in the regulation of enzymes *in vitro* by other mechanisms (21). In the earlier experiments, it was found that cyclic AMP had no direct effect on succinate oxidation in mitochondria. Verification of the possible effects of biogenic amines on succinate oxidation was therefore attempted (Table IV). Norepinephrine, epinephrine, and serotonin

TABLE IV
STIMULATION OF HEPATIC MITOCHONDRIAL SUCCINATE OXIDATION *IN VITRO* BY AMINES

Preincubation	Concentration µmoles/mg protein	% Stimulation	
		PMS assay	NT assay
Norepinephrine	2.5	61	33
Epinephrine	2.8	114	49
Serotonin	1.3	48	48
Tryptamine	2.5	21	56
Tyramine	2.9	7	50
Histamine	2.7	26	20
Dopamine	1.6	0	0

About 1 mg of freshly isolated mitochondrial protein was preincubated at 37° for 7 min with 100 µmoles of phosphate buffer (pH 7.6) and the amines at the concentrations indicated in a total volume of 1.0 ml. Suitable aliquots were used for determining enzyme activities.

showed activation of the reduction of both PMS and NT by succinate. Of particular interest is the appreciable stimulation of succinate-NT reductase by tyramine and tryptamine, which are known to be thermogenic (22). Dopamine and histamine showed only small changes. The above stimulation seemed to be part of the Q-activation system itself since addition of exogenous Q raised the activity to the same maximum in all cases. It would therefore appear that these amines accomplish their effect by redistribution of the Q in the membranes, thus making more of it available at the site of succinate-NT reductase. Also, the increase in the activity of succinate-PMS reductase seems to be mediated by QH_2, since the stimulation by amines resembled that of QH_2 with respect to the stability to washing of such mitochondria. On testing it was found that these amines are capable of reducing the mitochondrial Q and their oxidation by mitochondria has already been demonstrated (23,24).

COMPETITION BETWEEN OXIDATIONS OF NADH AND SUCCINATE

In the foregoing discussion, SDH (PMS assay) increased in most conditions considered to be thermogenic. This has direct implication on the regulation of the total electron flow through conservative or shunt pathways. This can be appreciated when the following two aspects are considered.

1. Both succinate and NADH dehydrogenases compete for the same cytochrome chain and the Q pool (25,27). Thus, if SDH increases, NADH oxidation decreases.
2. Oxidation of NADH is quantitatively far more significant than succinate. In terms of energy conservation there is a net difference of one ATP less for succinate (mole/mole), and the total number of moles of NADH per mole of succinate oxidized through each turn of the Krebs cycle is 4 and therefore a net difference of 4 ATP molecules.

Therefore, the activation of SDH would reduce NADH oxidation and indirectly lead to a decreased production of ATP by diverting part of the NADH away from the phosphorylating pathway.

ALTERNATE ROUTES OF OXIDATION OF NADH

Increased food intake and O_2 consumption would implicate increased generation of NADH as well. This has to be reoxidized for recycling NAD. Alternate, possibly thermogenic, shunt pathways have to be operative during cold exposure and similar conditions. The data in Fig. 4 show an interesting marked increase in the activity of mitochondrial NADH-NT reductase which deserves further study.

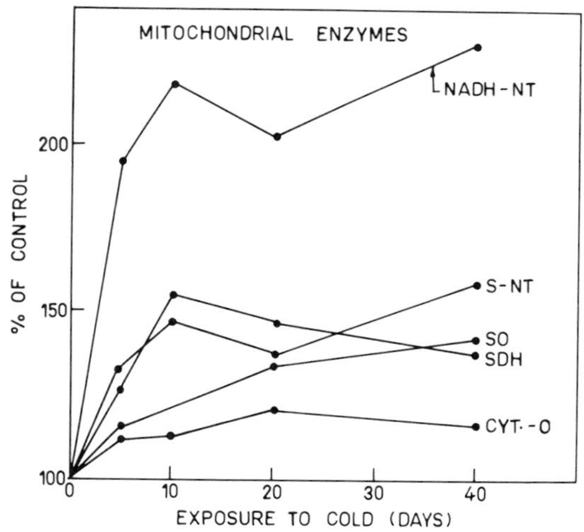

Fig. 4. Changes in the mitochondrial enzyme activities in livers of rats exposed to cold. NADH-NT reductase (NADH-NT); succinate-NT reductase (S-NT); succinate oxidase (SO); succinate dehydrogenase (SDH); cytochrome oxidase (Cyt-O).

It is well known that the reducing equivalents from NADH generated in the mitochondria can be transported out into the cytosol by a number of shuttle mechanisms. It is possible that these may also be activated under these conditions. Two powerful oxidative systems, NADH-cytochrome c reductase and NADH-NT reductase activities, present in the microsomes, can then reoxidize NADH at the rates required.

These microsomal enzyme systems are not coupled to phosphorylation and the energy is therefore transformed into heat. True enough, they increase progressively during cold exposure (Fig. 5). A touch of respectability will be added on when this function is assigned to these microsomal enzyme systems.

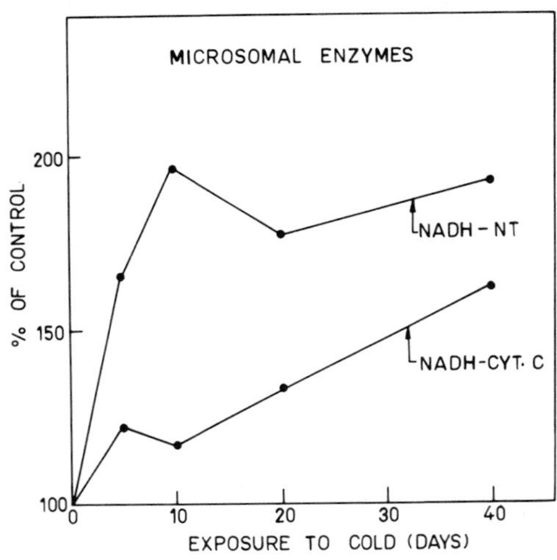

Fig. 5. Alterations in the hepatic microsomal enzyme activities in cold exposed rats.

THE NATURAL ACCEPTORS IN THE SHUNT PATHWAYS

The tetrazolium salts have been used to detect histochemically the dehydrogenases. NT had been introduced as an alternative to measure SDH and it gains importance in view of its unique activation by added Q and other features which indicate that it could represent a measure of the extra heat production. The natural acceptor reacting at this site is yet to be discovered. Any of the several redox components having an E_0' value of 0.05 or higher can fit in. Ascorbate can be considerd to be one such candidate and increased demand for ascorbate during cold exposure lends further support (28). The presence of an alternate route in mitochondria for the oxidation of NADH has already been shown which is insensitive to amytal and antimycin A and bypasses coupled phosphorylation (20). Recently it was shown that tetrahydrobiopterin can link up reduced pyridine nucleotides pools directly to O_2 via cytochrome c and cytochrome oxidase (30). Further work is needed to establish the bypass nature of these systems and thermogenic capabilities.

A MODEL FOR CHEMICAL THERMOGENESIS

The following features have to be explained by the model: increased uptake of oxygen and oxidation of substrates, oxidation without uncoupling of phosphorylation or damage of respiratory control, no increase in the ATP production, liberation of electron

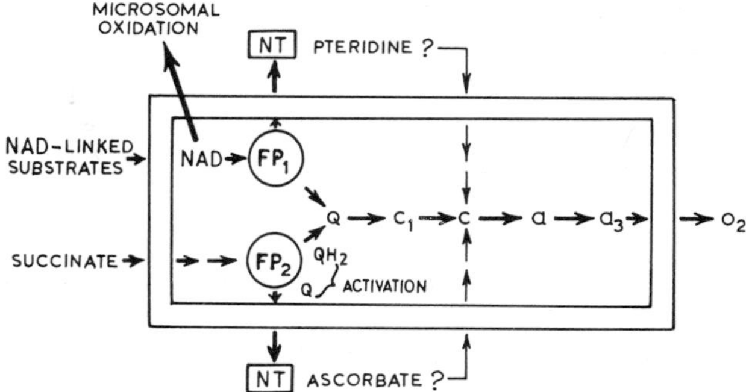

Fig. 6. A model for chemical thermogenesis at the level of mitochondrial oxidations. FP = Flavoprotein.

transport energy as heat. The model shown in Fig. 6 attempts to explain these features:

1. Q is mobilized to the appropriate site of SDH for activating the shunt pathway. This results in the bypass of the electrons at the second phosphorylating site and transfer to cytochrome c for further oxidation which generates one half of total ATP but without respiratory control.

Ascorbate may have a significant role in this bypass.

2. Part of the pool of Q will become reduced; this in turn activates SDH and by competition decreases NADH oxidation. By this process the overall electron transport through the phosphorylating pathway will be decreased.
3. Oxidation of NADH will occur by the shunt pathway represented by the pteridine bypass which eliminates the restriction of respiratory control and the first two phosphorylating sites. Minimum ATP should still be obtained in the third site as with succinate.
4. Further oxidation of NADH can also be obtained by increasing the shuttles to take the reducing equivalents out of mitochondria and utilizing the microsomal electron transport.

The functional role of membranes in the process will become significant first by the initiation of the mobilization of the lipid-quinone Q and alterations in its redox state, and secondly by possibly utilizing the large content of α-helical structure in membrane proteins. A suprahelical sequence of hydrogen-bonded π-electron system of peptide units exists (31) on α-helix which is conjectured to facilitate transport of electrons along the path of -CO..HN-CO..HN-. As illustrated in Fig. 7, if the energy of the electron is dropped during the transfer, the hydrogen bonds and the helix conformation may be broken and reversal thereof will release heat and this is already known to occur in reversible helix-coil transitions (32). This will provide a molecular framework of energy transduction to the form of heat.

It still remains to be seen what triggers the modifications, whether all the tissues behave thermogenically to a varying degree and what are the possible sites of action of known calorigenic substances. Some of these investigations are expected to illuminate on the wisdom of location of the energy transduction processes in membranes.

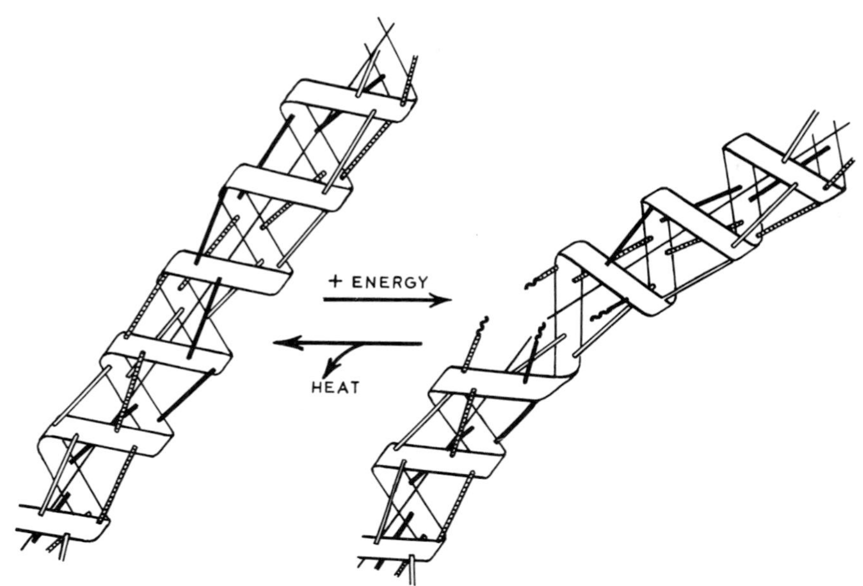

Fig. 7. A diagrammatic representation of possible thermogenesis through reversible transitions of hydrogen bonds in α-helix.

REFERENCES

1. DAVIS, T. R. A., JOHNSTON, D. R., BELL, F. C., AND CREMER, B. J. (1960) Am. J. Physiol. 198, 471.
2. STONER, H. B. (1963) Fed. Proc. 22, 851.
3. WILLIAMSON, J. R., JAKOB, A., AND SCHOLTZ, R. (1971) Metabolism 20, 13.
4. KLAIN, G. J., AND HANNON, J. P. (1969) Fed. Proc. 28, 965.
5. HANNON, J. P. (1963) Fed. Proc. 22, 856.
6. POTTER, V. R. (1958) Fed. Proc. 17, 1060.
7. LESTER, R. L., AND SMITH, A. L. (1961) Biochim. Biophys. Acta 47, 475.
8. GUTMAN, M., KEARNEY, E. B., AND SINGER, T. P. (1971) Biochemistry 10, 2726.

9. SUSHEELA, L., AND RAMASARMA, T. (1973) Biochim. Biophys. Acta 292, 50.
10. KROGER, A., AND KLINGENBERG, M. (1966) Biochem. Z. 344, 317.
11. GUTMAN, M., KEARNEY, E. B., AND SINGER, T. P. (1971) Biochemistry 10, 4763.
12. KEARNEY, E. B. (1957) J. Biol. Chem. 229, 363.
13. MONOD, J., WYMAN, J., AND CHANGEUX, J. P. (1965) J. Mol. Biol. 12, 88.
14. LAZDUNSKI, M. (1972) In Current Topics in Cellular Regulation (Horecker, B. L., and Stadtman, E. R., eds.) p. 267, vol. 6, Academic Press, New York and London.
15. AITHAL, H. N., AND RAMASARMA, T. (1971) Biochem. J. 123, 677.
16. BEYER, R. E., NOBLE, W. M., AND HERSCHFELD, T. J. (1962) Can. J. Biochem. Physiol. 40, 511.
17. AITHAL, H. N., JOSHI, V. C., AND RAMASARMA, T. (1968) Biochim. Biophys. Acta 162, 66.
18. AXELROD, J. (1964) In "Biogenic Amines," Progress in Brain Research (Himmich, H. E., and Himwich, W. A., eds.) p. 81, vol. 8, Elsevier, Amsterdam.
19. CARLSON, A. (1970 In New Aspects of Storage and Release Mechanisms of Catecholamines (Schumann, H. J., and Kroneberg, G., eds.) Bayer Symposium II, p. 223, Springer Verlag.
20. PLETSCHER, A., BROSSI, A., AND GEY, K. F. (1962) Int. Rev. Neurobiol. (Pfeiffer, C. C., and Smythies, J. R., eds.) p. 275, vol. 4, Academic Press, N.Y.
21. PITOT, H. C., AND YATVIN, M. B. (1973) Physiol. Rev. 53, 228.
22. HEROUX, O. (1969) In Physiology and Pathology of Adaptation Mechanisms (Bajusz, E., and collaborators, eds.) p. 347, vol. 27, Pergamon.
23. GREEN, D. E., AND RICHTER, D. (1937) Biochem. J. 31, 576.
24. PASTAN, I., HERRING, B., JOHNSON, P., AND FIELD, J. B. (1962) J. Biol. Chem. 237. 287.
25. RINGLER, R. L., AND SINGER, T. P. (1959) J. Biol. Chem. 234, 2211.

26. SINGER, T. P., GUTMAN, M., AND KEARNEY, E. B. (1971) FEBS Lett. 17, 11.
27. GUTMAN, M., AND SILMAN, N. (1972) FEBS Lett. 26, 207.
28. SMITH, R. E., AND HOIJER, D. J. (1962) Physiol. Rev. 42, 60.
29. LEHNINGER, A. L. (1951) In Phosphorus Metabolism (McElroy, W. D., and Glass, H. B., eds.) p. 344, vol. 1, Johns Hopkins Press, Baltimore.
30. REMBOLD, H., AND BUFF, K. (1972) Eur. J. Biochem. 28, 579.
31. RAMASARMA, T., AND VIJAYAN, M. (1974) FEBS Letters, in press.
32. CHOU, P. Y., AND SCHERAGA, H. A. (1971) Biophysics 10, 657.

COOPERATIVE INTERACTIONS IN MITOCHONDRIAL MEMBRANE FUNCTION

S. R. Panini and C. K. Ramakrishna Kurup

Homeostasis has long been recognized as one of the important characteristics of a living system. The dictates of homeostasis demand that the converging degradative pathways and the diverging biosynthetic pathways be so regulated as to provide effective coordination between the two. The catabolic processes provide not only the biosynthetic precursors but also the energy required to turn the anabolic machinery. Adenosine triphosphate, which acts as the link between catabolic and anabolic processes, has thus a key role to play in the maintenance of homeostasis. It is logical, then, to infer that an effective method of regulation of cellular processes *in toto* will be provided by the modulation of the output of energy by the mitochondrion. This is exemplified in the concepts of "adenylate energy charge" (1) and "phosphorylation potential" (2).

Since most of the energy requirement of the cell is met by the mitochondrion, it is obvious that both intra- and extra-mitochondrial control mechanisms must be operative in the generation and release of chemical energy. Many intra-mitochondrial control characteristics of the respiratory assembly have been detailed by Chance (3). The relative impermeability of mitochondria to electron transport components (pyridine nucleotides) and the existence of specific carriers for the transport of substrates as well as components of energy transduction (adenine nucleotides) are examples of control mechanisms external to the respiratory assembly.

During the course of our investigations on the inhibition of oxidative phosphorylation in rat liver mitochondria by the antihypercholesterolaemic compound chlorophenoxyisobutyric acid, we observed that the inhibition curves were sigmoidal. The "Dixon plots" (4), in which the reciprocal of reaction velocity (electron transport or phosphorylation) is plotted against inhibitor concentration, were non-linear. Such non-linearity is indicative of the interaction of more than one inhibitor molecule with the enzyme (5). Sigmoid kinetics are characteristic of allosteric interactions, which have been recognized as one of the most important control mechanisms in the regulation of metabolic processes (6). This raised the possibility that cooperative interactions may play an important role in the regulation of energy production in the mitochondrion. It may be recalled that the "conformation coupling theory" (7) envisages protein conformational changes as the primary event in the transduction of energy in the mitochondrion.

While with a single enzyme meaningful interpretation of sigmoid kinetics is possible, the question arises as to how such kinetics can be interpreted in a membrane-bound, multi-stepped and integrated process such as oxidative phosphorylation. That complex biological phenomena like cellular respiration, cell division, etc., which show cooperative interactions, are amenable to such kinetic analysis has been shown by Loftfield and Eigner (8). It was suggested that the Hill equation (9) could be applied to such phenomena. When applied to an enzyme system under the influence of an inhibitor, the Hill equation takes the form (6):

$$\log\left(\frac{V_o - V}{V - V_{sat}}\right) = n \log [I] + \log K$$

where V_o is the reaction velocity in the absence of the inhibitor and V and V_{sat} are the velocities in the presence of sub-optimal [I] and saturating concentrations of the inhibitor respectively. This equation forecasts a straight line graph when $\log[(V_o-V)/(V-V_{sat})]$ is plotted against $\log[I]$, the slope of the line representing "n", the Hill-coefficient. According to Atkinson (6) and Koshland (10) the Hill-coefficient is a measure of both the number of interacting functional binding sites and the strength of interaction between them. In a complex process the experimentally realizable value of "n" has been termed the "interaction coefficient" (λ). In such a process a linear Hill-plot would be an indication that the inhibitor interacts with a "single critical step" in the process. On the other hand, if two or more critical steps with different association constants are affected, it is to be expected that the slope of the Hill-plot would increase with increasing concentration of the inhibitor (8). In other words, the number of breaks in the Hill-plot would indicate the number of critical steps affected by the inhibitor and the slope of each segment a measure of the affinity of the inhibitor for that step. Based on the above assumptions, in this chapter we propose to apply the Hill-treatment to the data available in the literature on the action of inhibitors and uncouplers of oxidative phosphorylation.

Chlorophenoxyisobutyrate

This compound has been shown to act as an inhibitory uncoupler of coupled phosphorylation in rat liver mitochondria without any direct effect on the respiratory chain (11). The Hill-plots constructed for the inhibition of coupled phosphorylation by this compound gave two interaction coefficients irrespective of the number of coupling sites activated. The inhibition of succinate oxidation also showed a similar pattern. However, succinate oxidation in the presence of uncouplers was inhibited with a single interaction coefficient. Partial reactions of oxidative phosphorylation, such as ATPase (stimulation of latent ATPase, inhibition of uncoupler stimulated ATPase) and ATP-dependent reduction of NAD^+ by succinate through reversed electron transport were all affected yielding two interaction coefficients in each case (12). These results are summarized in Table I.

TABLE I
INTERACTION COEFFICIENTS (λ) FOR THE INHIBITION OF MITOCHONDRIAL ACTIVITIES BY CHLOROPHENOXYISOBUTYRATE

	Reaction	λ_1	λ_2
1	Phosphorylation with glutamate + malate	1.6	3.6
2	Phosphorylation with succinate	1.4	3.6
3	Phosphorylation with ascorbate + TMPD	2.0	3.6
4	Oxidation of succinate	0.6	1.8
5	Uncoupled oxidation of succinate	-	1.8
6	Latent ATPase (activation)	1.0	2.0
7	Uncoupler-stimulated ATPase	0.8	1.4
8	Submitochondrial particle ATPase	0.8	2.0
9	Reversed electron transport	1.0	2.0

Data summarized from Panini and Kurup (12).

The existence of two interaction coefficients would indicate that the compound interferes with two functionally important steps in the energy transfer pathway. This would support the earlier inference arrived at from a study of inhibition of uncoupler stimulated reactions in mitochondria by this compound (11). The interaction coefficients for the inhibition of ATP generating reactions were approximately 2 and 4 while those for ATP degrading reactions were 1 and 2. This may be taken to mean either that the synthetic and degradative pathways are different as suggested by Lardy *et al.* (13) or that the sensitivity of the forward and reverse reactions of energy transduction to inhibitors is different as observed by Danielson and Ernster (14).

Antimycin

The sigmoid nature of the inhibition of mitochondrial respiration by antimycin is well documented (15,16). In his pioneering study, Thorn (15) recognized the complexity of the inhibition curve (see inset in Fig. 1) but preferred to emphasize that the degree of inhibition was proportional to antimycin concentration if the initial part of the curve is ignored. Nevertheless, with commendable insight,

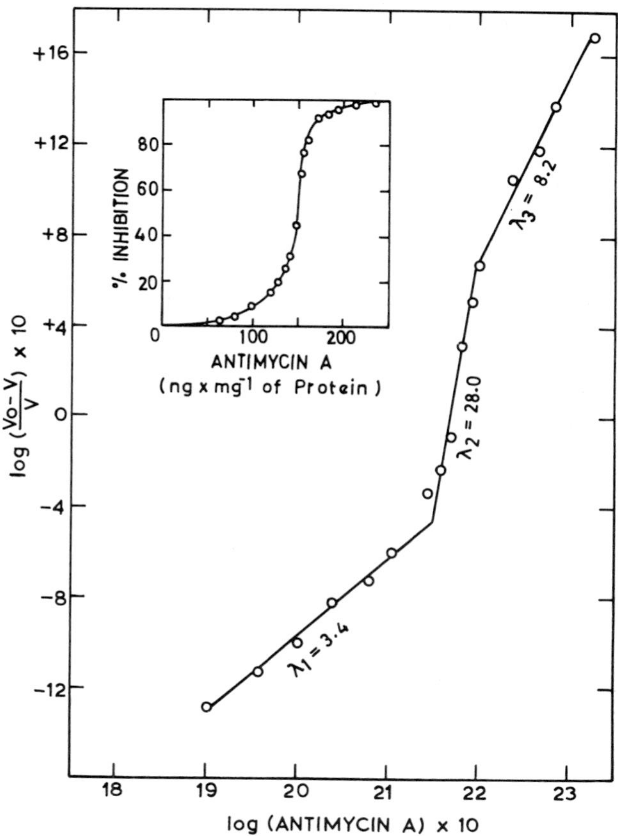

Fig. 1. Hill-plot of the inhibition of succinate oxidase activity of Keilin-Hartree particles by antimycin. Values recalculated from the data of Thorn (15).

he attributed the sigmoidicity of the curve to the organization and function of the electron transport system. Slater and his associates (16) in a detailed discussion of this sigmoidicity concluded that antimycin was an allosteric inhibitor of the respiratory chain. The Hill-plot constructed from the data of Thorn (15), presented in Fig. 1, confirms the conclusion of Slater. It also indicates that the compound inhibits at least two (possibly three) critical steps. This multi-site action of antimycin is indicated also by its oligomycin-like action at low concentrations (17), uncoupler-like action at high concentrations (18,19), inhibition of reactions other than succinate-cytochrome _c_ reductase (20), and its effect on the composition and structural organization of complex III, particularly cytochrome _b_ (21,22,23).

Rotenone and Piericidin A

In contrast to antimycin action, inhibition of mitochondrial respiration by both rotenone and piericidin A is confined to a single locus. This is clear from the linear Hill-plots obtained for the inhibition of glutamate + malate oxidation in rat liver mitochondria (24) presented in Fig. 2. Similar plots for the inhibition of NADH oxidase activity of electron transport particles made from the data of Singer _et al._ (25) were also linear (Fig. not given). In both cases the interaction coefficient for piericidin A was 2-3 times higher than that for rotenone, indicating greater affinity of the former to the binding sites. This may account for the observation of Singer _et al._ (25) that while piericidin could displace rotenone from the binding sites, neither rotenone nor serum albumin could displace piericidin.

Aurovertin

The inhibition of mitochondrial oxidation (State 3) by aurovertin is sigmoidal as is the binding curve (13,26). This has been interpreted (26) as an indication of cooperative interaction of aurovertin with

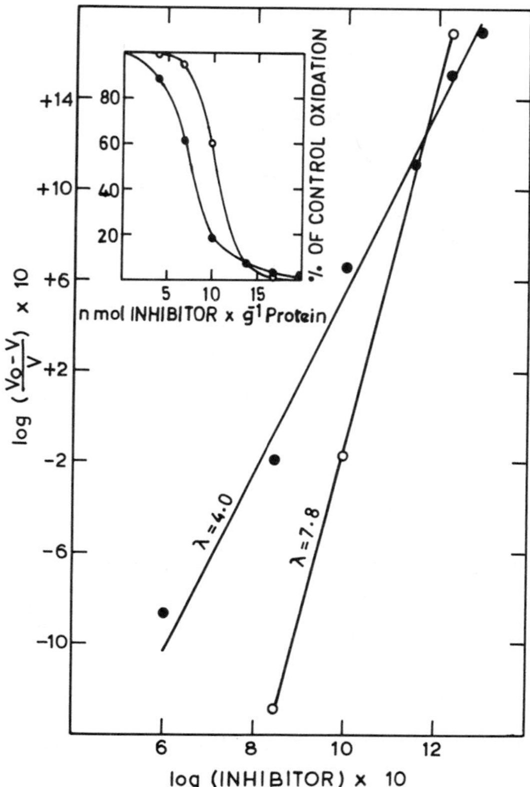

Fig. 2. Hill-plots of the inhibition of glutamate + malate oxidation in rat liver mitochondria by rotenone and piericidin A. Values recalculated from the data of Garland et al. (24). ●, Rotenone; o, Piericidin A.

the energy transfer pathway. The Hill-plot constructed from the data of Slater and his associates (26) shows two interaction coefficients (Fig. 3), suggesting interference with two functionally important steps in the energy transfer pathway. It is interesting to note that the analysis of the data according to the model of Monod et al. (27) also indicates a sharp increase in cooperativity at a certain concentration of aurovertin (26). This is consistent with the scheme of Lardy et al. (13). Aurovertin

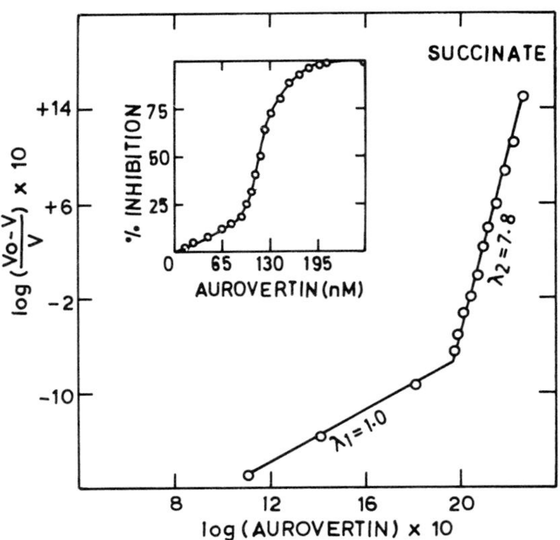

Fig. 3. Hill-plot of the inhibition of State 3 oxidation of succinate in rat heart mitochondria by aurovertin. Values recalculated from the data of Bertina et al. (26).

inhibits the activity of purified ATPase and two binding sites have been recognized on the protein (28,29).

Oligomycin

In contrast to aurovertin, oligomycin, which also inhibits active oxidation in a sigmoid fashion (30), shows a single interaction coefficient in the Hill-plot (Fig. 4). It is known that inhibition by oligomycin is not instantaneous, but shows a time lag which is not abolished by preincubation. During this period, the interaction coefficient increases from 4.0 to 7.0, indicating that as the system turns over the cooperativity is enhanced.

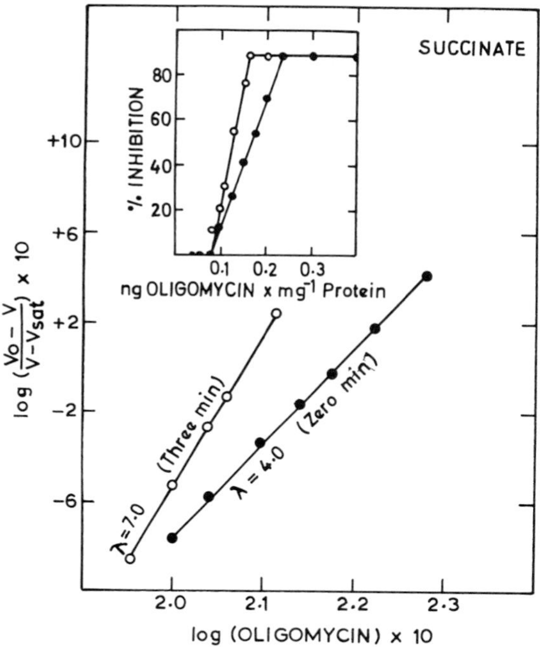

Fig. 4. Hill-plots of the inhibition of State 3 oxidation of succinate in rat liver mitochondria by oligomycin. Values recalculated from the data of Slater and ter Welle (30). ●, zero min; o, three min.

2,4-Dinitrophenol

It is clear from Fig. 5 that the stimulation of latent ATPase by dinitrophenol is a two-stepped process. The interaction coefficients compare favorably with those obtained for the same effect of chlorophenoxyisobutyrate on latent ATPase (Table I). It was observed by Katyare et al. (31) that the stimulation of ATPase activity by low concentrations of dinitrophenol was insensitive to fluoride while the stimulation at higher concentrations of the uncoupler was inhibited by fluoride. The Hill-plots in Fig. 5 reveal that fluoride effectively prevents interaction of the uncoupler with the second step. The stimula-

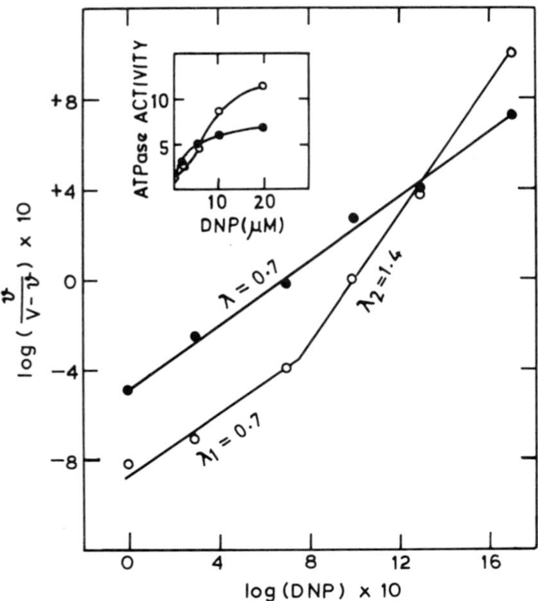

Fig. 5. Hill-plots of the stimulation of ATPase activity of rat liver mitochondria by dinitrophenol. Values recalculated from the data of Katyare et al. (31). For a positive modifier the Hill equation takes the form: $\log[v/(V-v)] = n \log[M] - \log K$, where v is the reaction velocity in the presence of modifier [M] and V is the maximal reaction velocity (6). o, no KF; ●, in the presence of KF.

tion of ATPase activity which is sigmoidal in nature is changed to the classical hyperbolic pattern in the presence of fluoride (Fig. 5 inset). Fluoride thus acts as a desensitizer.

Atractyloside and Bongkrekic acid

The application of the Hill treatment to the pattern of inhibition of oxidative phosphorylation by these two inhibitors of adenine nucleotide translocase system brings out subtle differences between their modes of action. The data of Henderson and

Lardy (32) on the effects of bongkrekic acid and atractyloside have been recalculated and plotted in Figs. 6 and 7 respectively. At low concentrations of ADP, the former exhibits two interaction coefficients (Fig. 6). Increasing concentrations of ADP abolish the first interaction coefficient and enhance the second. This would mean that of the two steps susceptible to attack by this inhibitor, only the first one is protected by ADP. In contrast to this, atractyloside shows a single interaction coefficient, the magnitude of which is decreased by increasing concentrations of ADP (Fig. 7).

In a recent publication Klingerberg and Buchholz (33) have discussed in detail the kinetics of bongkrekic acid binding to the adenine nucleotide carrier.

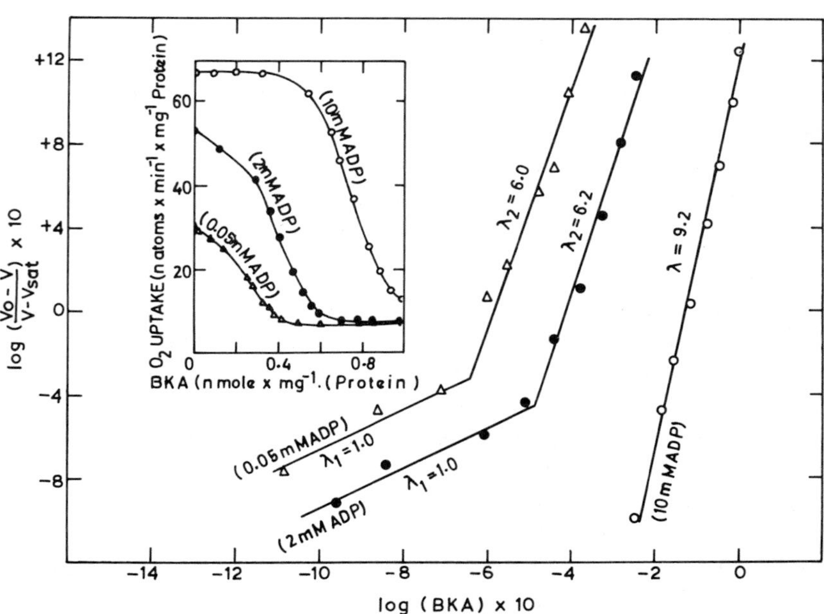

Fig. 6. Hill-plots of the inhibition of glutamate oxidation in rat liver mitochondria by bongkrekic acid. Values recalculated from the data of Henderson and Lardy (32). BKA, bongkrekic acid.

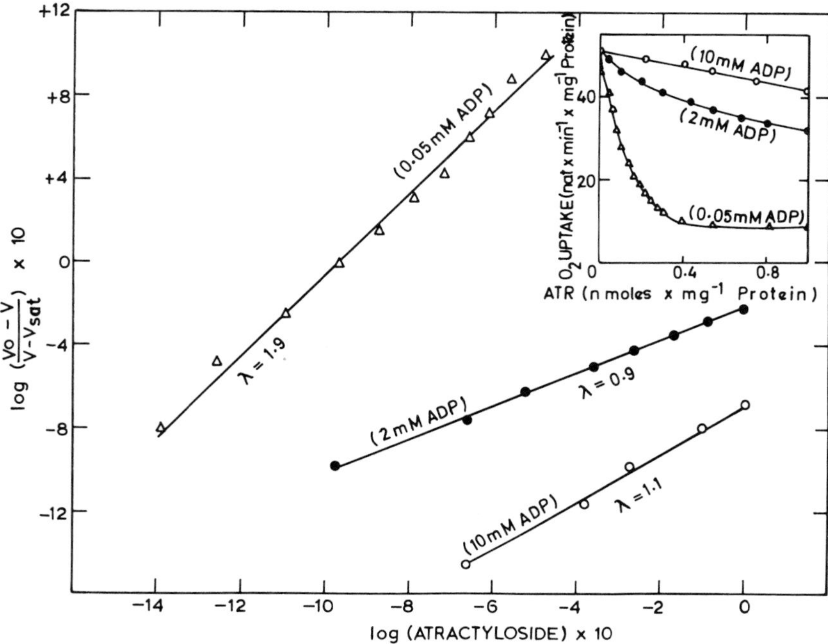

Fig. 7. Hill-plots of the inhibition of glutamate oxidation in rat liver mitochondria by atractyloside. Values recalculated from the data of Henderson and Lardy (32). ATR, atractyloside.

Their assumption that the modes of binding of bongkrekic acid and atractyloside to the carrier are different finds support from the Hill analysis of the inhibition effects. The two inhibitors have been shown to have opposite effects on calcium-stimulated State 4 oxidation (34). The observation (33) that ADP promotes bongkrekic acid binding is echoed by the increase in the value of the interaction coefficient for bongkrekic acid with increasing concentrations of ADP (Fig. 6). It may be mentioned, however, that Klingenberg and Buchholz (33) now do not favor the view that the sigmoidicity of bongkrekic acid effect reflects allosteric interactions. On the other hand, they postulate that the binding sites on the carrier are common to both adenine nucleotide and bongkrekic acid and that the sigmoidicity arises due to the binding of the inhibitor to sites other than those on the carrier.

DISCUSSION

In this communication we have mostly used data available in the literature on the action of various inhibitors and uncouplers to demonstrate the existence of cooperative interactions in mitochondrial oxidative phosphorylation. It is probable that small inaccuracies could have crept in during the computation of values from the figures in the literature. In spite of this limitation, the results presented here give ample indication that the Hill treatment could be profitably applied to analyze membrane-associated complex biological phenomena. It is pertinent to remark here that the possibility of operation of allosteric interactions between macromolecules organized into supramolecular structures has begun to gain recognition (35). The role of conformational alterations in the energy conservation mechanism has been emphasized by Green (7), Chance (36), and Slater (37).

It is noteworthy that the Hill-plots for most of these inhibitors show a break, indicating that at least two critical steps are being interfered with. This would indicate that the pathway of energy transduction is more intricate and may involve more steps than proposed in any mechanism at present. Our analysis has helped to bring out differences in the inhibitory patterns of compounds which are supposed to have similar mechanisms and same sites of action, the examples being rotenone and piericidin A, aurovertin and oligomycin, and bongkrekic acid and atractyloside.

A complete understanding of the significance of the interaction coefficient is difficult in complex systems. It should not be interpreted as a measure of the number of binding sites since even in the case of a single protein such as hemoglobin, which contains four binding sites for oxygen, the maximum value of the coefficient realized experimentally is only 3.0

(10). Our results indicate that the interaction coefficients generally give a comparative measure of the effectiveness of two similar modifiers. However, for a given modifier a change in the interaction coefficient under an altered set of conditions need not necessarily reflect a similar change in the effectiveness.

The most important and unique regulatory phenomenon occurring in the mitochondrion, namely "respiratory control," points to the existence of negative feedback control of energy on energy. Although two decades have passed since the demonstration of respiratory control in the laboratories of Lardy (38) and Chance (39), the mechanism by which mitochondrial respiratory activity is stimulated by ADP and inhibited by ATP is still far from clear. The favored interpretation invokes kinetic and equilibrium controls (2). The existence of cooperative interactions would indicate that "activity control" (2) may also play a role in the regulation of energy transduction.

SUMMARY

The kinetics of the effects of various inhibitors and uncouplers on oxidative phosphorylation and related reactions have been analyzed according to the Hill equation. The results indicate that the pathway of energy transduction is more intricate and may involve more steps than proposed in any mechanism at present. This analysis has helped to bring out differences in the inhibitory patterns of compounds which are supposed to have similar mechanisms and same sites of action. The existence of cooperative interactions in mitochondrial membrane function indicates that "activity control" may play an important role in the regulation of energy transduction.

REFERENCES

1. ATKINSON, D. E. (1970) *in* The Enzymes (Boyer, P. D., ed.) Vol. 1, p. 461-489, Academic Press, New York and London.

2. KLINGENBERG, M. (1965) in Control of Energy Metabolism (Chance, B., Estabrook, R. W., and Williamson, J. R., eds.) p. 149-155, Academic Press, N.Y.
3. CHANCE, B. (1965) in Control of Energy Metabolism (Chance, B., Estabrook, R. W., and Williamson, J. R., eds.) p. 9-12, 415-435, Academic Press, N.Y.
4. DIXON, M., AND WEBB, E. C. (1965) Enzymes, p. 328-330, Academic Press, London.
5. WEBB, J. L. (1963) Enzyme and Metabolic Inhibitors, p. 174, Academic Press, New York.
6. ATKINSON, D. E. (1966) Ann Rev. Biochem. 35, 85-124.
7. GREEN, D. E., AND JI, S. (1972) Bioenergetics 3, 159-202.
8. LOFTFIELD, R. B., AND EIGNER, E. A. (1969) Science 164, 305-308.
9. HILL, A. V. (1913) Biochem. J. 7, 471-480.
10. KOSHLAND, D. E., JR. (1970) in The Enzymes (Boyer, P. D., ed.) Vol. 1, p. 341-396, Academic Press, N.Y. and London.
11. PANINI, S. R., AND KURUP, C. K. R. (1974) Biochem. J., in press.
12. PANINI, S. R., AND KURUP, C. K. R. (1974) Biochim. Biophys. Acta (communicated).
13. LARDY, H. A., CONNELLY, J. L., AND JOHNSON, D. (1964) Biochemistry 3, 1961-1968.
14. DANIELSON, L., AND ERNSTER, L. (1963) in Energy-Linked Functions of Mitochondria (Chance, B., ed.) p. 157-180, Academic Press, N.Y.
15. THORN, M. B. (1956) Biochem. J. 63, 420-436.
16. BRYLA, J., KANIUGA, Z., AND SLATER, E. C. (1969) Biochim. Biophys. Acta 189, 317-326.
17. HOWLAND, J. L. (1968) Biochim. Biophys. Acta 153, 309-311.
18. LOW, H., AND VALLIN, I. (1963) Biochim. Biophys. Acta 69, 361-374.
19. HASS, D. W. (1964) Biochim. Biophys. Acta 92, 433-439.

20. TEETER, M. E., BAGINSKY, M. L., AND HATEFI, Y. (1969) Biochim. Biophys. Acta 172, 331-333.
21. PUMPHREY, A. M. (1962) J. Biol. Chem. 237, 2384-2390.
22. BAUM, H., SILMAN, H. I., RIESKE, J. S., AND LIPTON, S. H. (1967) J. Biol. Chem. 242, 4876-4887.
23. RIESKE, J. S., LIPTON, S. H., BAUM, H., AND SILMAN, H. I. (1967) J. Biol. Chem. 242, 4888-4896.
24. GARLAND, P. B., CLEGG, R. A., LIGHT, P. A., AND RAGAN, C. I. (1969) in Inhibitors-Tools in Cell Research (Bücher, Th., and Sies, H., eds.) p. 217-246, Springer-Verlag, Berlin.
25. SINGER, T. P., HORGAN, D. J., AND CASIDA, J. E. (1968) in Flavins and Flavoproteins (Yagi, K., ed.) p. 192-213, University of Tokyo Press, Tokyo.
26. BERTINA, R. M., SHRIER, P. I., AND SLATER, E. C. (1973) Biochim. Biophys. Acta 305, 503-518.
27. MONOD, J., WYMAN, J., AND CHANGEUX, J. P. (1965) J. Mol. Biol. 12, 88-118.
28. CHANG, T. M., AND PENEFSKY, H. S. (1973) J. Biol. Chem. 248, 2746-2754.
29. LARDY, H. A., AND LAMBETH, D. (1972) in Energy Metabolism and Regulation of Metabolic Processes in Mitochondria (Mehlman, M. A., and Hanson, R. W., eds.) p. 287-288, Academic Press, N.Y. and London.
30. SLATER, E. C., AND TER WELLE, H. F. (1969) in Inhibitors-Tools in Cell Research (Bücher, Th., and Sies, H., eds.) p. 258-278, Springer-Verlag, Berlin.
31. KATYARE, S. S., FATTERPAKER, P., AND SREENIVASAN, A. (1971) Arch. Biochem. Biophys. 144, 209-215.
32. HENDERSON, P. J. F., AND LARDY, H. A. (1970) J. Biol. Chem. 245, 1319-1326.
33. KLINGENBERG, M., AND BUCHHOLZ, M. (1973) Eur. J. Biochem. 38, 346-358.
34. OUT, T. A., KEMP, A., AND SOUVERIJN, J. (1971) Biochim. Biophys. Acta 245, 299-304.

35. CHANGEUX, J. P., AND THIERY, J. (1968) *in* Regulatory Functions of Biological Membranes (Jarnefelt, J., ed.) p. 116-138, Elsevier Publishing Co., Amsterdam.
36. CHANCE, B., LEE, C. P., AND MELA, L. (1967) *Federation Proc.* *26*, 1341-1354.
37. SLATER, E. C. (1969) *in* Mitochondria - Structure and Function (Ernster, L., and Drahota, Z., eds.) p. 205-217, Academic Press, London.
38. LARDY, H. A., AND WELLMAN, H. (1952) *J. Biol. Chem.* *195*, 215-224.
39. CHANCE, B., AND WILLIAMS, G. R. (1956) *Advan. Enzymol.* *17*, 65-134.

FLUORESCENCE PROPERTIES OF CHLOROPLASTS *IN VIVO* AT LOW TEMPERATURES

P. V. Sane, V. G. Tatake and T. S. Desai

INTRODUCTION

Spinach leaf segments subjected to different light and chemical treatments were used to study the changes in fluorescence yield of chlorophylls during warm up from 77°K to 300°K. Light of 436 or 480 nm was used for excitation of fluorescence which was monitored at 682 nm. Fluorescence yield of leaf frozen in dark and illuminated at 77°K with white light, when excited with very low intensity light, showed the following changes with temperature:

(a) after an initial rise in fluorescence due to induction at 77°K there was a decrease as the temperature rose from 77°K to 95°K,

(b) a further rise in temperature resulted in an increase in fluorescence yield with a peak at 180°K followed by a decrease until a temperature of 235°K was reached,

(c) continued warming of the leaf resulted in another peak in fluorescence yield at 250°K.

The rise in fluorescence yield peaking at 180°K will be explained by proposing an additional quencher of chlorophyll fluorescence Q_1. The quencher is reduced by Q and is probably situated on the electron transport chain prior to the site of DCMU action. The results will also be discussed in terms of donors of electrons to PS II at low temperatures.

Abbreviations: cyt - cytochrome; DCMU - 3-(3',4'-dichlorophenyl)-1,1-dimethylurea; NADP - nicotinamide adenine dinucleotide phosphate; PS - photosystem.

Previously in our laboratory we had observed that when chloroplasts or spinach leaf discs frozen in white light to liquid nitrogen temperature were subsequently heated at a slow rate in the dark they emitted light at different temperatures (Fig. 1).

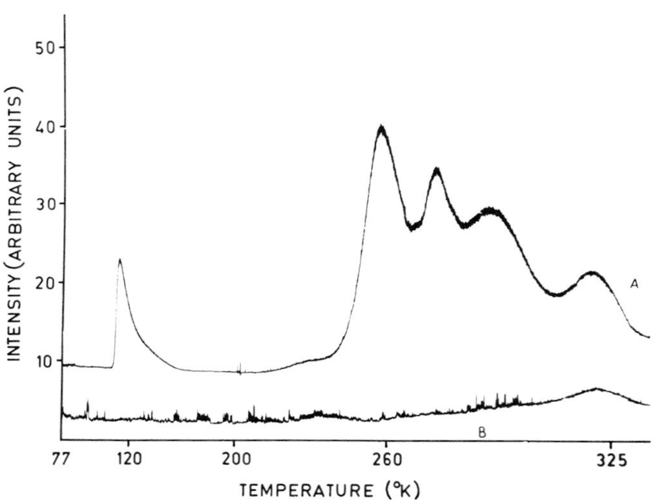

Fig. 1. Glow curves of a spinach leaf frozen to 77°K. A, Leaf frozen under 2.8×10^3 ergs·cm^{-2}·sec^{-1} illumination for two minutes. B, Relaxed leaf frozen in dark.

We were interested in relating the appearance of these thermoluminescence (TL) peaks to known photochemical reactions of chloroplasts. It is known that the primary acceptor of PS II (Q) can be reduced at liquid nitrogen temperature. We therefore thought it very likely that the reduced Q could act as a primary source of electrons for different electron transport reactions induced by gradual rise in the temperature of the preilluminated leaf. We therefore were interested in studying the oxidation of Q at different temperatures. Since fluorescence yield of chlorophylls is an indicator of the redox state of Q we have studied fluorescence characteristics of spinach leaves at low temperatures. Although the results did not establish a firm relation between oxidation of Q and appearance of all the TL peaks the data have provided evidence for the existence of an additional quencher of chlorophyll fluorescence in spinach leaves.

MATERIALS AND METHODS

In all the experiments fresh spinach leaves cut into strips of 8 × 20 mm were used. The strip was placed into a cuvette and supported by an alluminum strip. The leaf strip placed in the cuvette was frozen to 77°K by dipping the cuvette in the liquid nitrogen either in light or in dark as desired. The cuvette containing liquid nitrogen was transferred to the solid state assembly of the Aminco Bowman Spectrophotofluorometer. The leaf temperature was monitored with an iron-constantan thermocouple placed just above the area of excitation. The fluorescence was excited with 436 nm and monitored at 682 nm using R 446 photomultiplier. Excitation with 480 nm was also tried but it gave essentially the same results as excitation with 436 nm. The intensity of the exciting beam was reduced when desired by inserting a neutral density filter and Kodak cut-off filters. The incident energy on the leaf strip was measured with a precalibrated thermopile and has been indicated in figures. The temperature of the leaf remained at 77°K for 8 seconds after its transfer to the solid state assembly. During this time there was induction of fluorescence and the fluorescence reached a high level until the temperature started rising. This has not been shown in the figures.

Chloroplasts were isolated according to Sane, Goodchild and Park (1). The suspension of chloroplasts was smeared on a 8 × 20 mm glass slide and

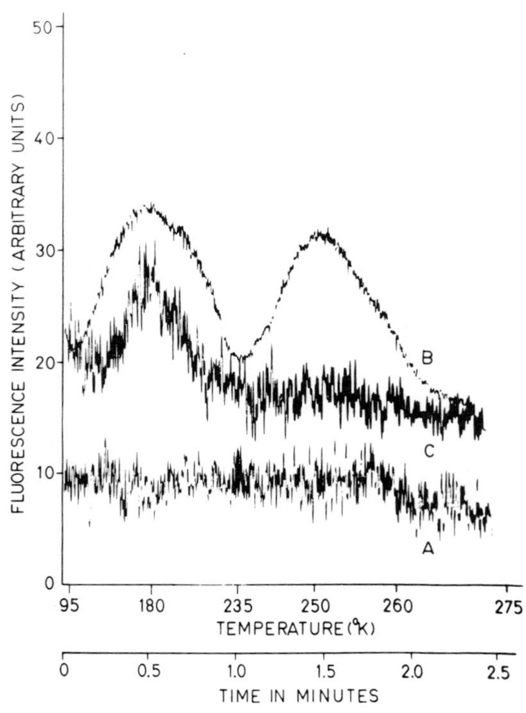

Fig. 2. Changes in fluorescence yield of a leaf with increase in temperature at different monitoring intensities of excitation. A, Excitation intensity less than 2 ergs$\cdot$cm$^{-2}\cdot$sec^{-1}, fluorescence amplifier gain 1, leaf frozen in dark. B, Excitation intensity 10 ergs$\cdot$cm$^{-2}\cdot$sec^{-1} fluorescence amplifier gain 1/20 of A, leaf frozen in dark. C, Leaf illuminated with 7.5×10^5 ergs $\cdot$cm$^{-2}\cdot$sec^{-1} for two minutes at 77°K after freezing in dark, excitation intensity and amplifier gain same as that for A.

this slide replaced the leaf strip in experiments in which chloroplasts were used.

RESULTS AND DISCUSSION

Since our major objective was to study the oxidation of Q during warming we wanted to choose such a low intensity of exciting light that would not reduce Q. In Fig. 2 we have compared two intensities of light. When a leaf frozen in the dark was studied for the changes in fluorescence yield during warming it was observed that the yield was very low and remained unchanged if the exciting light intensity was less than 2 $ergs \cdot cm^{-2} \cdot sec^{-1}$. If, however, the intensity of exciting beam was 10 $ergs \cdot cm^{-2} \cdot sec^{-1}$ the following changes were observed:

1. At first there is a decrease in fluorescence yield as the temperature increases from 77°K to 95°K.
2. This is followed by a rise in the yield, giving a peak at 180°K. A further rise in temperature from 180°K to 235°K results in the lowering of the fluorescence yield.
3. A subsequent rise in temperature results in a second increase in fluorescence yield which reaches a maximum at 250°K after which it declines to a lower level.

From Fig. 2C it is observed that the very low intensity of exciting beam also shows the changes mentioned in 1 and 2 above, provided the leaf frozen in the dark was illuminated with white light at 77°K. These data indicate that when Q is reduced by exposing a leaf at 77°K to white light even the very low intensity of exciting light shows changes in fluorescence yield as the leaf temperature rises. Since the very low intensity of exciting light did not show any changes with rise in temperature in the case of a leaf frozen in the dark, we believe that this intensity of light is unable to reduce Q by itself. One can now explain the decrease and increase in fluorescence yield observed by very low intensity of exciting beam in terms of Duysens and Sweers' (2) proposal. According to them a component Q in an oxidized state quenches the chlorophyll fluorescence whereas in the reduced state it increases the fluorescence yield of chlorophylls. On this basis one could explain the observations mentioned in 1 and 2 above as follows:

An increase in temperature from 77°K to 95°K which resulted in a decrease in fluorescence yield must be due to oxidation of Q during this rise in temperature. The subsequent rise in fluorescence yield giving a peak at 180°K must be due to reduction of Q as the temperature rose from 95°K to 180°K. Whereas in the case of Fig. 2B one may propose that the exciting intensity could cause the reduction of Q, such a proposal is unacceptable in the case of Fig. 2C. This is because we know that this intensity of light is unable to reduce Q by itself as mentioned earlier.

An alternate explanation could be that the changes observed are due to some other physical factors such as the changes in phase transitions of ice crystals during warming as suggested by Cho and Govindjee (3). If this was true we would have expected changes similar to those mentioned in 1, 2 and 3 irrespective of pretreatment of leaf. In Fig. 3A we have shown the changes in fluorescence yield of a leaf frozen in light. Here we do not see the peak appearing at 180°K but instead a high yield of fluorescence is observed until a temperature of 250°K is reached, after which it declines. It may be mentioned here that although the exciting intensity used in this case was 10 ergs$\cdot$cm$^{-2}\cdot$sec^{-1}, a similar result is obtained with a very low intensity of light, viz., less than 2 ergs$\cdot$cm$^{-2}\cdot$sec^{-1}. We have in fact observed that characteristics of a peak appearing at 180°K can be studied either by using 10 ergs$\cdot$cm$^{-2}\cdot$sec^{-1} exciting light or by less than 2 ergs$\cdot$cm$^{-2}\cdot$sec^{-1} after the Q is reduced. We, therefore, do not think that the observed fluorescence changes could be due to the phase transitions of the ice crystals.

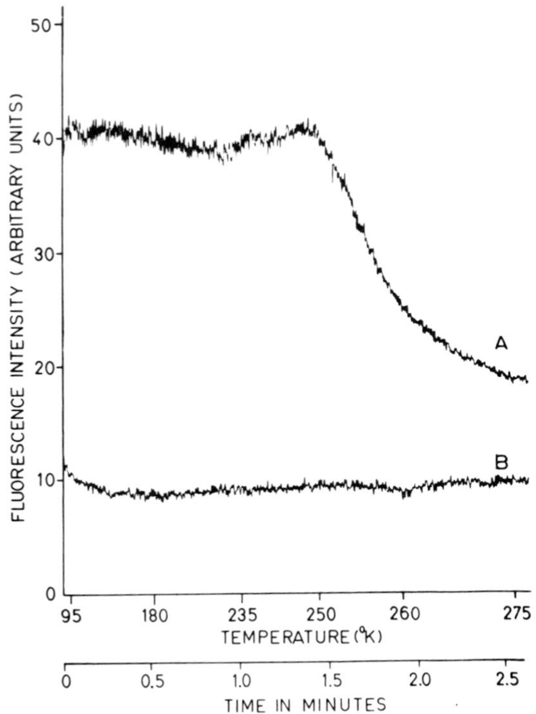

Fig. 3. Changes in fluorescence yield of leaf with increase in temperature, excitation intensity 10 ergs $\cdot cm^{-2} \cdot sec^{-1}$. A, Leaf frozen under 7.5×10^5 ergs $\cdot cm^{-2} \cdot sec^{-1}$ illumination. B, Leaf heated at 90°C for 3 minutes prior to freezing.

One may also explain the rise in fluorescence yield from 95°K to 180°K by considering the reduction of Q by a hypothetical compound Q' which could be reduced at 77°K by white light. If this were true we would have expected a similar reduction of Q and hence a rise in fluorescence yield peaking at 180°K even in a leaf frozen in limiting light intensity and illuminated at 77°K. However, under these conditions a curve similar to Fig. 3A is observed and hence this explanation has to be abandoned.

Although the rise in fluorescence yield can be explained by considering the reduction of Q, it is not possible at all to explain its reduction again under the conditions where the measuring beam is unable to do so. Further, the reduction of Q at 180°K can also not be explained by assuming a reductant formed during the preillumination at 77°K. Thus the rise in fluorescence yield giving a peak at 180°K cannot be explained by involving Q as the only quencher of chlorophyll fluorescence. We, therefore, propose an additional quencher Q_1 which also governs the fluorescence yield of chlorophylls. We further propose that Q is able to reduce Q_1 at temperatures higher than 95°K. According to our proposal the decrease in fluorescence yield observed between 77°K and 95°K is due to oxidation of Q by a compound A. The increase in fluorescence yield bewteen 95°K and 180°K is due to reduction of Q_1 by Q through the compound A. This reduction of Q_1 results in more or less bringing back the yield of chlorophyll fluorescence to the same level as observed prior to the oxidation of Q. We believe that Q_1 has an effect on chlorophyll fluorescence identical to Q. The decrease in fluorescence yield beyond 180°K is due to oxidation of Q_1 and in sequence of Q as a result of increase in temperature. We will explain the maintenance of fluorescence yield at a high level in a leaf frozen in light by considering the fact that in this leaf the components of the electron transport chain between Q and PS I (including Q_1) are in a reduced state. This prevents oxidation of Q until a temperature of 250°K is reached. The decrease in fluorescence yield after 250°K is probably due to a back reaction of Q with Z, the electron donor to PS II. This has been discussed by us elsewhere.

Figure 3B shows the changes in fluorescence yield in a leaf heated to 90°C for 3 minutes. In this case we do not see the peaks appearing at 180°K or 250°K. The heating of a leaf to 90°C destroys photochemical reactions of the chloroplasts. This experiment, therefore, confirms that the changes observed in fluorescence yield are due to photochemical reactions.

Since we believe that the proposed quencher Q_1 is probably situated on the electron transport chain from Q to PS I we studied its position with respect to the site of DCMU action. From Fig. 4A it is clear that the peak due to Q_1 appearing at 180°K is observed even

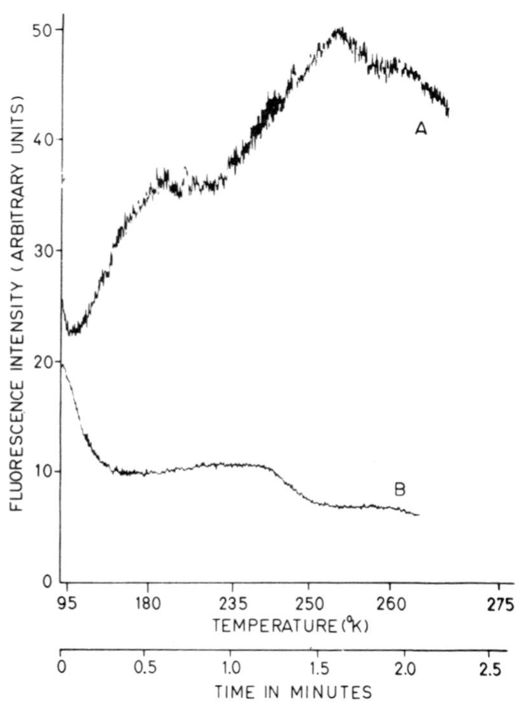

Fig. 4. Changes in fluorescence yield of leaf with increase in temperature, excitation intensity 10 ergs $\cdot cm^{-2} \cdot sec^{-1}$. A, Leaf treated with DCMU. B, Chloroplasts.

in a leaf vacuum infiltrated with 10^{-5} M DCMU. We, therefore, conclude that Q_1 is situated prior to the site of DCMU action.

On the basis of our proposal we would predict that if the rise kinetics of fluorescence are studied at 180°K for a leaf frozen in the dark we shall see a biphasic rise curve due to reduction of Q and Q_1 in sequence. However, if we study the rise kinetics in a leaf which was frozen in light we should see a monophasic rise curve as the Q_1 would already be in a reduced state as stated earlier. Figure 5 shows that

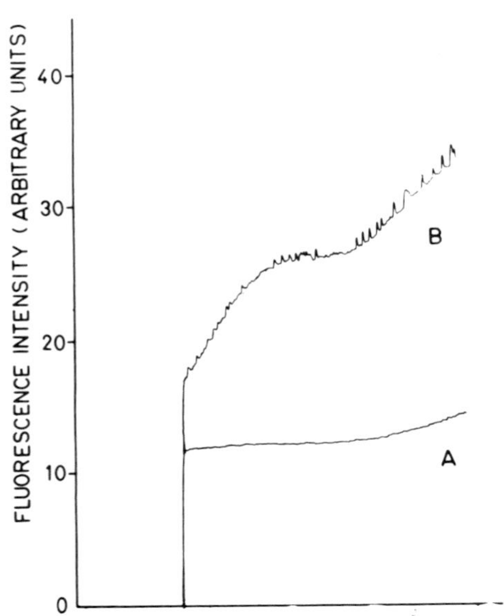

Fig. 5. Rise kinetics of fluorescence at 180°K, fluorescence excitation intensity 5×10^3 ergs·cm^{-2}·sec^{-1}. A, Leaf frozen in light. B, Leaf frozen in dark.

our predictions are experimentally confirmed and hence lend support to our proposal. The rise kinetics of fluorescence have been previously studied by Thorne and Boardman (4). However, our results cannot be compared with theirs because they used isolated chloroplasts instead of leaf. In their system the terminal electron acceptor, viz., NADP, was absent and in addition they may have lost a soluble component of the electron transport chain between Q and PS I. We tried to observe the peak due to Q_1 in isolated chloroplasts. However, we have not yet been able to see it (Fig. 4B).

The increase in fluorescence yield peaking at 180°K can also be considered in terms of donors of electrons to PS II at low temperatures. In recent studies conducted by Butler and his coworkers (5,6,7) as well as by Vermeglio and Mathis (8), it has been clearly shown that at low temperatures a high potential cytochrome b_{559} and an unknown compound Z act as donors of electrons to PS II. Okayama and Butler (6) have proposed that when chloroplasts are illuminated at 77°K the following changes take place:

$$\text{Cyt } b_{559} \cdot P_{680} \cdot C\text{-}550 \xrightarrow{h\nu} \text{Cyt } b_{559} \cdot P^{+}_{680} \cdot C^{-}\text{550}$$

Here P_{680} is the reaction center chlorophyll of PS II and C-550 is the primary acceptor of PS II and probably equivalent to Q. In a subsequent reaction Cyt b_{559} is oxidized with the formation of the following products resulting in high fluorescence yield:

$$\text{Cyt } b_{559} \cdot P^{+}_{680} \cdot C^{-}\text{550} \rightarrow \text{Cyt } b^{+}_{559} \cdot P_{680} \cdot C^{-}\text{550}$$

Thus it is considered that the donor cyt b_{559} is involved in controlling the fluorescence yield of chlorophylls at 77°K. In our studies we had illuminated leaf at 77°K prior to measurement of chlorophyll fluorescence (Fig. 2C). We therefore had started with an oxidized cyt b_{559} and hence it could not be acting as a donor of electrons to PS II during subsequent warming. In fact it has been shown that cyt b_{559} is the donor of electrons to PS II only at 77°K (7,8). With increase in temperature another donor Z comes into the picture and it is the most efficient donor at -100°C. This temperature corresponds to the peak temperature of 180°K observed by us for the maximum fluorescence yield. It is conceivable that, in our studies, the rise in fluorescence from 95°K to 180°K may be due to the gradual reduction of oxidized chlorophylls by Z. We could assume that the donors and acceptors of PS II in a leaf illuminated at 77°K are in the following redox states:

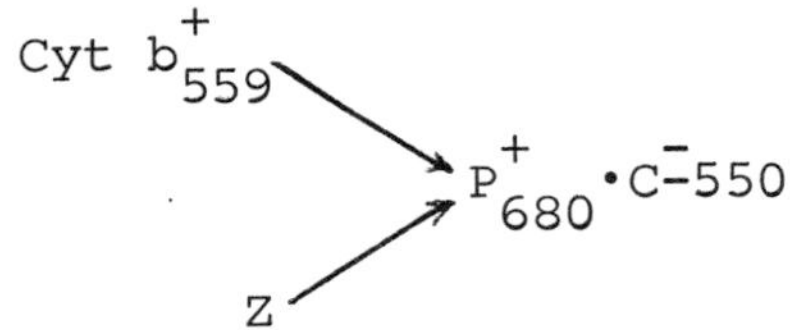

An increase in temperature from 77°K to 95°K may result in partial oxidation of reduced C-550 resulting in a decrease in fluorescence yield. A subsequent rise in temperature may result in gradual oxidation of Z and reduction of P_{680} yielding the following product:

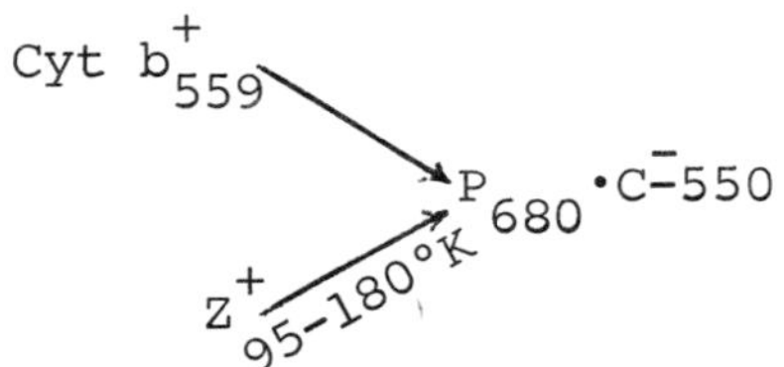

The P_{680} • C^--550 complex could be responsible for giving us higher yield of fluorescence.

The maintenance of high fluorescence yield in a leaf frozen in light can be explained on the basis of this proposal. We could assume that the redox states of donors and acceptors in the case of a leaf frozen in light are as follows:

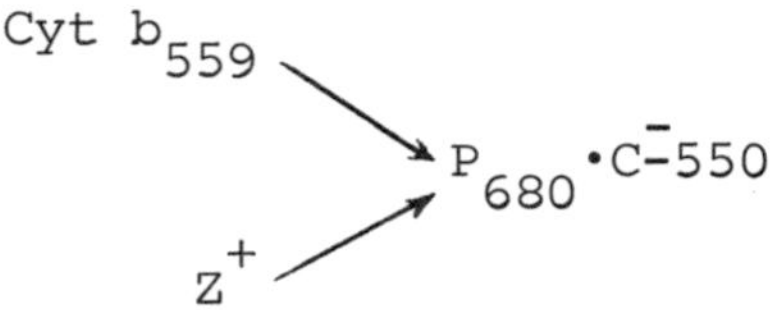

As mentioned above the complex $P_{680} \cdot C\bar{}$-550 would be responsible for high fluorescence yield which will be maintained until oxidation of C-550 takes place. It has already been shown by Erixon and Butler (5) that in chloroplasts frozen in light C-550 is in a reduced state but cyt b_{559} is not.

Although it seems that our results can be explained on the basis of donors of electrons to PS II at low temperatures, two observations made by us go against it. One is that we have not yet observed the increase in fluorescence yield peaking at 180°K in isolated chloroplasts. It is known that in isolated chloroplasts both the donors mentioned above are present and that they function normally. The other is that we do not see the changes in fluorescence yield in a leaf which was frozen and thawed only once. This has been shown in Fig. 6. It is known that chloroplasts isolated from frozen and thawed leaves are quite active in PS II reactions and that donors cyt b_{559} and Z are not destroyed by just one thawing. In view of this we tend to favor our proposal involving an additional quencher Q_1 which seems to be a very delicate component of the chloroplast membrane to explain our results. In recent years quenchers in addition to Q have been proposed by Govindjee *et al.* (9), Cramer and Bohme (10), and Delosme (11). The characteristics of Q_1 proposed by us distinguish it from quenchers proposed by Govindjee *et al.* and Cramer and Bohme but it seems to resemble that proposed by Delosme.

We have studied the appearance of the peak due to Q_1 in a greening leaf. The results obtained are shown in Fig. 7. It is observed that this peak is absent in an etiolated leaf but starts appearing with the greening of the leaf. Its appearance also corres-

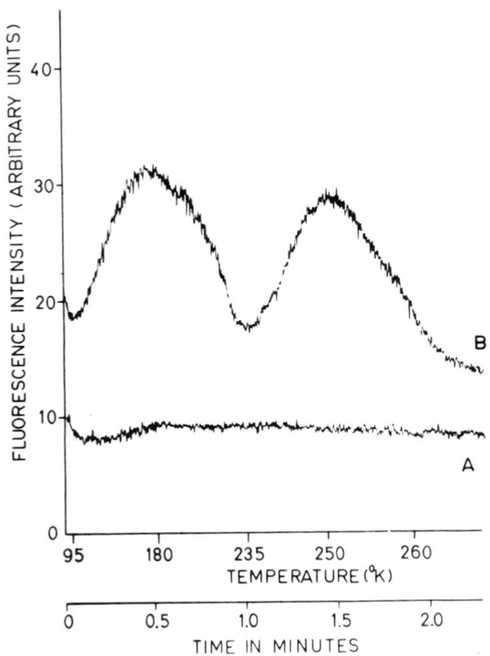

Fig. 6. Changes in fluorescence yield of a leaf frozen in dark (B) and refrozen after warming to 273°K (A), excitation intensity 10 ergs$\cdot$cm$^{-2}\cdot$sec^{-1}.

ponds with the shift of the excitation maximum of the fluorescence from 440 nm to 480 nm. These results are consistent with the assumption that the changes observed are due to photochemical reactions of PS II.

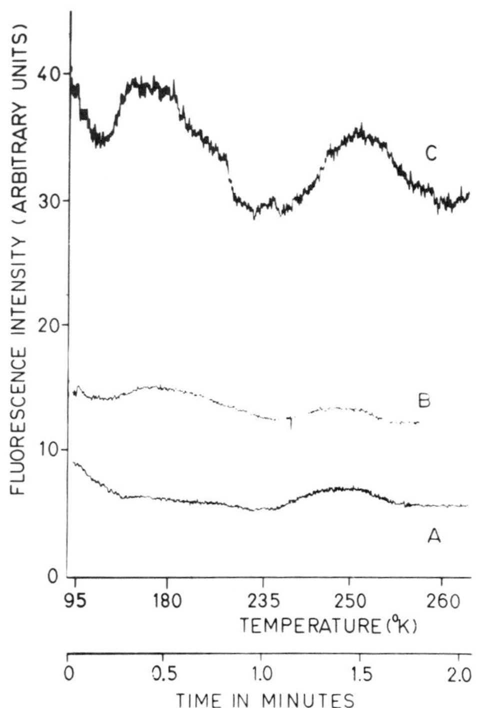

Fig. 7. Appearance of fluorescence peak due to reduction of Q_1 during greening of an etiolated leaf exposed to white light for different periods. A, after 1.5 hours of exposure. B, after 5 hours of exposure. C, after 24 hours of exposure; amplifier gain for C was half that for A and B.

REFERENCES

1. SANE, P. V., GOODCHILD, D. J., AND PARK. R. B. (1970) Biochim. Biophys. Acta 216, 162.
2. DUYSENS, L. N. M., AND SWEERS, H. E. (1963) in Studies on Microalgae and Photosynthetic Bacteria (Miyachi, S., ed.) p. 353, University of Tokyo Press, Tokyo.
3. CHO, F., AND GOVINDJEE (1970) Biochim. Biophys. Acta 205, 371.
4. THORNE, S. W., AND BOARDMAN, N. K. (1971) Biochim. Biophys. Acta 234, 113.
5. ERIXON, K., AND BUTLER, W. L. (1971) Biochim. Biophys. Acta 234, 381.
6. OKAYAMA, S., AND BUTLER, W. L. (1972) Biochim. Biophys. Acta 267, 523.
7. BUTLER, W. L., VISSER, J. W. M., AND SIMONS, H. L. (1973) Biochim. Biophys. Acta 292, 140.
8. VERMEGLIO, A., AND MATHIS, P. (1973) Biochim. Biophys. Acta 292, 763.
9. GOVINDJEE, R., GOVINDJEE, LAVOREL, J., AND BRIANTAIS, J. M. (1970) Biochim. Biophys. Acta 205, 361.
10. CRAMER, W. A., AND BOHME, H. (1972) Biochim. Biophys. Acta 256, 358.
11. DELOSME, R. (1967) Biochim. Biophys. Acta 143, 108.

THE ROLE OF MEMBRANES IN RADIATION DAMAGE

B. B. Singh

INTRODUCTION

The concept of targets and hits for cell killing by high energy radiation was historically introduced to explain dose-effect relationships (1). Earlier studies have indicated the possibility of cellular genome to be one of the most probable targets. Consequently, enormous amounts of effort have been made to identify the nature of radiation chemical alterations induced in DNA. These include chemical modifications on bases and sugar moieties and formation of breaks in one or both strands of sugar-phosphate backbone. The repair of such damage has also been investigated and has been postulated to play an important role in radiation sensitivity of cells (2,3). Notwithstanding this, results have been accumulating which indicate the membrane to be another equally important target for radiation killing of cells. Of particular mention are the observations on oxygen effect and its relation to LET of radiations which strongly support this view (4). In fact, it would be inappropriate to assign the role of these two targets independently considering the various results which suggest an interdependence of cellular processes on DNA and the membrane. For instance, the association of DNA with bacterial membrane or with the nuclear membrane in higher cell systems has been clearly demonstrated and the DNA replication is now believed to be initiated at the membrane in bacteria (5). In addition, membrane bound polysomes containing 35 to 50% of total cellular RNA are supposed to be the major sites for the bulk synthesis of proteins (6). The DNA polymerase activity has also been shown to be associated with bacterial membrane (7). It would thus appear most appropriate to suggest a "cooperative target" with DNA and membrane as its two components playing an interrelated role in realization of cell killing by radiation. The damaging events could occur in either of these components and the repair of one could be achieved with the cooperation of the other. This is schematically represented in Fig. 1.

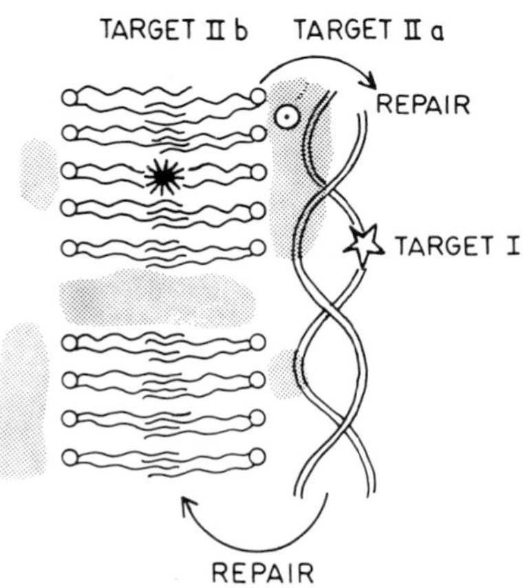

Fig. 1.

In this figure the membrane is shown as a combination of lipid-bilayer and proteins and the damage to the cooperative target has been subdivided into three different types. Type I refers to DNA damage in terms of base alterations or strand breaks; type IIa refers to protein damage resulting in the inhibition of enzymatic processes; and type IIb denotes damage to lipids causing structural disorganization in the membrane. In the present chapter the significance of Type IIa and Type IIb targets will only be discussed with particular reference to the modification of radiation lethality of cells by chemicals.

TARGET IIa

Iodination of proteins by irradiation

The radiation lethality of cells has been shown to increase on irradiation in the presence of iodine compounds like iodoacetamide (8) iodoacetic acid, and iodides (9). Using I^{131} labelled iodoacetic acid and sodium iodide it was observed that radioactivity was incorporated in *E. coli* B/r cells on irradiation with gamma rays. On chemical fractionation of these cells, over 83 percent of the total radioactivity was found to be present in proteins which were associated with the membrane (10). Gel electrophoretic studies made at pH 2.3 and 9.5 revealed that the radioactive iodine was present in all protein bands indicating non-specificity of the iodination reaction (11). Furthermore, paper chromatographic analysis of iodinated proteins revealed the presence of iodine in all amino acid spots. As a consequence, membrane bound enzymes such as ATPase and succinoxidase were inactivated to a greater extent when cells or isolated membrane were irradiated in the presence of iodine compounds (11). In addition, post-irradiation syntheses of DNA and proteins were also considerably inhibited under these conditions.

Chemically iodinated cells and radiation repair

Even without irradiation bacterial cells could be iodinated using Fenton's reaction and sodium iodine. In *M. radiodurans* this resulted in lethality of cells as well as in considerable reduction in DNA and protein synthesis after gamma-irradiation. In addition, the large shoulder on the dose survival curve of *M. radiodurans* was absent if cells were chemically iodinated before irradiation, which has been interpreted in terms of inactivation of the DNA repair system (12). It was further supported by the observation that the u.v. survival curves of pre-iodinated M. radiodurans also lacked the shoulder which is specifically attributable to the repair of DNA damage. The presence of iodine containing sensitizers per se during u.v. irradiation, however, neither caused sensitization nor gave rise to incorporation of iodine in cells. These observations, therefore, clearly demonstrated that radiosensitization by iodine compound was caused due to the inhibition of DNA repair (Target-I) on account of permanent damage induced in membrane proteins (Target-IIa) in the form of iodination of various amino acid residues.

TARGET IIb

Procaine hydrochloride, which is a well-known membrane-specific drug, has been demonstrated to induce many changes in various membrane characteristics (13,14).

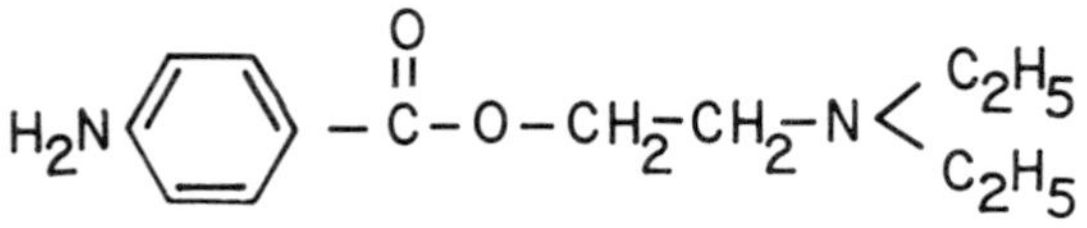

$$H_2N-C_6H_4-\overset{O}{\overset{\|}{C}}-O-CH_2-CH_2-N\begin{matrix} C_2H_5 \\ C_2H_5 \end{matrix}$$

NMR studies on procaine in aqueous dispersions of phosphatidylserine and lecithin showed broadening of proton signals nearest to the two nitrogen nuclei, clearly indicating ionic interactions between procaine and negatively charged phosphate groups on lipids. Our NMR studies have further demonstrated absence of any such interaction between procaine and bovine serum albumin, thus indicating this drug to be specifically localized in the lipid milieu of membranes.

Using E. coli B/r cells as the test system it was observed that procaine HCl at concentrations varying between 0.25 and 25 mM enhanced the lethality of cells if irradiations were carried out under anoxia (Table I). Since the presence of nutrient

TABLE I
EFFECT OF PROCAINE HCl ON POST-IRRADIATION
LETHALITY OF E. COLI B/r CELLS

Medium	Sensitizer (mM)	Conditions of irradiations	D_1 (krads)	DMF
Buffer	-	N_2	74.0±1.4	-
	0.25	N_2	40.0±0.6	0.54
	25.0	N_2	30.9±1.0	0.42
	25.0	Sensitizer washed off before irradiation	74.0	1.00
	25.0	Sensitizer washed off after irradiation	31.2±1.3	0.42
	25.0	Irradiated anoxic cells treated with unirradiated sensitizer	38.5±1.5	0.52
Nutrient broth	-	N_2	87.0±7.0	-
	25.0	N_2	43.5±1.5	0.52
	-	Air	28.0±2.0	0.32
	25.0	Air	28.0±2.0	1.00

D_1 = Dose required to give 1% survival.
DMF = D_1 in presence of sensitizer/D_1 for control cells.

broth or scavengers for radiolytic transients of water could not modify this effect (15), the role of any radiolytic products of procaine in the sensitizing process could be ruled out. This was further supported by the fact that enhanced lethality was noticed when cells irradiated under anoxia were subsequently exposed to unirradiated procaine solution. However, if procaine was removed from the cell suspension before irradiation, no effect was noticed but the removal of procaine after irradiation did not reduce the sensitizing effect. These results can be satisfactorily interpreted to indicate that irradiation leads to certain temporary structural alterations in the lipid milieu of the bacterial membrane. Such alteration could be fixed by procaine if it is present with cells during irradiation or allowed to interact with cells even after irradiation within a certain interval of time before the damage could be repaired. Our preliminary results have indicated that procaine is effective up to 3 hours of incubation at 37°C in phosphate buffer following irradiation of cells under anoxia. This period being sufficient to allow complete DNA repair, the process of repair of membrane damage would seem to follow the repair of DNA. Similar non-repairable permanent structural alterations due to formation of lipid peroxides may be induced on irradiation in the presence of oxygen. Under these circumstances, therefore, procaine would be ineffective in modifying the radiation damage to cells as observed in the present studies.

REFERENCES

1. ZIMMER, K. G. (1961) Studies on Quantitative Radiation Biology, p. 10, Oliver & Boyd, London.
2. HARIHARAN, P. V., AND CERUTTI, P. A. (1972) J. Mol. Biol. 66, 65.
3. HOWARD-FLANDERS, P. (1968) Ann. Rev. Biochem. 36, 175.
4. ALPER, T. (1973) In Effects of Neutron Irradiation upon Cell Function, IAEA Symposium Neuherberg FRG, 22-26 October.
5. JACOB, F., BRENER, S., AND CUZIN, F. (1963) Cold Spring Harbor Symp. Quant. Biol. 28, 329.
6. GODSON, G. N., HUNTER, G. D., AND BUTLER, J. A. V. (1964) Biochem. J. 81, 59.
7. BILLEN, D. (1962) Biochem. Biophys. Res. Commns. 7, 179.
8. ALEXANDER, P., LETT, J. T., AND DEAN, C. J. (1965) Prog. Biochem. Pharmacol. 1, 22.
9. MULLENGER, L., SINGH, B. B., ORMEROD, M. G., AND DEAN, C. J. (1967) Nature (Lond.) 216, 372.
10. SHENOY, M. A., SINGH, B. B., AND GOPAL-AYENGAR, A. R. (1968) Science 160, 999.
11. SHENOY, M. A., JOSHI, D. S., SINGH, B. B., AND GOPAL-AYENGAR, A. R. (1970) Adv. Biol. Med. Phys. 13, 255.
12. SHENOY, M. A., SINGH, B. B., AND GOPAL-AYENGAR, A. R. (1971) Ind. J. Expl. Biol. 9, 518.
13. HAUSER, H., PENKETT, S. A., AND CHAPMAN, D. (1969) Biochim. Biophys. Acta 183, 466.
14. PAPAHADJOPOULAS, D. (1972) Biochim. Biophys. Acta 265, 169.
15. SHENOY, M. A., SINGH, B. B., AND GOPAL-AYENGAR, A. R. (1974) Nature (Lond.) in press.

BICARBONATE STIMULATION OF OXYGEN EVOLUTION IN CHLOROPLAST MEMBRANES

Alan Stemler and Govindjee

INTRODUCTION

The ability of bicarbonate ion to stimulate (5-10 fold) Hill reaction in isolated chloroplast membrane fragments is now well documented (1-6), though very little is known about its precise role in the process. Bicarbonate ion, rather than CO_2, appears to be the active species (7). A more extensive investigation (7-9) of this phenomenon has yielded surprising results and indicates that HCO_3^- has a more critical role in oxygen evolution than previously suspected. Our recent experiments and conclusions (7-9) will be summarized below.

INCREASE IN HCO_3 DEPENDENCE WITH PREILLUMINATION

Isolated oat (*Avena sativa*) chloroplast membranes (see ref. 7 for details), not previously depleted of HCO_3^-, were placed on the surface of a platinum rate electrode that was polarized at +0.7 volt relative to the Ag/AgCl electrode so as to measure ferricyanide production (see ref. 9 for details). The chloroplast preparation (to be referred to as chloroplasts for brevity) was preilluminated with saturating white light in the presence of ferricyanide for a variable length of time while HCO_3^- free solution passed over the membrane holding the chloroplasts to the surface of the platinum (see legend of Fig. 1 for the composition of the suspension medium). After this period

---

Abbreviations: DCPIP - 2,6-dichlorophenol indophenol; DPC - diphenyl carbazide; DCMU - 3-(3,4-dichlorophenyl)-1,1-dimethylurea.

of illumination several half-second assays were conducted. The chloroplasts were then provided HCO_3^- containing solution and several more half-second assays were performed.

The ability of preillumination to enhance dependence of the Hill reaction on HCO_3^- is shown in Fig. 1.

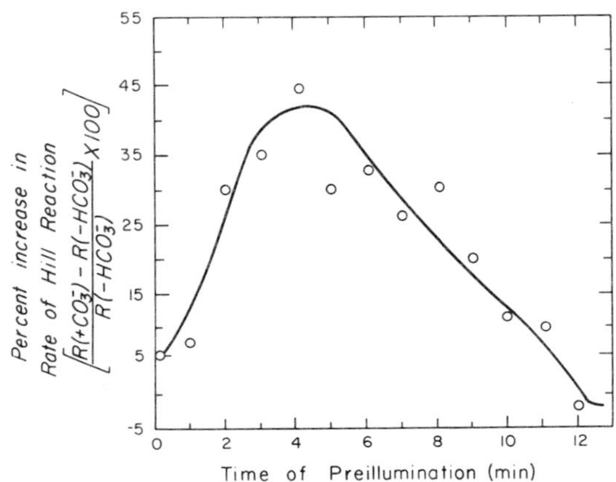

Fig. 1. Percent increase in the rate of ferricyanide reduction with added HCO_3^- as a function of preillumination time. The solution passing over the electrode membrane contained 0.25 M NaCl, 0.04 M Na acetate, 0.05 M phosphate pH 6.8, 0.5 mM potassium ferricyanide ±0.01 M $NaHCO_3$. Saturating white light (10^3 W/m^2) used during preillumination and half-second assays. Anaerobic conditions. Oat chloroplast membrane fragments (43 µg chlorophyll/ml in stock). Average of two series [after Stemler and Govindjee (9)].

Progressive increase in HCO_3^- dependence continues for the first four minutes of preillumination. Dark controls, and chloroplasts preilluminated in the absence of ferricyanide, showed no change in dependence on HCO_3^-. It should be mentioned that the decline in HCO_3^- dependence after 4 minutes is associated with a decline in the overall activity, i.e., it reflects photoinactivation.

Since $H^{14}CO_3^-$ was not incorporated (A. Stemler, unpublished) into a stable compound during Hill reaction, and thus "used up", increased dependence on HCO_3^- with time of illumination suggests that this ion is initially bound, perhaps ionicly to reaction centers directly, or exists in some complexed condition. As Hill reaction proceeds, it may become unbound or otherwise free (possibly as CO_2?).

BICARBONATE EFFECT AS A FUNCTION OF LIGHT INTENSITY

Oat chloroplast fragments, again not previously depleted of HCO_3^-, were placed on the platinum rate electrode and allowed to perform Hill reaction in the

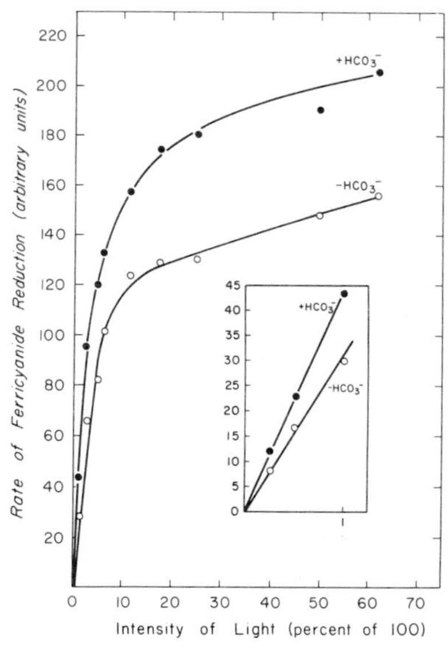

Fig. 2. Rate of ferricyanide reduction with and without 0.01 M $NaHCO_3$ as a function of light intensity. Oat chloroplast fragments preilluminated 4 min. in saturating white light. Ligh intensity was varied with calibrated neutral density filters. Average of 3 series. Other conditions as in Fig. 1 [after Stemler and Govindjee (9)].

absence of exogenous HCO_3^- for 4 minutes to induce a degree of dependence on HCO_3^- as shown in Fig. 1. Half-second assays were then conducted without, and then with, HCO_3^- as described above. In this case light intensity was varied by means of calibrated neutral density filters. Under these conditions the bicarbonate effect is independent of light intensity as shown in Fig. 2. Even at the lowest intensity where ferricyanide reduction could be accurately measured (see insert), HCO_3^- stimulated the activity to the same degree as at saturating intensities, or about 35 percent at all intensities. We conclude from the results presented in Fig. 2 that HCO_3^- is involved in early photochemical reactions of photosystem II rather than purely enzymatic reactions somewhat removed from the reaction centers.

COMPARISON OF THE EFFECT OF HCO_3^- ON OXYGEN EVOLUTION AND FERRICYANIDE REDUCTION MEASURED SIMULTANEOUSLY

Maize (Zea mays) was grown and chloroplast membrane fragments isolated from them as described earlier (7). Chloroplasts, to be depleted of HCO_3^-, were suspended in a solution containing 0.25 M NaCl, 0.04 M Na acetate, and 0.05 M Na phosphate, at pH 5.0, (see reference 7 for further details). The chloroplasts remained in this solution for 30 minutes at room temperature while N_2 gas was bubbled through the suspension to remove CO_2. Both high salt content and low pH were necessary for developing maximum HCO_3^- dependence. Ferricyanide reduction and oxygen evolution were measured as described in ref. 9.

When maize chloroplast fragments were depleted of HCO_3^-, over 90% of their oxygen evolving ability was suppressed while at the same time their ability to reduce ferricyanide was suppressed less than 80%. This is shown in Fig. 3. If HCO_3^- was present, however, equal μ equivalents of oxygen and ferricyanide were produced, at least during the first several minutes of illumination. That is, in the presence of HCO_3^-, for every molecule of oxygen evolved, four electrons are transferred to ferricyanide.

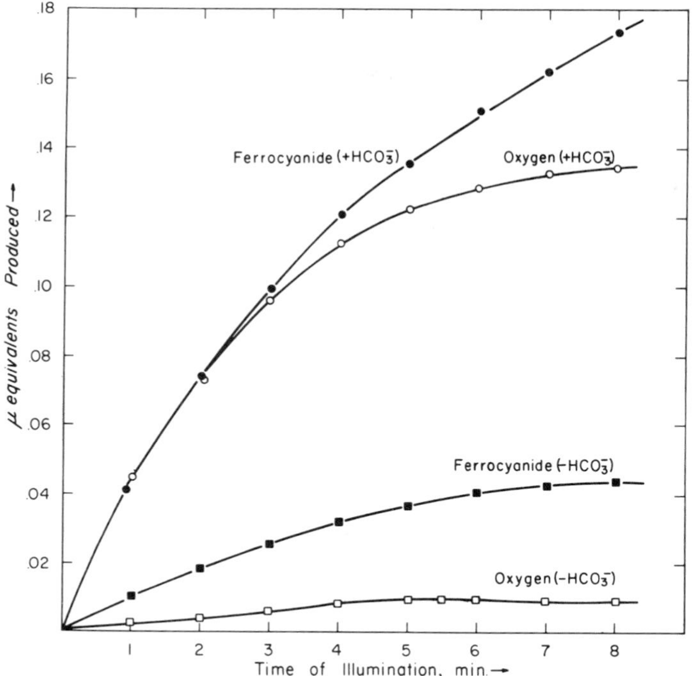

Fig. 3. Comparison of oxygen evolution and ferricyanide reduction in the presence and absence of HCO_3^-. Reaction mixtures contained 0.25 M NaCl, 0.04 M Na acetate, 0.05 M phosphate, pH 6.8, 1 mM potassium ferricyanide ±0.01 M $NaHCO_3$ and 15 μg chlorophyll/ml of maize chloroplast membrane fragments suspension. The light intensity was 500 W/m^2. Corning C.S. 3-71 (yellow) cut-off filters were used [after Stemler and Govindjee (9)].

These results can be explained by assuming the presence of a fairly substantial amount of some endogenous electron donor capable of reducing ferricyanide without evolving oxygen (see section 4 below). This assumption is consistent with the data of Kahn (10) and more recently of Huzisige and Yamamoto (11) who showed residual ferricyanide reduction in the absence of oxygen evolution in chloroplast particles. However

TABLE I
RATE OF DCPIP REDUCTION WITH AND WITHOUT BICARBONATE

Treatment	Rate $-NaHCO_3$	Rate $+0.01$ M $NaHCO_3$	$+HCO_3^-$ / $-HCO_3^-$
	μmoles DCPIP reduced/mg Chl/hr		Ratio
None	15.8 ± 3.8	70.0 ± 6.2	4.4
Heat	0.00	Trace	--
Heat + 0.05mM DPC	50.0 ± 8.9	56.0 ± 11.6	1.12
Heat + 0.5mM DPC + 50 μM DCMU	0.0	0.0	--

Initial rate of DCPIP reduction in normal and heat treated (maize) chloroplast fragments with DPC as electron donor with and without added bicarbonate. The reaction mixture contained 0.25 M NaCl, 0.04 M Na Acetate, 0.05 M phosphate buffer, pH 6.8, 39 μM DCPIP, and 15 μg chlorophyll/ml. Saturating red light was 2×10^3 W/m^2. The data are the average of five experiments [after Stemler and Govindjee (7)].

the identity of this endogenous donor, or its possible role in "normal" electron flow, is not known. In any case, HCO_3^- appears to direct electron flow away from this donor and "couples" it rather to oxygen evolution.

EFFECT OF BICARBONATE ON AN ARTIFICIAL ELECTRON DONOR SYSTEM

To test whether HCO_3^- acts near the oxygen evolving site or farther along the electron transport chain, an artificial electron donor to photosystem II, diphenyl carbazide (DPC), was given to maize chloroplast membranes made unable to evolve oxygen by heat treatment. It was reasoned that if HCO_3^- acted between the oxygen evolving site and the site of electron donation by DPC, no effect of bicarbonate would be observed on the rate of electron flow from DPC to the electron acceptor (the Hill oxidant). The acceptor chosen for the present study was DCPIP, measured spectrophotometrically as described in ref.7.

Table 1 indicates that while normal (unheated) chloroplast fragments show a large HCO_3^- effect (4.4 fold increase with HCO_3^-) when DCPIP reduction is coupled to the natural electron donor, no significant effect is seen when DPC is the electron donor (1.12-fold increase with HCO_3^-).

We conclude from the data in Table I that at least one site of action of HCO_3^- is at, or very near, the oxygen evolving mechanism.

THE EFFECT OF BICARBONATE ON CHLOROPHYLL *a* FLUORESCENCE TRANSIENTS

Chlorophyll *a* fluorescence transients were measured as described in ref. 8. Figure 4 shows the fast fluorescence transient observed in isolated maize chloroplast fragments exposed to $80W/m^2$ blue-green light. While constant fluorescence (F_0 or "O" level) is not affected in HCO_3^- depleted chloroplasts (see ref. 8), the initial fluorescence rise from the first recorded level to "I" is rapid, but the I → D

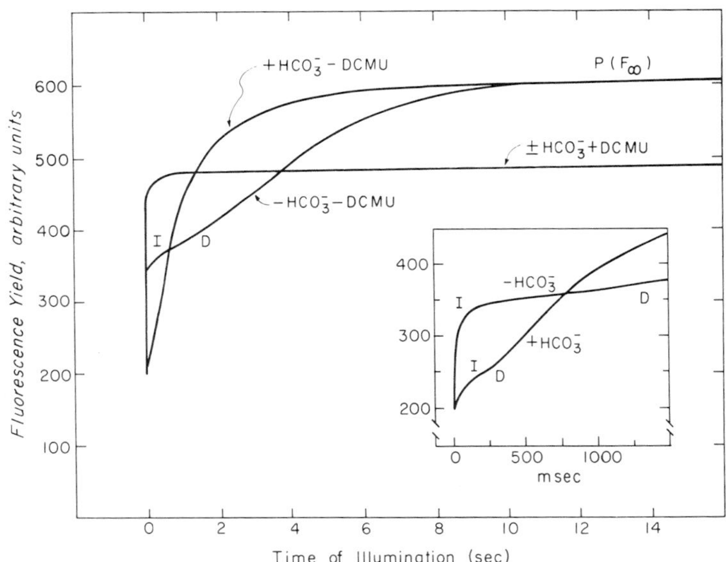

Fig. 4. Fluorescence yield of chlorophyll a at 685 nm as a function of time of illumination in the presence and absence of 0.01 M $NaHCO_3$. Maize chloroplast fragments previously depleted of HCO_3^- were suspended in 0.25 M NaCl, 0.04 M Na acetate, 0.05 M phosphate buffer pH 6.8, ±10 μM DCMU; 10 μg chlorophyll/ml. Blue actinic light, 80 W/m^2. [After Stemler and Govindjee (8).]

(the inflection) and the D → P (peak F_∞) phases are slow. In the presence of 10 mM HCO_3^-, the initial fast rise to I appears slower and is depressed, the level D is not clear and occurs earlier (see insert Fig. 4), and the D → P rise is much more rapid with a half rise time of about 1 vs 4s in bicarbonate-depleted chloroplasts. The P level fluorescence, like that of the O level, is insensitive to HCO_3^-.

To explain Chl a fluorescence transients, it is generally assumed (see Duysens and Sweers, 12) that the yield of variable fluorescence reflects the redox state of Q, the primary electron acceptor of photosystem II. HCO_3^- depletion may block electron flow either before or after Q. However, if the block occurs after Q, preventing the reoxidation of Q by intersystem intermediates (A pool), variable fluorescence should be at all times greater in the absence of HCO_3^-. Since this is not the case, we feel that the effect of HCO_3^- on fluorescence transients provides further evidence that HCO_3^- is acting on the oxygen evolving side of photosystem II (for further analysis, see ref. 8).

THE EFFECT OF BICARBONATE ON DELAYED LIGHT EMISSION

Delayed light emission (DLE) measurements were made as described in ref. 8. Maize chloroplast fragments were illuminated for 60 sec with low intensity (0.4 W/m^2) blue light. When observed in the time period starting about 0.5 sec after the cessation of illumination, the greatest amount of DLE results when HCO_3^- was resupplied during illumination (see Fig. 5.) Relatively less (about 70% of the maximum) DLE is observed when DCMU is also present; the omission of HCO_3^- causes an even greater decline in DLE to about 50% of the maximum. If HCO_3^- is omitted and DCMU is present, DLE is almost completely quenched, only about 5% of the maximum remaining.

Since DLE, in the time period observed here, is usually considered to be due to back reactions (i.e. charge recombinations) in photosystem II reaction centers the effect of HCO_3^- appears puzzling. HCO_3^- is known to stimulate oxygen evolution (1-4) and ferricyanide and DCPIP reduction (5-7) indicating greater photochemical efficiency of photosystem II. At the same time back reactions leading to increased DLE are also stimulated, arguing for less overall efficiency. To render these observations compatible, we refer to the kinetic model of oxygen evolution of Forbush _et al._ (13) and propose that HCO_3^- permits the formation of higher oxidation states ($>S_1$) of the photosystem II reaction centers. Such states are not only necessary for oxygen evolution, but are also the only states with significant propensity to decay (back react) yielding much more DLE (14, 15) (for more detailed discussion of HCO_3^- effects on DLE, see ref. 8.)

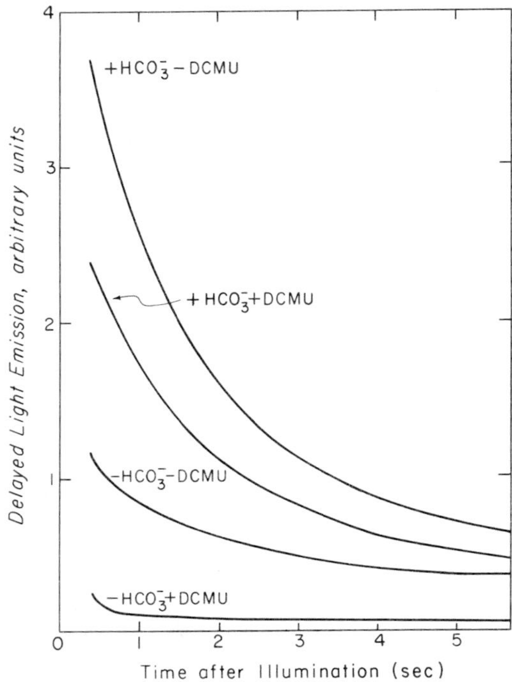

Fig. 5. Delayed light emission decay from maize chloroplast fragments, previously depleted of HCO_3^-, illuminated for 60 s. in weak blue light (0.4 W/m^2). Chloroplast fragments were suspended in 0.25 M NaCl, 0.04 M Na acetate, 0.05 M phosphate buffer pH 6.8, ±0.01 M $NaHCO_3$ ± 10 μM DCMU; 15 μg chlorophyll/ml. [After Stemler and Govindjee (8).]

CONCLUSIONS

From the evidence presented here and elsewhere we conclude that HCO_3^- has some important role in the oxygen evolving process. One possible mode of action has been eliminated. HCO_3^- does not in any way change the spillover of excitation energy between photosystem II and photosystem I as inferred from the absence of any effect of HCO_3^- on the fluorescence emission spectra both at room and liquid nitrogen temperatures (T. Wydrzynski and Govindjee, unpublished.)

Many other possible mechanisms of action remain. Otto Warburg (1) cited this phenomenon as proof of his theory that oxygen is evolved as a result of splitting the CO_2 molecule and not from splitting H_2O. There are, of course, other possible modes of action. Metzner (16) has published a scheme suggesting that the bicarbonate ion, bound to an acceptor, may function as the electron donor of photosystem II. HCO_3^- may be simply an allosteric effector operating on the oxygen evolving enzyme. It may be causing conformational changes in membrane protein or influencing membrane potential in some way. We now have some evidence (Stemler et al., in preparation (17)) that HCO_3^- is, in some manner, facilitating the formation of higher oxidation states of the photosystem II reaction centers. Since none of the above possibilities have been completely ruled out we are continuing the investigation of this phenomenon.

SUMMARY

Dependence of ferricyanide reduction (Hill reaction) by broken chloroplast membranes on bicarbonate ion increases with time of illumination. The stimulation caused by HCO_3^- is clearly observed even at low light intensities, suggesting an involvement of this anion in early photochemical reactions. The greater dependence of oxygen evolution than of ferricyanide reduction on HCO_3^-, as reported here, is interpreted to indicate the presence of an endogenous, non-oxygen evolving electron donor. Electron flow from the artificial electron donor diphenyl carbazide (DPC) to dichlorophenol indophenol (DCPIP) through photosystem II, in contrast to normal flow, is insensitive to HCO_3^-, indicating that HCO_3^- acts at or near the oxygen evolving site. Effects of HCO_3^- on chlorophyll *a* fluorescence and delayed light emission also suggest that this ion acts on the oxygen evolving side of photosystem II. We conclude that HCO_3^- has an important and a critical role in the oxygen evolution steps of photosynthesis.

REFERENCES

1. WARBURG, O., AND KRIPPAHL, G. (1960) Z. Naturforschg. 156, 367.
2. STERN, B. K., AND VENNESLAND, B. (1960) J. Biol. Chem. 235, 51.
3. VENNESLAND, B., OLSON, E., AND AMMERAL, R. N. (1965) Fed. Proc. 24, 873.
4. GOOD, N. E. (1963) Plant Physiol. 38, 294.
5. WEST, J., AND HILL, R. (1967) Plant Physiol. 42, 819.
6. BATRA, P., AND JAGENDORF, A.T. (1965) Plant Physiol. 40, 1074.
7. STEMLER, A., AND GOVINDJEE (1973) Plant Physiol. 52, 119.
8. STEMLER, A., AND GOVINDJEE (1973) Photochem. Photobiol. 19, 227.
9. STEMLER, A., AND GOVINDJEE (1974) Plant Cell Physiol., in press.
10. KAHN, J. (1964) Biochim. Biophys. Acta 79, 234.
11. HUZISIGE, H., AND YAMAMOTO, Y. (1972) Plant and Cell Physiol. 13, 477.
12. DUYSENS, L. N. M., AND SWEERS, H. E. (1963) in Studies on Microalgae and Photosynthetic Bacteria (Ashida, J., ed.) p. 353-372, University of Tokyo Press, Tokyo.
13. FORBUSH, B., KOK, B., AND McGLOIN, M. (1971) Photochem. Photobiol. 14, 307.
14. JOLIOT, P., JOLIOT, A., BOUGES, B., AND BARBIERI, G. (1971) Photochem. Photobiol. 14, 287.
15. BARBIERI, G., DELOSME, R., AND JOLIOT, P. (1970) Photochem. Photobiol. 12, 197.
16. METZNER, H. (1966) Naturwiss. 53, 141.
17. STEMLER, A., BABCOCK, G. T., AND GOVINDJEE (1974) Plant Physiol. Annual Supplement US ISSN 0079-2241, June, 1974, abstract #91, p. 34.

CHLOROPLAST MEMBRANES: LIPID, PROTEIN, AND CHLOROPHYLL INTERACTIONS

Marcia Brody

Benson and co-workers (1,2,3) have demonstrated that the mono- and digalactosyl diglycerides (containing the most unsaturated fatty acids) constitute the major portion of the non-pigmented lipids of the chloroplast (in this respect, differing from most other membranes), and have suggested that these surfactant materials (by virtue of their possessing both hydrophilic and hydrophobic groups, as does chlorophyll itself) act to determine the molecular interfacial absorption properties, as well as the structure and stability of oriented molecular leaflets which occur in the chloroplast. They further suggested interaction of these lipids, via their water soluble galactosyl groups, with porphyrin, and via their hydrophobic, "liquid-like" (because of the two tri-unsaturated C_{18} fatty esters of linolenic acid) groups with similarly hydrophobic sections of adjacent protein molecules.

In 1964 Sastry and Kates (4) first reported galactolipase activity in extracts of runner-bean primary leaves (*Phaseolus multifloris*); they reported that approximately 70% of the enzymatic activity was associated with a chloroplast-rich fraction.

In 1969 M. Brody *et al*. (5) observed that time- and temperature-dependent changes in fluorescence emission occured either with isolated chloroplasts of the castor-oil plant, *Ricinus communis*, or with cross-reaction systems, i.e., chloroplasts from other sources (algae or higher plants) treated with a crude, but "active protein"-containing cell-free extract of *Ricinus* leaves. In Fig. 1 may be seen fluorescence emission spectra excited at 435 nm, from

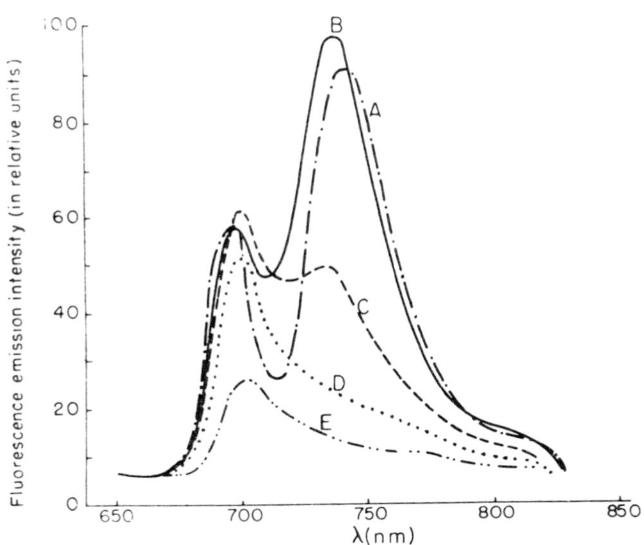

Fig. 1. Fluorescence emission spectra, excited at 435 nm, from Ricinus chloroplasts (at -196°C) as a function of time of incubation at 37°C. Curves A, B, C, D and E are for 0, 5, 10, 20, and 60 min., respectively. Figure from M. Brody et al. (5).

Ricinus chloroplasts (at 77°K), as a function of time of incubation at 37°C. With increasing time of incubation it was observed that (a) the (low temperature) band with maximum at 685 nm (F_{685}) disappeared as a discrete entity, (b) the band with maximum at 735 nm (F_{735}) markedly decreased in intensity, and (c) the band with maximum at 698 nm (F_{698}) intensified to such a degree that an essentially one-banded emission spectrum was eventually obtained. F_{685} has been ascribed to bulk or antennae chlorophyll monomers of (Photo) system II, F_{735} has been associated with aggregated chlorophyll of system I, and F_{698} has been assigned to special reaction center monomers of system II [see, for example, reviews of Butler (6) and Brody, M. (7). Simultaneous changes in (low temperature) fluorescence action spectra were noted as a function of time of incubation; there occurred, for example, in the spectrum for exciting long wavelength fluorescence (F_{735}), decreased participation by bands ascribed to aggregated chlorophyll, but enhanced participation by bands ascribed to monomeric chlorophyll.

Preliminary studies with the electron microscope and experiments on oxygen evolution, made concommittantly with these fluorescence observations, suggested that disruptions in normal lamellar structure and electron transport were also occurring.

Attempts (8) to duplicate the above-cited changes in steady-state fluorescence were made with a number of substances, e.g., several detergents, reducing agents, oxygen evolution inhibitors and lytic enzymes (including pronase, trypsin and wheat-germ lipase); only when pancreatin or scarlet runner bean leaf extract [the latter prepared according to Sastry and Kates (4)] were tested did there seem to occur fluorescence changes similar to those brought about by *Ricinus* leaf extract. Because both pancreatin and scarlet runner bean leaf extract are rich in galactolipases, Cohen *et al*. investigated the influence of exogenous fatty acids on plant material, additionally motivated by earlier reports of the inhibitory effects of fatty acids on Hill Reaction (9-14), and pyocyanin-catalyzed cylic photophosphorylation, as well as the uncoupling effects of fatty acids on ferricyanide-mediated non-cyclic phosphorylation (11). Significantly, Molotkovsky and Zheskova (12) had additionally reported that inhibition was accompanied by light scattering changes--interpreted by these workers as having their origin in swelling of chloroplasts.

Cohen *et al*. (8) showed, not only that exogenous long-chain (C_{18}) unsaturated fatty acids (linolenic, linoleic, oleic or ricinoleic) serve as good models for the action of the protein fraction of *Ricinus* leaf extract on chloroplasts (whole or fragmented) in regard to the above described fluorescence changes, but also in regard to changes in chloroplast ultrastructure. From Fig. 2, for example, it may be seen

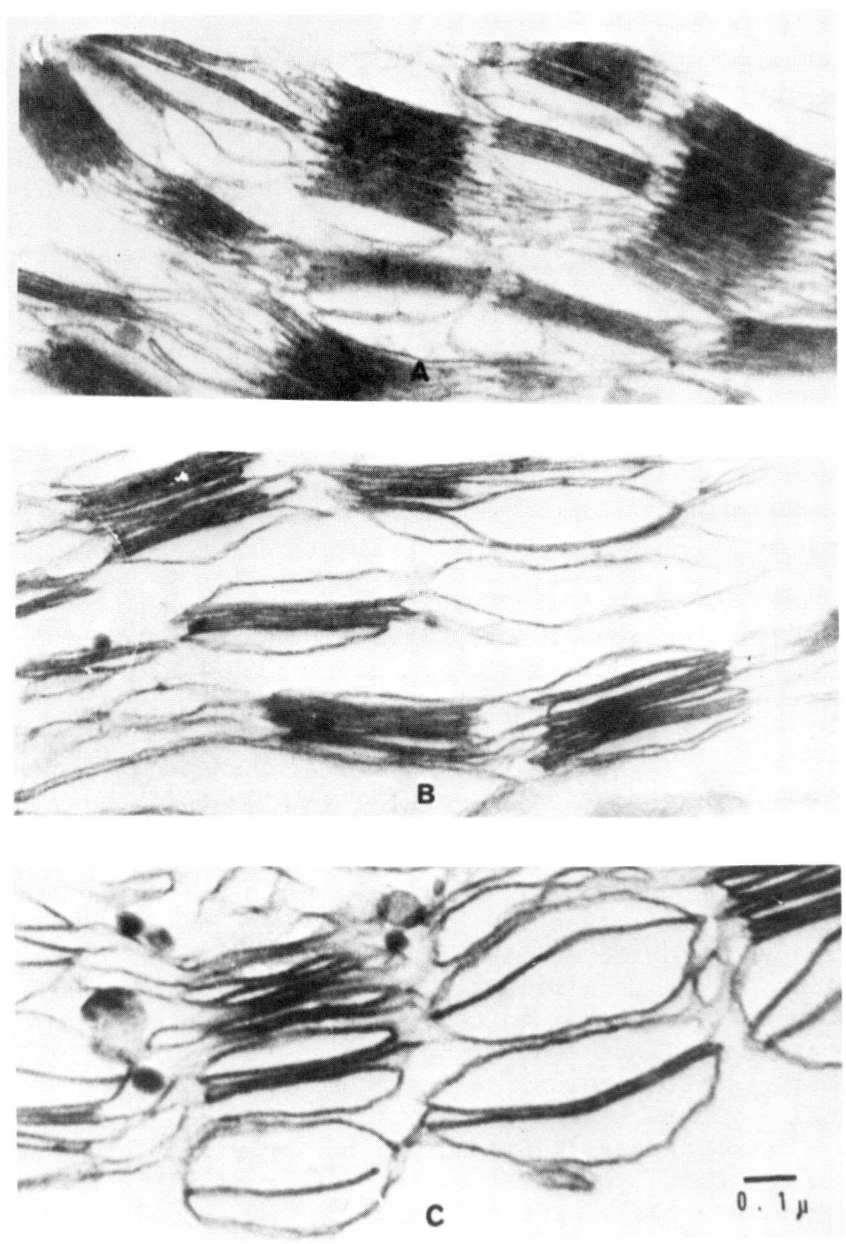

Fig. 2. (Legend on opposite page)

that with increasing concentration of exogenous linolenic acid, i.e., in the mole to mole ratio of fatty acid to chlorophyll ("FA/CHL") range of 0.001 to 0.5--whole spinach chloroplasts progressively become naked lamellae until, at concentration of FA/CHL ≃ 1, no intact chloroplasts are observed. Occurring concommittantly with these changes in chloroplast envelope are other configurational changes in lamellar membranes. Sequentially, the stromal (intergranal) membranes (Fig. 2A) are the first to separate (FA/CHL of 0.001 to 0.01); at intermediate concentrations (FA/CHL of 0.01 to 0.1) granal thylakoids also begin to swell (Fig. 2B), while at higher concentrations (FA/CHL of 0.1 to 1.0) there is observed (Fig. 2C) what seems to be the beginning of separation of the fusion layers of the grana thylakoids. At concentrations of FA/CHL of about 10, chloroplast lamellae become distorted "myelin-like" figures. Essentially the same (in this case, time-dependent) configurational changes result when chloroplasts are treated with the protein fraction of *Ricinus* leaf extract. That exogenous long chain unsaturated fatty acids serve as good models for the action of this leaf extract was determined for several other photosynthesis-associated parameters (8), including light-induced absorption changes of chlorophylls a_{II} and a_{I} [for greater detail, and a consideration of how some of these changes relate to effects on thylakoid membranes, see S. Brody *et al.* (15) and M. Brody *et al.* (16); Hill activity (compare Table I with unpublished results shown in Table II); and fluorescence induction, see Table I, Fig. 3, and for greater detail, M. Brody (18)]. (In addition, it was demonstrated by Cohen *et al.* that there occurs sequential inhibition of system II- and system I-

---

Fig. 2. Electron micrographs of whole spinach chloroplasts incubated in linolenic acid. A, separation of stromal thylakoids (FA/CHL = 0.001 to 0.01). B, swelling of granal thylakoids (FA/CHL ≃ 0.1. C, separation of granal thylakoids (FA/CHL ≃ 1). Figure from Cohen *et al.* (8).

TABLE I

THE EFFECT OF LINOLENIC ACID CONCENTRATION ON VARIOUS CHLOROPLAST PARAMETERS[a,b]

Fatty Acid/ chlorophyll (FA/CHL)	F_{735}/F_{698} (-196°C) (R)	Variable Part of Fluorescence Induction	Light-induced absorption changes		Dye Reduction	
			Chl a_I	Chl a_{II}	Sys. I	Sys. II
2-3	100	50	100	50	150	50
10	50	0	(-)[c]	0	50	0

[a] Experiments with _Spinacia_ and _Zea mays_ chloroplasts
[b] Data expressed as percent of control
[c] Slight retardation in decay of absorption at 705 nm.

Table from Cohen, _et al_. (8)

TABLE II
THE EFFECT OF RICINUS CHLOROPLAST PROTEIN ON NADP REDUCTION[a]

Protein Concentration (μg/ml)	Electron donor H_2O	DCPIP-Ascorbate + DCMU
0	100	100
58	46[b]	90
100	16	80

[a] Spinach chloroplasts were incubated (at 26°C) in 25 mM Tricine-NaOH, pH 8.3 + 6.6 mM $MgCl_2$. The control rates of NADP reduction were (μ moles reduced/mg chlorophyll/hr): from H_2O = 18, from ASC-DCPIP = 7.
[b] Data are expressed as percent of control.
Table from Cohen (17).

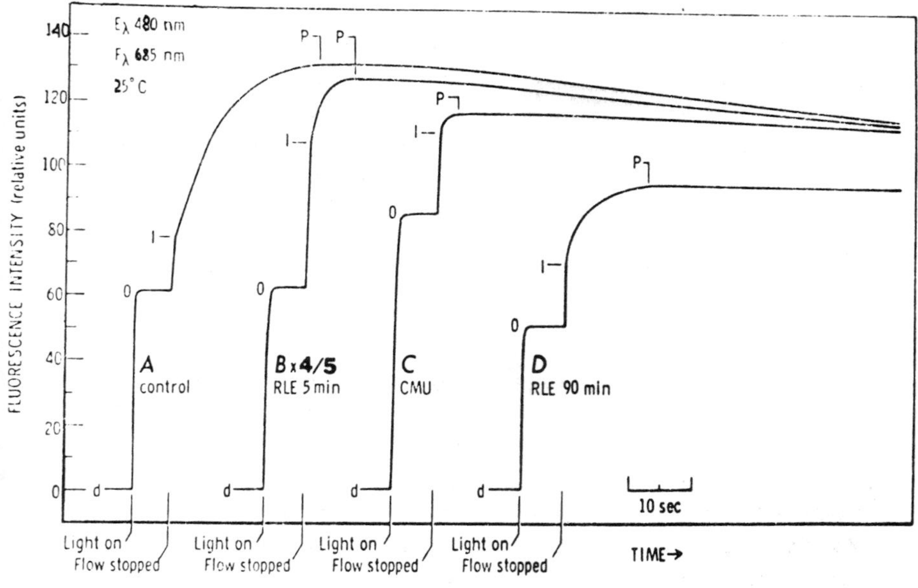

Fig. 3. (Legend on following page)

associated electron flow as a function of increasing concentration of exogenous fatty acids; see Table I.)

Cohen *et al*. (8), on the basis of their works, expressed the opinion that fatty acids may bring about direct effects on chlorophyll pigments, and also suggested that the temporal dependence of the action of *Ricinus* leaf extract "may be interpreted as indicating release, from the chloroplast membrane, of substances similar to fatty acids" [in spite of the fact that "protection" of chloroplasts by defatted bovine serum albumin (BSA) was observed only in the case of treatment with exogenous fatty acids, and not in the case of treatment with *Ricinus* leaf extract].

In extension of the above described earlier works on fluorescence changes, Nathanson and M. Brody (19) studied, in detail, the time- and temperature-dependent changes in emission (at 77°K and room temperature) which occur upon incubation of isolated *Ricinus* chloroplasts, or treatment of chloroplasts from other plants (cross reaction systems) with *Ricinus* leaf extract. In correlation with room temperature absorption and low temperature fluorescence excitation studies of chloroplasts undergoing such emission changes, these workers suggested that the action of the (proteinaceous component of the crude) extract results in a deaggregation of the long wavelength fluorescing forms of chlorophyll (F_{735}) into a monomeric form that contributes to fluorescence at 698 nm. (That the low temperature changes in these bands is

---

Fig. 5 (figure on preceding page)
Fluorescence induction of *Zea mays* chloroplasts. Curve A, control; curves B and D, chloroplasts incubated in RLE for 5 min. and 1-1/2 hr, respectively; curve C, CMU-treated chloroplasts (8×10^{-5} M). Curves A to D are tracings of recordings; curve B has been multiplied by 4/5 for ease of comparison with other curves. Fluorescence excited at 480 nm and monitored at 685 nm. Abscissa time units indicated by scale of 10 sec. Figure from M. Brody (18).

absolute, and not only relative, was shown by M. Brody and S. Brody (20), in experiments in which there was, co-suspended with spinach chloroplasts, a fluorescence standard -- that is, a plastic-embedded dye -- inert to the action of added linolenic acid.) In consideration of the configurational changes induced by treatment of chloroplasts with exogenous fatty acids or *Ricinus* leaf extract (recall the works of Brody *et al.* (5), and Cohen *et al.*, (8)), and in recognition that such configurational changes reflect conformational changes [Murakami and Packer (21)], Nathanson and Brody proposed (19) that the active protein of *Ricinus* leaf extract interacts with lipoprotein chloroplast lamellae, either to (1) bring about direct conformational changes (thus deaggregating chlorophyll aggregates into monomers) or to (2) bring about the liberation of a smaller molecule (from the extract itself, or from the lipoprotein chloroplast lamellae), which could then produce conformational changes and/or directly convert the 735 nm fluorescing form of aggregated chlorophyll, into a monomeric form contributing to fluorescence at 698 nm.

Recent work in my laboratory has been concerned with a) the isolation, characterization, and purification of the active protein of *Ricinus* leaf extract, with the aims of determining whether similar proteins exist in all plants, and whether these proteins exert regulatory influences on the structure and/or function of the photosynthetic apparatus, and b) the mechanism(s) of interaction(s) of fatty acids with the various components of chloroplast membranes (and fractions derived therefrom).

I will turn my attention first to the work done on *Ricinus* leaf extract. Although attempts to isolate the active factor soon indicated its proteinaceous nature (e.g. heat lability, NH_4SO_4 precipitability, inactivation by pronase), largely unrewarding results were obtained when classical methods for protein purification were employed; however, G-25 Sephadex exclusion chromatography (18) resulted in obtaining turbid protein fractions exhibiting 2- to 5-fold increases in specific activity.

TABLE III
PREPARATION OF RICINUS CHLOROPLAST PROTEIN

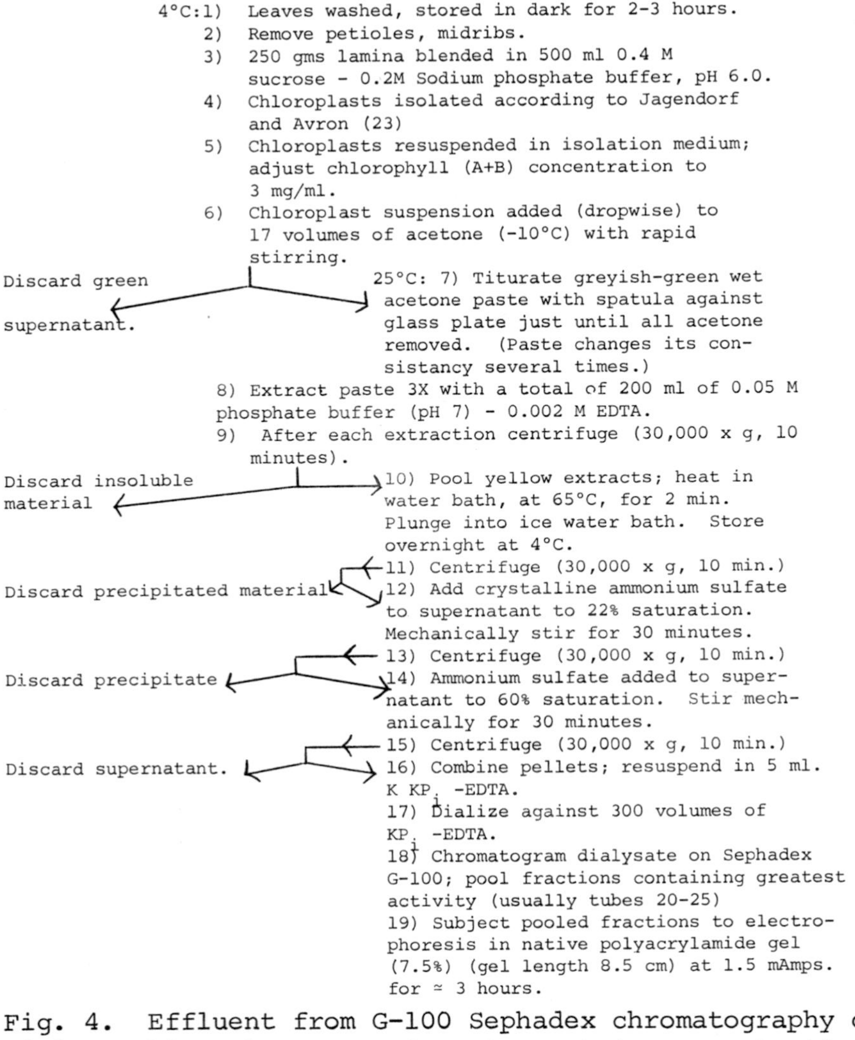

4°C:1) Leaves washed, stored in dark for 2-3 hours.
2) Remove petioles, midribs.
3) 250 gms lamina blended in 500 ml 0.4 M sucrose - 0.2M Sodium phosphate buffer, pH 6.0.
4) Chloroplasts isolated according to Jagendorf and Avron (23)
5) Chloroplasts resuspended in isolation medium; adjust chlorophyll (A+B) concentration to 3 mg/ml.
6) Chloroplast suspension added (dropwise) to 17 volumes of acetone (-10°C) with rapid stirring.

Discard green supernatant.

25°C: 7) Titurate greyish-green wet acetone paste with spatula against glass plate just until all acetone removed. (Paste changes its consistancy several times.)
8) Extract paste 3X with a total of 200 ml of 0.05 M phosphate buffer (pH 7) - 0.002 M EDTA.
9) After each extraction centrifuge (30,000 x g, 10 minutes).

Discard insoluble material

10) Pool yellow extracts; heat in water bath, at 65°C, for 2 min. Plunge into ice water bath. Store overnight at 4°C.
11) Centrifuge (30,000 x g, 10 min.)

Discard precipitated material

12) Add crystalline ammonium sulfate to supernatant to 22% saturation. Mechanically stir for 30 minutes.
13) Centrifuge (30,000 x g, 10 min.)

Discard precipitate

14) Ammonium sulfate added to supernatant to 60% saturation. Stir mechanically for 30 minutes.
15) Centrifuge (30,000 x g, 10 min.)

Discard supernatant.

16) Combine pellets; resuspend in 5 ml. K KP_i -EDTA.
17) Dialize against 300 volumes of KP_i -EDTA.
18) Chromatogram dialysate on Sephadex G-100; pool fractions containing greatest activity (usually tubes 20-25)
19) Subject pooled fractions to electrophoresis in native polyacrylamide gel (7.5%) (gel length 8.5 cm) at 1.5 mAmps. for ≃ 3 hours.

Fig. 4. Effluent from G-100 Sephadex chromatography of *Ricinus* chloroplast protein. A, protein concentration as a function of fraction number. B, inhibition of FeCN-Hill activity and change in *R* (F_{735}/F_{698}) as functions of fraction number; control rates of FeCN reduction (μmoles/mg chl./hr) = 70; control value of R = 2.44. Incubation temperature of spinach chloroplasts 30°C. Figure from Cohen (17).

Since fluorescence changes could be observed both with *Ricinus* chloroplasts (except when isolated in a medium of 0.4 M sucrose - 0.02 M potassium phosphate buffer; pH 6.0) and with other chloroplasts suspended in *Ricinus* leaf extract, it was suspected that the source of the active factor(s) was the *Ricinus* chloroplast (as reported also by Sastry and Kates (4), for galactolipases from runner-bean primary leaf). Dr. R.E. McCarty of Cornell University suggested a (unpublished) method for the preparation of chloroplast proteins from *Ricinus* chloroplasts, rather than from leaf homogenates. McCarty was using this method for the preparation of chloroplast proteins from the primary leaves of *Phaseolus vulgaris* (pole bean). A summary of the procedure [see also Anderson *et al.* (22)] is shown in flow chart form in Table III. In Fig. 4A,

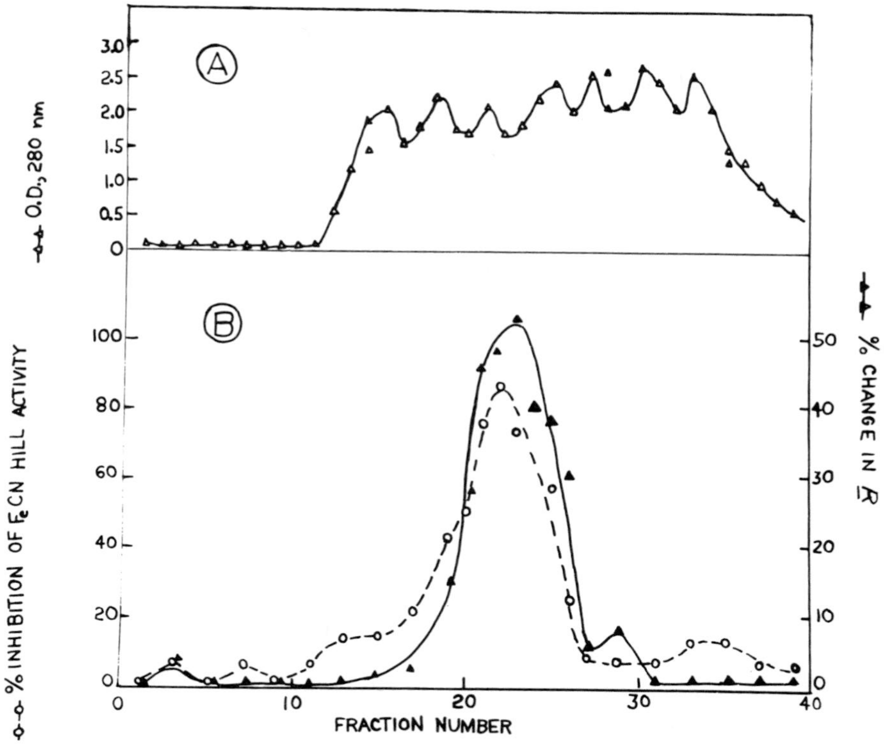

Fig. 4. (legend on previous page)

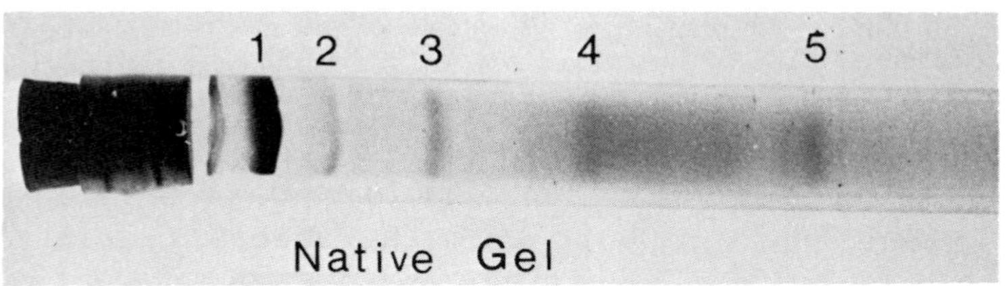

Fig. 5. Results of electrophoresis of Ricinus chloroplast protein (prepared according to Table III) on 7.5% polyacrylamide native gel.

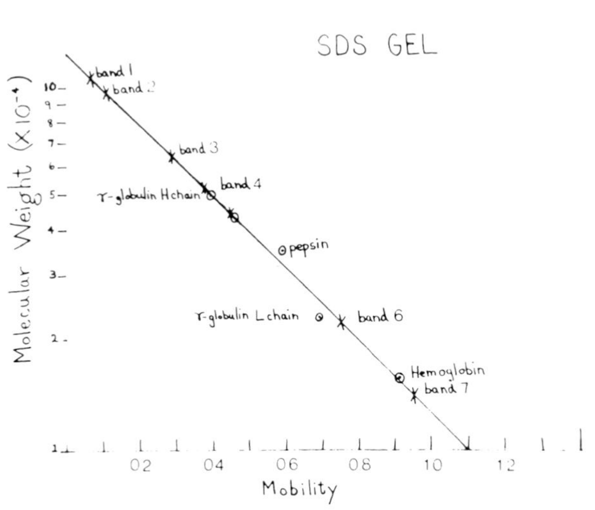

Fig. 6. (Legend on opposite page)

one may see the elution patterns from G-100 Sephadex column (the latter constitutes the final step in the protein extraction procedure); in Fig. 4B are plotted ferricyanide Hill activity and percent change in R(F735/F698) as functions of fraction number. It may be observed that the fractions of protein (tubes 20-25) which inhibit the Hill reaction also seem to be responsible for the fluorescence changes. Since G-100 Sephadex columns exclude proteins of molecular weight 150,000, and since the active protein is somewhat retarded on the column (being eluted at approximately 25% of the bed volume), all one could conclude was that the *Ricinus* chloroplast protein consists mainly of molecules larger than 10,000 M.W. and smaller than 150,000 M.W. Purification through these steps results in a three order of magnitude increase in specific activity compared to crude *Ricinus* extract.

Proceeding according to Brewer and Ashworth (24), polyacrylamide gel electrophoresis (native gel, 7.5%) of the purified *Ricinus* protein yields five discrete bands and perhaps an equal number of faint bands [see Fig. 5 (unpublished results of M. Brody and V. Tsao)]; similar results were obtained when electrophoresis was done on 5% native gel. Preliminary experiments indicate that all five bands manifest some degree of activity; specific activity determinations are in progress.

To remove the charge differences on the known and unknown proteins (and, therefore, to have mobility primarily a function of molecular weight) *Ricinus* chloroplast protein was co-subjected to electrophoresis on SDS (Sodium Dodecyl Sulfate) gel, according to the method of Weber and Osborn (25). From Fig. 6 (unpublished work of Brody and Tsao) it may be observed that electrophoresis of *Ricinus* chloroplast protein

---

Fig. 6. Comparison of molecular weights of *Ricinus* chloroplast proteins [prepared according to Table III (bands 1 through 7)] and four polypeptides of known molecular weight, with their electrophoretic mobilities on sodium dodecyl sulfate gel.

on such gel yields seven bands of different polypeptide lengths, having molecular weights of between 14 and 106×10^3. [The relationships of these molecules (e.g. sub-unit relationships) is presently being investigated.]

Using essentially the same procedure he communicated to us (for extraction of _Ricinus_ chloroplast protein) McCarty and co-workers (22) recently isolated and purified, from the primary leaves of _Phaseolus vulgaris_, a lipase which hydrolyzes purified substrate monogalactosyl diglyceride (from spinach leaves) at a rapid rate, and (even more rapidly) liberates fatty acids from spinach subchloroplast particles.

Since chloroplasts from bean leaves are notoriously unstable (photo- and bio-chemically), in our laboratory we attempted to determine whether "stable" chloroplasts contain proteins similar in action to those of _Ricinus_ chloroplasts. To this end, chloroplasts of spinach were utilized for extraction. All data -- those of electrophoretic characteristics -- as well as those of fluorescence emission and circular dichroism (see below) changes (induced in isolated chloroplasts, including those of spinich) indicate the presence of such protein, albeit of considerably lower activity (unpublished results of M. Brody and V. Tsao). There is some evidence which suggests that the activity requirement for 1 to 2 orders of magnitude more spinach protein than _Ricinus_ protein results from the presence of a tightly coupled inhibitor in the former case.

It has not escaped the author's attention (see ref. 18) that the presence, in chloroplasts, of molecules similar in action to those of _Ricinus_ provides a possible _in vivo_ monitor for controlling or regulating both the partition of energy and electron transfer between system II and system I of photosynthesis.

I will now consider some of the experiments which were done (in conjunction with Dr. B. Nathanson, while she was a doctoral student in my laboratory) on the mechanism(s) of interaction(s) of fatty acids with derivatives of chloroplast membranes. Since with chlorophyllous systems of various levels of organization (from intact chloroplasts to chlorophyll in solution), spectral changes induced by exogenous fatty acids or Ricinus protein were found to be alike, it was postulated that these changes resulted from a deaggregation to monomeric chlorophyll (19). While direct action on the chlorophyll chromophore could be exclusively invoked to explain the (postulated) deaggregation process, the (above described) configurational changes which obtain upon treatment of chloroplasts with Ricinus protein or exogenous fatty acids prompted consideration of the possibility that there also occurs indirectly-induced deaggregation, in which fatty acids act on the protein and/or lipid environment of the chlorophyll molecule, resulting in conformational changes capable of altering the states of chlorophyll.

Since aggregate and monomer bands of chlorophyll may be resolved separately in circular dichroism (CD) spectroscopy (26,27), this tool (as well as infra-red spectroscopy), was used to investigate the possible occurrence of conformational changes (28). In the case of chlorophyll in solution (CCl_4 or Cl_4-hexane) de-aggregation was found to be by direct action on the chromophore. In the case of system I chlorophyll-protein-lipid complexes, exogenous fatty acids also induced deaggregation; essentially similar results were obtained upon treatment of lipid-free-HP700 reaction center chlorophyll-protein complex. In the case of the latter preparation, incubation in urea (or exogenous linolenic acid) brought about changes in the ultraviolet C.D. region, at concentrations lower than those which yielded deaggregation in the visible (red) region of the spectrum--see Fig. 7 (29). Such changes in near ultraviolet C.D. bands (which have been attributed to conformational changes in cases of other proteins) were found (albeit, not always consistently) to reverse upon partial removal of the urea; concommittantly, reaggregation was observed to occur in the visible C.D. region.

In their C.D. study, M. Brody and Nathanson (28) pointed out that the additional structural array,

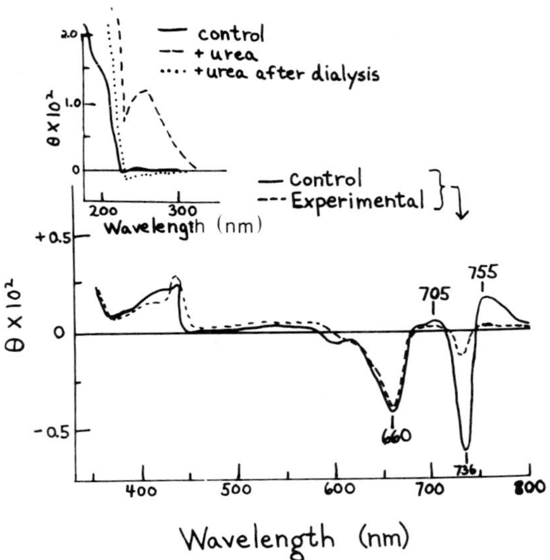

Fig. 7. C.D. spectra of a) control HP700 chlorophyll-protein complexes suspended in STN buffer, pH 8.0 (solid line), and b) HP700 chlorophyll-protein complexes in STN buffer to which had been added 8 M urea (broken line). Insert: Ultraviolet C.D. of same particles, plus b), as above, dialyzed 1/2 hour (given by dotted line). Chlorophyll concentration = 2.5×10^{-5} M; path length 1 cm. Figure from Nathanson (29).

brought about by organization (of chlorophyll-protein-lipid associations) into membranes (digitonin-isolated particles containing either system I or systems II and I, subchloroplast particles obtained by means of sonication, and specially prepared intact chloroplasts), conferred an orientation of visible C.D. bands opposite in sign to those of corresponding C.D. bands in non-membranous systems. Deaggregation of chlorophyll in membranous systems results primarily from conformational changes in protein. In Fig. 8 (20) it may be observed that exogenous linolenic acid converts the ultraviolet C.D. bands of control digitonin-isolated particles (containing systems II and I) from α-helix to random coil (compare with Fig. 9, re-drawn from Gratzer and Cowburn, (30); this conversion is accompanied by de-

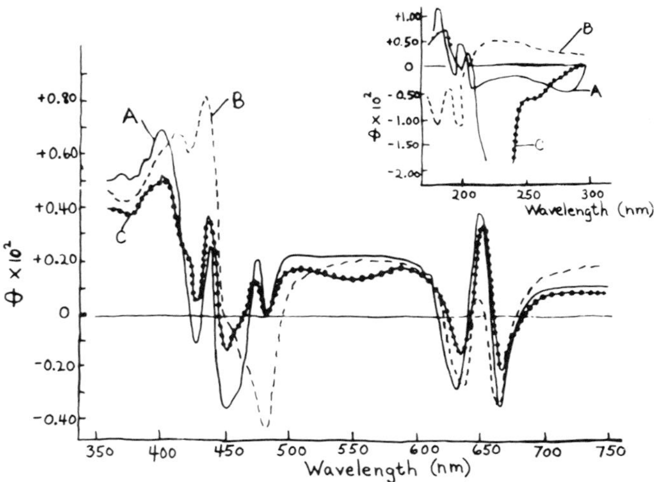

Fig. 8. C.D. spectra of a) control digitonin-isolated particles (chlorophyll a + b concentration 1×10^{-5} M) containing systems I and II, suspended in 0.01 M KCl, 0.05 M Tris-HCl buffer pH 7.8 (solid line), b) particles after 1 min of incubation in buffer to which had been added 2×10^{-4} M linolenic acid (broken line), and c) particles after 1 min of incubation in buffer to which had been added 2×10^{-4} M linolenic acid, followed by 1 min incubation in the presence of 20 mg/ml defatted BSA (dotted line). Insert: Ultraviolet C.D. spectra of aforementioned particles. Figure from Nathanson (29).

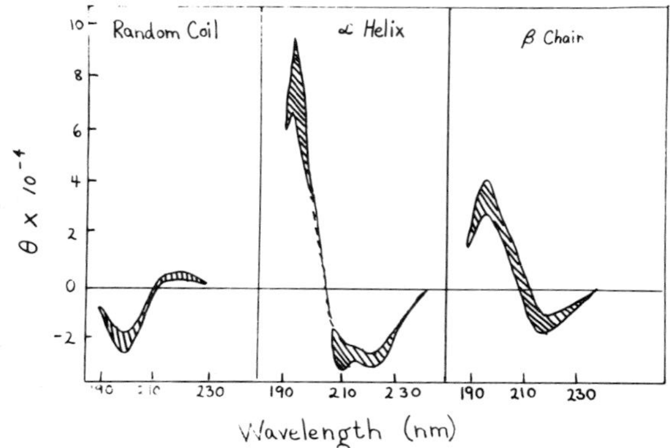

Fig. 9. (Legend on following page)

aggregation of chlorophyll. Post treatment of such particles with defatted BSA restores helical (probably β-extended) structure in the ultraviolet and brings back aggregate bands of chlorophyll in the visible C.D. It is interesting to note (in this case, with spinach chloroplasts), that post treatment with defatted BSA not only (largely) restores visible C.D. to the control situation (Fig. 10), but also restores the variable part of fluorescence induction [see Fig.

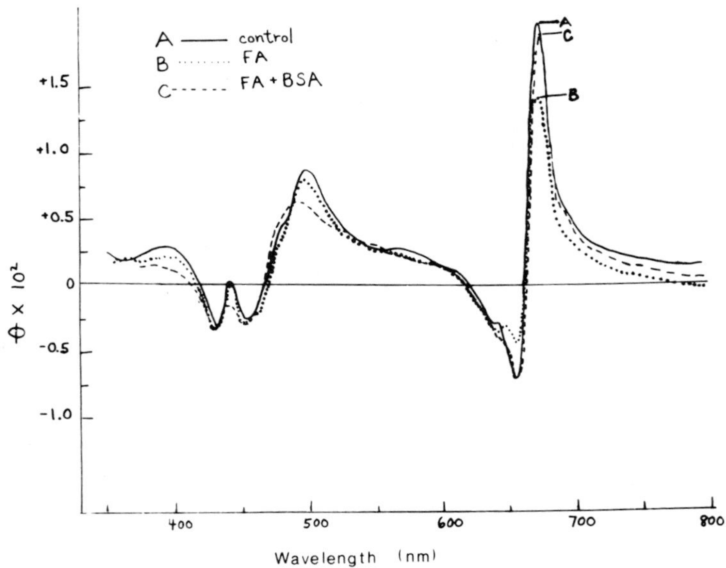

Fig. 10. C.D. spectra of spinach chloroplasts (chl a+b conc. 2.5×10^{-5} M). a) control; b) treated with linolenic acid at FA/CHL = 10; c) as in b), additionally post-treated with BSA (de-fatted) at a final concentration of 20 mg/ml. (See accompanying fluorescence induction experiment in Fig. 11.)

Fig. 9. (Figure on preceding page)
Ultraviolet C.D. spectra for random coil, α-helix and for β-extended chain of polyglutamic acid in aqueous solution. [Curves re-drawn from Gratzer and Cowburn (30).]

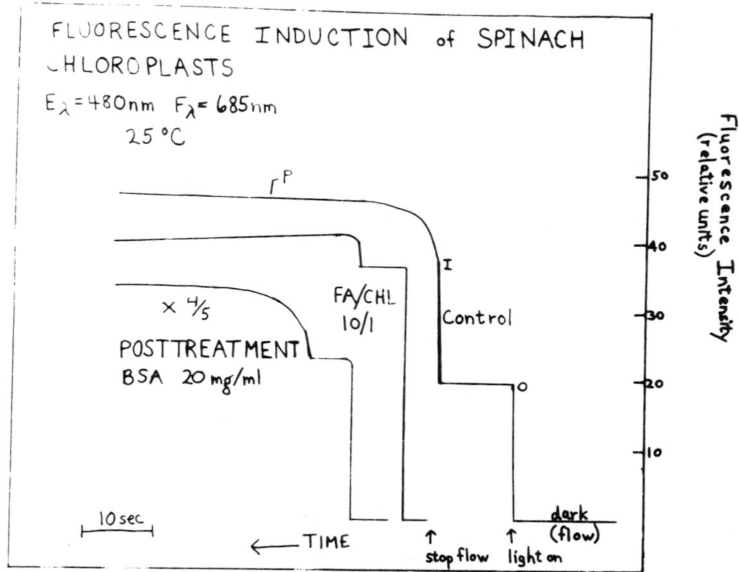

Fig. 11. Fluorescence induction of spinach chloroplasts; experimental conditions as described in Fig. 10. Curves are tracings of recordings; the Post Treatment curve has been multiplied by 4/5 for ease of comparison with other curves. Abscissa time units indicated by scale of 10 seconds.

11 (unpublished data), and compare with Fig. 3].

In closing, it might be well to recall that the role of the uniquely preponderant (long-chain, unsaturated fatty acid-rich) galactosyl diglycerides of chloroplast membranes has never been fully understood. As Benson (2) long ago suggested, it is likely that galactolipids and linolenic acid in chloroplasts exist in dynamic equilibrium between the free and bound state, and, as M. Brody and Nathanson (28) pointed out, in their free state (the degree of which may be critically regulated) these molecules may well act to bring about conformational changes in the proteins to which chlorophyll is attached, thus causing shifts both in electron transport patterns and in populations of system I and system II pigments.

REFERENCES

1. BENSON, A. A. (1963) *in* Photosynthetic Mechanisms of Green Plants, Publ. 1145, p. 571, Nat'l. Acad. Sci.-Nat'l. Res. Council, Washington, D.C.
2. BENSON, A. A. (1964) *Ann. Rev. Plant Physiol.* *15*, 1.
3. WEIR, T. E., AND BENSON, A. A. (1967) *Am. J. Bot.* *54*, 389.
4. SASTRY, P., AND KATES, M. (1964) *Biochem.* *3*, 1280.
5. BRODY, M., NATHANSON, B., AND COHEN, W. (1969) *Biochim. Biophys. Acta* *172*, 340.
6. BUTLER, W. (1966) *in* The Chlorophylls (Vernon, L., and Seely, G., eds.) p. 343, Academic Press, New York City.
7. BRODY, M. (1969) *in* The Biology of *Euglena* (Buetow, D. E., ed.) vol. 2, p. 215, Academic Press, New York.
8. COHEN, W. S., NATHANSON, B., WHITE, J. E., AND BRODY, M. (1969) *Arch. Biochem. Biophys.* *135*, 21.
9. SPIKES, J. LUMRY, R., AND RIESKE, J. (1954) *Arch. Biochem. Biophys.* *55*, 25.
10. KROGMANN, D., AND JAGENDORF, A. (1959) *Arch. Biochem. Biophys.* *80*, 421.
11. McCARTY, R., AND JAGENDORF, A. (1965) *Plant Physiol.* *40*, 725.
12. MOLOTKOVSKY, Y., AND ZHESKOVA, I. (1966) *Biochim. Biophys. Acta* *112*, 170.
13. CONSTANTOPOULOS, G., AND KENYON, C. (1968) *Plant Physiol.* *43*, 521.
14. KATOH, S., AND SAN PIETRO, A. (1968) *Arch. Biochem. Biophys.* *128*, 378.
15. BRODY, S., BRODY, M., AND DÖRING, G. (1970) *Z. Naturforsch.* *25b(4)*, 367.

16. BRODY, M., BRODY, S., AND DÖRING, G. (1970) Z. Naturforsch. 25b(8), 862.
17. COHEN, W. S. (1970) Ph.D. Thesis, Hunter College, C.U.N.Y., New York City.
18. BRODY, M. (1971) Biophys. J. 11(2), 189.
19. NATHANSON, B., AND BRODY, M. (1970) Photochem. Photobiol. 12, 469.
20. BRODY, M., AND BRODY, S. (1971) Photochem. Photobiol. 13, 293.
21. MURAKAMI, S., AND PACKER, L. (1970) Biochim. Biophys. Acta 180, 420.
22. ANDERSON, M. M., McCARTY, R. E., AND ZIMMER, E. A. (1974) Plant Physiol., in press.
23. JAGENDORF, A., AND AVRON, M. (1958) J. Biol. Chem. 231, 277.
24. BREWER, J., AND ASHWORTH, R. (1969) J. Chem. Ed. 46, 41.
25. WEBER, K., AND OSBORN, M. (1969) J. Biol. Chem. 244, 4406.
26. DRATZ, E., SCHULTZ, A., AND SAUER, K. (1966) Brookhaven Symp. Biol. 19, 303.
27. HOUSSIER, C., AND SAUER, K. (1970) J. Am. Chem. Soc. 92, 779.
28. BRODY, M., AND NATHANSON, B. (1972) Biophys. J. 12, 774.
29. NATHANSON, B. (1973) Ph.D. Thesis, Hunter College, C.U.N.Y., New York City.
30. GRATZER, W., AND COWBURN, D. (1966) Nature 222, 426.

INTERACTION BETWEEN CHLOROPLASTS AND MITOCHONDRIA IN ISOLATED LEAF CELLS

A. Gnanam, A. G. Govindarajan, and M. Vivekanandan

INTRODUCTION

The chloroplasts and the mitochondria are the two major organelles of the eucaryotic cells in which distinct electron transport systems and the other structural components are distributed for energy transduction. They show a remarkable degree of compartmentalization of the metabolic pools, the mediating enzymes and other substances. As we know today, photosynthesis takes place in the chloroplasts; the Embden-Meyerhof pathway, the hexose monophosphate pathway and the related enzymes are located in the cytoplasm; and the tricarboxylic acid cycle together with its associated electron transport systems are situated in the mitochondria. However, one can visualize a great deal of interaction between these two organelles since the starting products of photosynthesis and the end products of respiration are similar, and also because the end products of photosynthesis provide substrates for respiration. Besides both seem to require substances that are common to energy transducing processes such as inorganic phosphate, pyridine nucleotides, the adenylates, 3-phosphoglyceric acid and triose phosphate.

More definitive evidence for their actual interaction comes from the discovery of a pyridine nucleotide

---

Abbreviations: DCMU - Dichlorophenyl-dimethylurea; NADP - Nicotinamide adenine dinucleotide phosphate; $NADPH_2$ - Nicotinamide adenine dinucleotide phosphate reduced; PGA - 3-phosphoglyceric acid; Pi - Orthophosphate; TCAC - Tricarboxylic acid cycle; NAD - Nicotinamide adenine dinucleotide; $NADH_2$ - Nicotinamide adenine dinucleotide reduced.

transhydrogenase in chlorophyll containing cells. It is suspected to play an important controlling role in the breakdown of the respiratory substrates by promoting the formation of DPNH, mediated by the photosynthetic pyridine nucleotide reductase on illumination (1). Further, the demonstration of ion and organic acid fluxes and the attendant structural (osmotic) changes of these organelles under light and dark conditions (2) along with the fact that the photosynthetically fixed carbon is traceable in the mitochondrial intermediates within minutes after illumination are indicative of the transport of the common intermediates among these systems (3). In this paper we report the results of our efforts in understanding the interactions at structural and functional levels of these two major energy transducing systems using isolated cells from plant sources and the possible effect on interfering preferentially with one organelle on the other and on the overall energy turnover of the cell in which they reside.

MATERIALS AND METHODS

Materials: The plant materials used in this study, *Dolichos lab lab* L. and *Canna edulis* Ker., were grown in the Botanic Garden. *Euglena gracilis*, strain Z, was axenically cultured in autotrophic conditions at 27°C using modified Hunter's medium (4).

Isolation of mesophyll cells: Fresh and fully expanded leaves were harvested and the mesophyll cells were isolated according to the procedure described by one of us (5).

O_2 exchange studies: The O_2 exchange rates by cells were measured at 30°C using Clark oxygen electrode (Yellow Springs Instruments Co., Ohio) connected to Heath Servo recorder Model EU-20 B. Photosynthetic evolution of oxygen was measured in 3 ml reaction mixtures containing cells (4.5×10^6), sucrose 400 mM; orthophosphate (50 mM), pH 7.0 and $MgCl_2$ 10 mM under saturating light intensity from a bank of light source. Unless otherwise indicated no sodium bicarbonate was added. Photosynthetic O_2 evolution was carried out with an initial oxygen content of 80-120 nmoles/ml of the reaction mixture. Respiratory O_2 uptake was measured under the same conditions but in the absence of light.

CO_2 fixation: The cells were aerated for 15-30 min. under illumination supplied by 500 W projector reflector lamp. After achieving steady state of photosynthetic rate the aeration was stopped and quickly ^{14}C-bicarbonate was added. ^{14}C-bicarbonate fixation by cells was carried out essentially as described by Bassham and Calvin (6). The reactions were followed at 25°C for 15 min. at the end of which the cells were quickly chilled and the suspension was acidified to remove the residual bicarbonate. They were then extracted with acidified alcohol and aliquots of this extract were counted for radioactivity in a gas flow proportional counter (Electronic Corporation of India, India).

Phosophorylation: The whole cell phosphorylation rates were measured using ^{32}Pi under the same conditions as those described for ^{14}C-bicarbonate fixation studies. The cells were pre-illuminated for 20 min. before the addition of ^{32}Pi. The reactions were carried out for 10 min., after which cold TCA (final conc. 4%) was added to stop the reactions. Aliquots were drawn from deproteinized supernatant for determining Pi esterification. The organic fraction was separated from the supernatant by isobutanol:benzene extraction as described by Nobel (7).

Isolation of aplastic cells and estimation of terpenoid compounds: Aplastic cells were derived from "albino" leaves (5) of Canna plants to which 10 mM aminotriazole was injected in the pseudostem of the plant. The plastoquinone and ubiquinone were extracted and estimated by the method described by Crane (8) and Crane and Barr (9) respectively. The amount of β-Carotene was estimated according to Goodwin (10) and sterols by the method of Abell *et al.* (11).

RESULTS

Oxygen exchange reactions - the photosynthetic O_2 evolution and respiratory O_2 uptake - of bean cells were comparatively lower than those of Euglena cells. The rate of O_2 evolution on illumination was 5 and 10-fold higher in bean and Euglena cells as compared to their respective O_2 uptake in the dark. Since this ratio of O_2 exchange rates was consistent over several experiments, it was concluded that the observed lower rate of O_2 exchange is intrinsic to the bean cells. The effect of various mitochondrial or chloroplast specific inhibitors on the rate of O_2 exchange is shown in Table I. At the concentration level used, amytal, antimycin, oligomycin, potassium cyanide, rotenone, sodium azide and DCMU inhibited completely the photosynthetic O_2 evolution in both Euglena and bean cells,

TABLE I
ABSOLUTE VALUES OF OXYGEN EXCHANGE REACTIONS[a]

Inhibitor	Conc. (mM)	System	nmoles $O_2/10^8$ cells/min			
			Control		Treated	
			P	R	P	R[b]
DCMU[c]	0.05	Bean	630	130	0(--)	208(160)[d]
	0.05	Euglena	1810	160	0(--)	240(150)
Rotenone	0.17	Bean	410	70	0(--)	56(81)
	0.34	Euglena	1500	140	0(--)	49(35)
Amytal	0.7	Bean	480	120	0(--)	50(42)
	1.4	Euglena	1800	160	650(36)	80(50)
Malonate	2.0	Bean	400	80	350(88)	53(66)
	4.0	Euglena	1630	150	1560(96)	70(47)
Sodium azide	0.3	Bean	600	120	0(--)	59(50)
	0.6	Euglena	2030	200	0(--)	186(93)
Oligomycin	0.015	Bean	710	130	0(--)	64(50)
	0.030	Euglena	2040	200	0(--)	102(50)
Antimycin	0.012	Bean	460	90	0(--)	37(41)
	0.024	Euglena	1740	160	0(--)	112(70)
Potassium cyanide	1.0	Bean	660	150	0(--)	0(--)
	1.0	Euglena	2000	190	0(--)	0(--)

[a]Composition of reaction mixtures is as given under "Materials and Methods."

[b]P denotes photosynthetic O_2 evolution and R the respiratory O_2 consumption.

[c]Cells were incubated with inhibitors for 5 min. before measuring O_2 exchange rates.

[d]Figures in parentheses represent respective per cent control values.

though some of the compounds are known to be specific only to mitochondrial reactions.

In order to understand whether the observed complete inhibition of O_2 evolution by these compounds is due to their interference at the sensitive water splitting site prior to electron transport system or due to their interference at the electron transport chain _per se_, ferricyanide was added to the inhibited cells and their O_2 evolution rate was measured. Ferricyanide reversed the inhibition of antimycin completely and of rotenone partially, and to a much lower extent of the inhibition caused by oligomycin and sodium azide (Table II). This indicates that they are really interfering with the electron transport chain.

TABLE II
EFFECT OF POTASSIUM FERRICYANIDE[a] ON OXYGEN EXCHANGE REACTIONS[b]

	Additives	Conc. (mM)	nmoles $O_2/10^8$ cells/min	
			Photosynthesis	Respiration
I	None		1030	225
	Rotenone	0.17	0	180
	+FeCN		600(58)[c]	-
II	None		1030	360
	Antimycin	0.012	0	135
	+FeCN		1350(131)	-
III	None		710	130
	Oligomycin	0.015	0	64
	+FeCN		90(13)	-
IV	None		600	120
	Sodium azide	0.3	0	59
	+FeCN		85(14)	-

[a]$K_3Fe(CN)_6$ was added to 5 mM final concentration in reaction mixtures.

[b]System used was bean cells.

[c]Figures in parentheses represent respective per cent values.

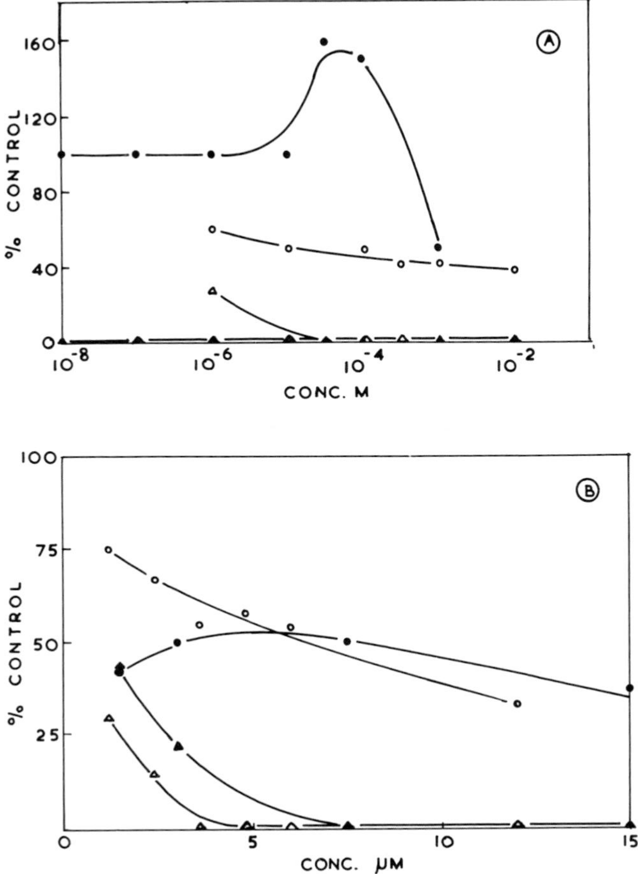

Fig. 1. Effect of increasing concentrations of inhibitors on O_2 exchange rates in bean cells.
A: o——o, amytal O_2 uptake; ●——●, DCMU O_2 uptake; Δ——Δ, amytal O_2 evolution; ▲——▲, DCMU O_2 evolution.
B: o——o, antimycin O_2 uptake; ●——●, oligomycin O_2 uptake; Δ——Δ, antimycin O_2 evolution; ▲——▲, oligomycin O_2 evolution.

Since we used a whole cell system, concentration levels chosen for the various inhibitors were about tenfold higher than that routinely used for in vitro chloroplast or mitochondrial studies to obtain optimal inhibition of either electron transport or other reactions. In order to check whether or not this high concentration was responsible for non-specific inhibition of both the organelle functions and to see the extent of the specificity of these substances to the organelles, a detailed measurement of the inhibition of O_2 exchange reactions at various concentrations of a selected few inhibitors were made. It can be seen from Figure 1 that even at low concentration levels amytal, antimycin, oligomycin and DCMU blocked photosynthetic O_2 evolution to a greater extent as compared to respiratory O_2 uptake. Apparently, chloroplasts and associated reactions are the primary targets to most of the externally added compounds. The complete inhibition of mitochondrial electron transport could never be obtained even at the highest concentrations used in our studies.

Further to confirm that loss of O_2 evolution is not due to depletion of the terminal oxidant pool or other intermediates NADP, PGA, sodium acetate and sodium bicarbonate were added and were found to have no effect in reversing the inhibition caused by these compounds.

The effect of few inhibitors on the phosphorylation rate of the whole cells under light and dark conditions are given in Table III. The phosophorylation in the dark was not affected to any significant level by the compounds tried. In fact rotenone was found to enhance the Pi esterification marginally. On the other hand photophosphorylation, as measured by the difference in the level of Pi esterification between light and dark conditions, was inhibited to an extent of nearly 50% by DCMU and rotenone. Arsenite appeared to be a potent inhibitor of light dependant phosphorylation.

The bicarbonate fixation by the bean cells was also inhibited to a large extent by all the substances

TABLE III
PHOSPHORYLATION BY BEAN CELLS[a]
Effect of Photosynthetic and Respiratory Inhibitors[b]

Inhibitors	Conc. (mM)	Pi esterified nmoles/10^8 cells/min					
		Control			Treated		
		Dark	Light	L-D	Dark	Light	L-D[c]
DCMU	0.05	1170	3100	1930	1000(85)	2130(69)	1130(59)[d]
Rotenone	0.17	300	2170	1870	370(123)	1350(62)	980(52)
Arsenite	0.50	930	3260	2330	800(86)	1030(32)	330(14)

[a]The reaction mixtures (4.0 ml) contained cells (6 × 10^6); Tris-HCl (50 mM), pH 7.8; NaCl 35 mM; $MgCl_2$ 10 mM and 200,000 cpm/ml carrier-free orthophosphate.

[b]Cells were incubated for 5 min. with the inhibitors before the addition of ^{32}Pi.

[c]Denotes light minus dark values.

[d]Figures in parentheses represent respective per cent control values.

TABLE IV
^{14}C-BICARBONATE FIXATION BY BEAN CELLS[a]
Effect of Photosynthetic and Respiratory Inhibitors[b]

Inhibitors	Conc. (mM)	μmoles $^{14}CO_2/10^8$ cells/hr		
		Control	Treated	% Control
DCMU	0.05	70.4	2.56	4
Rotenone	0.17	83.2	9.60	12
Amytal	0.7	96.0	44.80	47
Malonate	2.0	96.0	108.00	113
Sodium azide	0.3	96.0	25.60	27
Potassium cyanide	1.0	96.0	38.40	40

[a]The reaction mixtures contained cells (3 × 10^6); phosphate (50 mM); pH 7.8; NaCl 35 mM; $MgCl_2$ 5 mM, and $NaH^{14}CO_3$ (75nc/μmole), 3 μc in a final volume of 3.0 ml.

[b]Cells were incubated for 5 min. with the inhibitors before the addition of ^{14}C-bicarbonate.

tried except for malonate (Table IV), which had no effect on the chloroplast mediated O_2 evolution as well (Table I).

The comparative studies on the levels of the terpenoid compounds of green and aminotriazole bleached canna leaf cells are given in Table V. Though the

TABLE V
COMPARATIVE STUDIES ON THE LEVELS OF TERPENOID COMPOUNDS OF GREEN AND AMITROLE BLEACHED CANNA LEAF CELLS

Terpenoid compounds	Green cells (control)	Bleached cells (treated)	% Control
	$\mu g/10^8$/cells		
Chlorophyll	1851	0	--
β-Carotene	61	0	-
Sterols	905	1050	116
	$\mu mole/10^8$ cells		
Plastoquinone	0.32	0	-
Ubiquinone	0.04	0.047	118

chlorophylls, carotenes and plastoquinone are completely absent in the bleached cells indicating the absence of chloroplast structure, the ubiquinone and total sterols of the cells are practically the same in both the aplastic and green cells.

Other biochemical composition of amitrole bleached and normal green cells are compared in Table VI. The differences observed in the levels of nucleic acids, proteins and lipids could be attributed to the absence of chloroplasts in the bleached cell, rather than any derangement on their normal function as seen from the levels of free amino acids and carbohydrates.

The respiratory levels of both the types of cells are compared in terms of O_2 uptake. The mitochondrial functions appear to be normal in aplastic cells and quite comparable with green cells (Table VII).

TABLE VI
BIOCHEMICAL COMPOSITION OF GREEN AND AMITROLE BLEACHED CANNA LEAF CELLS

	Green cells (control)	Bleached cells (treated)	% Control
	mg/10^8 cells		
Nucleic acids (total)	0.98	0.69	70
RNA	0.73	0.49	67
Free amino acids	1.53	4.21	275
Protein	19.0	1.36	71
Starch	3.01	4.37	145
Soluble sugars (total)	7.15	3.96	55
Lipid (total)	242.0	73.0	30
	µmole/10^8 cells		
Phospholipid	5.65	1.70	30

TABLE VII
COMPARATIVE RESPIRATORY STUDIES ON NORMAL AND AMITROLE BLEACHED CANNA LEAF CELLS

Additives	Conc. (mM)	nmoles O_2/10^8 cells/min	
		Green cells (control)	Bleached cells (treated)
None		670(100)	580(100)[a]
Glucose	10.0	670(100)	770(133)
Succinate	2.0	870(130)	890(153)
Malic acid	2.0	820(122)	720(124)
Citric acid	2.0	700(104)	710(122)

[a]Figures in parentheses represent respective per cent control values.

DISCUSSION

Interaction between chloroplasts and mitochondria in a green cell could best be studied only under illumination since none of the chloroplast functions could be activated in the absence of light. On the contrary, there is compelling evidence to show that mitochondrial reactions are also operative under illumination. The ^{18}O exchange studies and the existence of ambient CO_2 compensation point for green tissues are some of the evidence for such a conclusion. However, it is not clear whether or not all the following major mitochondrial reactions, *viz.*, 1) TCAC with the attendant reduction of NAD and the release of CO_2 through decarboxylation reactions, 2) the re-oxidation of $NADH_2$ *via* the electron transport using molecular oxygen, and 3) the coupling of oxidative phosphorylation to the electron transport, proceed unabated since each one could be independently controlled and they in turn can regulate the corresponding TCAC reactions of mitochondria. For example, it is known that TCAC reactions can metabolize photosynthetically fixed carbon compounds during illumination (12). Similarly $NADH_2$ produced by TCAC reactions in mitochondria can donate electrons to photosystem I (13). Likewise, the phosphorylation rate of one organelle can regulate that of the other by controlling the relative levels of ADP and Pi. It is with this view various mitochondrial or chloroplast specific electron transport and/or phosphorylation inhibitors were employed in this study to see if any of the above possible interactions can be demonstrated by preferentially interfering with the partial reactions of one organelle. From the data presented it can be seen that their effect on the mitochondrial electron transport is only partial. Though one cannot equate the measured O_2 uptake in its entirety to a measure of mitochondrial electron transport, the observed partial inhibition of O_2 uptake in our experiments can still be attributed to mitochondrial respiration as none of the substances, except KCN, seem to have any effect on several oxidases presumably operating in whole cells. However, the extent to which the observed residual O_2 uptake can be ascribed either to mitochondrial respiration or other oxidases is hard to derive from the present data. The susceptibility of the chloroplast to all the substances tried may be due to the fact that the volume and surface area of the chloroplasts inside the cell is enormous as compared to those of mitochondria and hence chloroplasts form the primary targets of these inhibitors. Alternatively, the order of magnitude of chloroplast susceptibility to the inhibitors should be several-fold higher than that required for inhibition of mitochondrial activity. In fact, this is exactly the situation as can be seen from Figure 1.

The well-documented specificity of the compounds to either chloroplast or mitochondria under in vitro conditions impells one not to overlook their specificity in whole cells also. Notwithstanding the total inhibition of chloroplast partial reactions by these substances, their specificity, save DCMU, towards mitochondrial functions is corroborated by our data inasmuch as the extent of inhibition recorded by us goes well with those obtained for mitochondria in vitro. Similarly DCMU even at concentrations as low as 1×10^{-8} M abolishes the chloroplast O_2 evolution with least interference on mitochondrial O_2 uptake (Figure 1). Hence claim for their specificity to organelles in situ is not too far fetched and the results from such studies can still be meaningful. The reasons for their cross-over effects should be looked for elsewhere, and our studies are incomplete to delineate this aspect. The resumption of O_2 evolution by ferricyanide indicates the site of action of the inhibitors to be away from that of DCMU. The ferricyanide restoration of O_2 evolution is sensitive to DCMU. The inhibition of $^{14}CO_2$ fixation as a consequence of block in O_2 evolution falls much in line with DCMU action. This may be due to lack of turnover of NADPH although ATP level not being limiting as the cell can still switch over to cyclic photophosphorylation. The fact that even when the O_2 evolution and bicarbonate "fixation" is completely supressed the light mediated phosphorylation is only partially inhibited suggests that the chloroplast is switching over to more stable photosystem I which mediates cyclic photophosphorylation in the whole cells. This is what we see in the case of DCMU and rotenone treated cells, in which dark phosphorylation is not inhibited to any significant level. Arsenite appears to inhibit specifically the photosphosphorylation and this specificity indicates that it is not inhibiting the Pi esterification competitively. This is in contrast to the report by Whatley et al. (14). The inhibition of $^{14}CO_2$ fixation by arsenite (15) is thus accounted for by its effect on photophosphorylation. That the dark phosphorylation remains unaffected is explained by the low concentration (0.5 mM) used by us, 2-20 fold lower than that generally used for isolated mitochondria (16). It is interesting to note here that it is the chloroplast function again that is inhibited in spite of the low concentration of the inhibitor used at which level it would have no effect on mitochondrial function.

Another approach to the study of organelle interaction is to prevent one of the organelles from normal development. Using a chloroplast specific herbicide, 3-amino 1,2,4-triazole (17), we were able to derive such a system where it is possible to study the changes in the composition and function of mitochondria by the exclusive elimination of the chloroplast as a structure in these aplastic but otherwise normal cells. Though most of the tissues in a plant are having aplastic cells with normal mitochrondrial function, one can anticipate some alteration of their functions by the presence or absence of an additional energy transducing system, viz., chloroplast in the leaf tissue as the leaves were developmentally attuned to perform a lot of energy dependent processes other than CO_2 fixation like phloem loading and the active transport of Cl^- and other ions. However, from the data presented here it is seen that there is apparently not much change in the biochemical composition by the complete loss of chloroplasts. Also, there is no increase in either the number or function of the mitochondria in the aplastic cells, if we can take the level of ubiquinone as a marker.

In summary, our present studies do not seem to suggest a synergistic interaction between the chloroplasts and mitochondria either at the structural or functional level, although one would be tempted to hypothesize so. The fact that the basal cell function remains unaltered even when one of the organized organelle systems is either rendered deficient functionally or totally absent vouches for the above point.

SUMMARY

The effect of either chloroplast or mitochondrial specific electron transport and/or phosphorylation inhibitors on the cellular energy transducing systems was studied in intact leaf cells. The light dependent O_2 evolution and the CO_2 fixation of the whole cells were the first steps to be completely blocked by amytal, antimycin, oligomycin, potassium cyanide, rotenone and sodium azide. The mitochondrial electron transport and oxidative phosphorylation were only partially inhibited. The chloroplast specific inhibitor DCMU promoted mitochondrial O_2 exchange significantly without enhancing the phosophorylation. While ferricyanide partially restored the O_2 evolution, the addition of the terminal photo-oxidants like NADP, PGA, sodium acetate or sodium bicarbonate did not restore the O_2 evolution. Most of the inhibitors seem to interfere more with the absolute level of photophosphorylation than oxidative one in intact cells.

Using a chloroplast specific herbicide aminotriazole, aplastic but otherwise normal leaf cells were derived developmentally and their mitochondrial functions were studied. The absence of chloroplast did not appear to confer any special limitations on the leaf cells. Neither was there any increase in the number nor in the function of mitochondria to compensate for the lack of the other energy transducing organelle.

REFERENCES

1. GIBBS, M. (1962) in Physiology and Biochemistry of Algae (Lewin, R. A., ed.), p. 61, Academic Press, N. Y.
2. PACKER, L., MURAKAMI, S., AND MEHARD, C. W. (1970) Ann. Rev. Plant Physiol. 12, 271-304.
3. RAVEN, J. A. (1972) New Phytol. 71, 227-247.
4. PRICE, C. A., AND VALLEE, B. L. (1962) Plant Physiol. 37, 428-433.
5. GNANAM, A., AND KULANDAIVELU, G. (1969) Plant Physiol. 44, 1451-1456.
6. BASSHAM, J. A., AND CALVIN, M. (1957) in The Path of Carbon in Photosynthesis, Prentice-Hall, Inc., Englewood Cliffs, N.J.
7. NOBEL, P. S. (1967) Plant Physiol. 42, 1389-1394.
8. CRANE, F. L. (1959) Plant Physiol. 34, 128-131.
9. CRANE, F. L., AND BARR, R. (1971) Methods Enzymol. 18, 149-151.
10. GOODWIN, T. W. (1955) in Modern Methods of Plant Analysis (Paech, K., and Tracey, M. V., eds.), Vol. 3, p. 272, Springer, Heidelberg.
11. ABELL, L. L., BRODIE, B. B., AND KENDALL, F. E. (1952) J. Biol. Chem. 195, 357-366.
12. LEECH, R. M. (1966) in Biochemistry of Chloroplasts (Goodwin, T. W., ed.) Vol. 1, p. 65, Academic Press, N.Y.
13. HEALEY, F. P., AND MYERS, J. (1971) Plant Physiol. 47, 373-379.
14. WHATLEY, F. R., ALLEN, M. B., TREBST, A. V., AND ARNON, D. I. (1960) Plant Physiol. 35, 188.
15. SCHACTER, B., ELEY, J. H., JR., GIBBS, M. (1968) Plant Physiol. 43, S-30.
16. VAN DEN BERGH, S. G., MODDER, C. P., SOUVERIJN, J. H. M., AND PIERROT, C. J. M. (1969) in Mitochondria Structure and Function (Ernster, L., and Drahota, Z., eds.) Vol. 17, p. 137, Academic Press, N.Y.
17. VIVEKANANDAN, M. AND GNANAM, A. (1972) Biochem. J. 128, 55 p.

PART IV

DIFFERENTIATION IN MEMBRANES

Biomembranes show a complex hierarchy of structural and functional organization. The complexities can be shown to some degree by comparisons of the chemical, functional, and structural organization between membranes. For example, compare (a), the simple lipid bilayers produced by lung surfactin, the natural membranes across which gas exchange occurs, and which are almost devoid of protein, to membranes having greater amounts of protein, but little functional differentiation; to (b), the myelin membrane with little functional specialization ana approximately 20% protein composition; to (c), more complex membranes with greater protein concentration, such as those found in the retina or sarcoplasmic reticulum, where one protein dominates all others in occurrence and upon which the functional basis of this membrane rests, e.g., light reception in the visual process or the unidirectional calcium translocation; to (d), more complicated membranes where complex energy-transductions occur, such as those of the bacterial envelope membrane and inner membranes of mitochondria and chloroplasts where the ratio of protein to lipid composition is large and which possess complex functions and hundreds of different proteins. Differentiation in membranes takes many forms and involves the developmental process - it is to this latter context that several of the chapters selected in this section have been devoted. Differentiation in membranes is one of the most important areas to which future volumes should give greater emphasis.

GLYCOSYLTRANSFERASES LINKED TO SUBCELLULAR MEMBRANES OF THE SLIME MOLD DICTYOSTELIUM DISCOIDEUM

H. J. Risse, H. Rogge, M. Rath, G. Rothe, and F. Platzek

The eucaryotic cellular slime mold *Dictyostelium discoideum* is generally accepted as a suitable model for biological and biochemical investigations on a differentiating system. The microorganism begins its development as a vegetative amoeba growing on bacteria as substrate. After consumption of the substrate the differentiation of the cells is induced. Following a precisely regulated time program the cells develop the ability to aggregate and to form a multicellular pseudo-plasmodium. Further development leads to the formation of sporangia, which represent the nutrient-independent form of the slime mold (1,2).

During the aggregation phase two developmentally controlled events cooperate: (a) The cells migrate towards an aggregation center and form large cell clusters; this process is known to be produced by the chemotactic activity of cyclic 3'5'-AMP (3); (b) The cells form solid and specific intercellular linkages.

Gerisch and coworkers showed (4,5) that cell surface antigens are responsible for the increase in cell affinity. New cell surface antigens are synthesized during the transition from the vegetative to the aggregated state. The results from Lüderitz *et al.* (6) as well as Gerisch's data suggest that these antigens consist of glycoproteins. Our own investigations on glycosyltransferases in Dictyostelium started with the hypothesis that the enzyme systems which synthesize these glycoproteins should be induced or activated after the induction of differentiation.

This chapter presents some results on the localization of glycosyltransferases in subcellular fractions

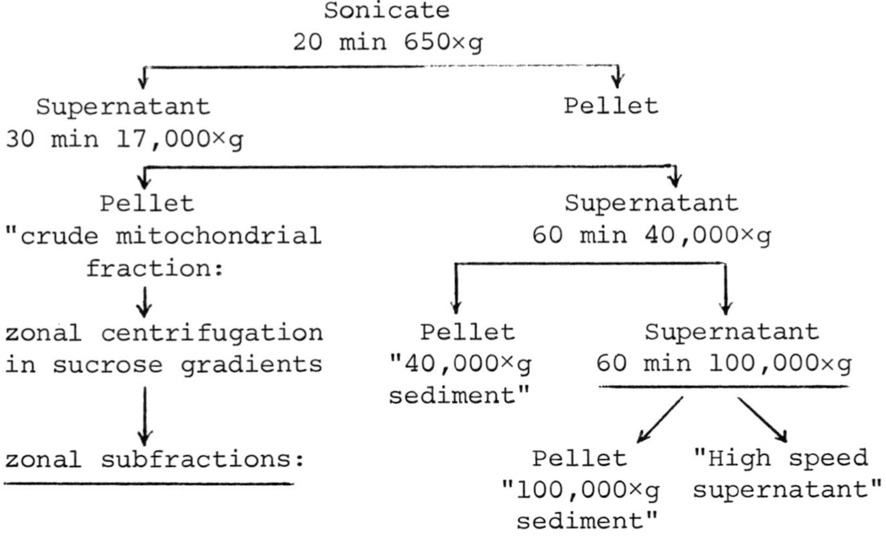

light endoplasmic fraction	0-15%	sucrose
heavy endoplasmic fraction	15-24%	"
mitochondria	37-42%	"
unidentified heavy fraction	47-50%	"

Fig. 1. Flow diagram of the fractionation of a Dictyostelium sonicate.

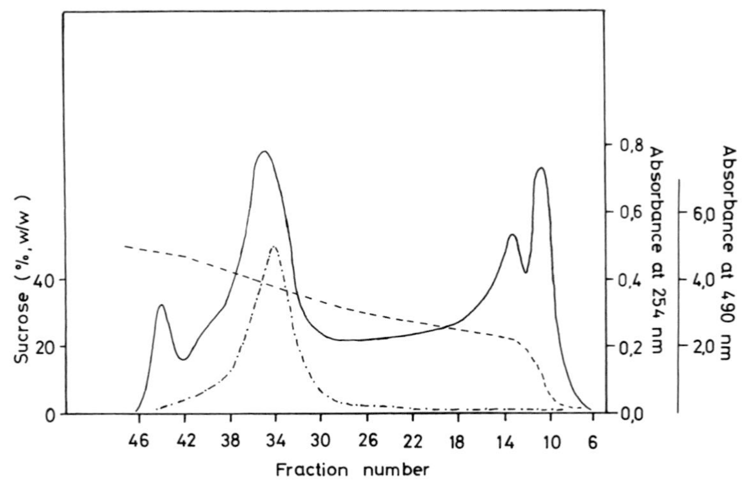

Fig. 2. (See legend on opposite page)

of Dictyostelium and their dependence on differentiation of the cells. The incorporation of the sugars D-mannose, L-fucose, D-glucose, and D-N-acetylglucosamine into high molecular weight products was examined. These sugars are constituents of the Dictyostelium glycoproteins.

All experiments were carried out with the "axenic mutant" described by Watts and Ashworth (7). The strain was cultivated on a synthetic medium with maltose as substrate. To induce the differentiation the cells were harvested under sterile conditions and transferred into nutrient-free phosphate buffer pH 6.0 as described in a previous paper (8). The cells were disrupted by very gentle sonication (Branson Sonifier B 12, output control 4, 4 sec) in 50 mM Tris/HCl pH 7.5 containing 10 mM glutathione (cell/buffer ratio 1/2 w/w). This treatment leaves about 20% of the cells intact. The mitochondria remain intact as shown by electron microscopy and by determination of the succinate reductase activity in a 17,000 ×g supernatant. The sonicate fractionation scheme for the separation of the subcellular particles is shown in Fig. 1.

The crude subcellular fractions were obtained by sedimenting the sonicate at 17,000×g, 40,000×g, and 100,000×g. The 17,000×g sediment was further fractionated by zonal centrifugation using a linear sucrose gradient from 20-50% sucrose (w/w). The effluents from the zonal centrifugation were measured at 254 nm wavelength in order to detect the particle containing fractions. Succinate reductase activity was used as a marker enzyme for mitochondria. A typical profile of a zonal separation of the crude 17,000×g fraction is shown in Fig. 2.

Fig. 2. Zonal centrifugation of the 17,000×g sediment in a linear sucrose gradient. ——— absorbance at 254 nm; -—- absorbance at 490 nm (succinate-reductase assay (14); ---- sucrose concentration (%, w/w). Centrifugation at 16,000 rpm, 20 min, fractions of 13 ml collected.

The electron-microscopic inspection of the zonal fractions confirmed that the mitochondria sediment quantitatively in the range between 35 and 42% sucrose (9). The fractions sedimenting slower than the mitochondria have in general the appearance of small vesicular membranes as usually found in preparations of the endoplasmatic reticulum. Ribosomal contamination of these fractions was minimal.

The fraction sedimenting faster than the mitochondria consists of ill-defined structures, the origin of which is not yet clear. Some analytical data suggest that it contains material from nuclei disintegrated during sonication.

The results of the glycosyltransferase assays in the fractions obtained by differential or zonal centrifugation are shown in Table I.

TABLE I

SPECIFIC ACTIVITIES OF GLYCOSYLTRANSFERASES IN SUBCELLULAR FRACTIONS OF *DICTYOSTELIUM*

Fraction	pMol Sugar/mg Protein × min			
	MAN	FUC	GLCNAC	GLC
Sonicate	2.6	0.110	0.37	0.94
Sediment 17,000 × g	2.9	0.130	0.38	0.21
Sediment 40,000 × g	5.5	0.275	2.14	4.13
Sediment 100,000 × g	4.9	0.230	2.90	3.01
Supernatant	less than 0.1 for all sugars			
Zonal fractions obtained from 17,000 × g sediment:				
Light endoplasmic fraction	37.0	0.229	n.d.	n.d.
Heavy endoplasmic fraction	39.8	0.226	1.19	1.24
Mitochondria	4.4	0.153	0.25	0.16

The values are given as pmoles sugar incorporated per mg protein and per minute.

The assays were performed by measuring the incorporation of radioactivity from labelled nucleotide sugars into trichloroacetic acid insoluble material. The assay mixtures were standardized from the concentrations of protein, substrate, divalent cations and for the pH. Replacing of the nucleotide sugar by the corresponding monosaccharide, the sugar phosphate, or replacing of the divalent cation abolished the sugar incorporation completely as shown in Table II. Moreover, the extraction of lipids from the active fractions destroyed the enzyme activity.

TABLE II
SUBSTRATE AND COFACTOR REQUIREMENTS FOR THE TRANSFER OF MANNOSE IN AN ENDOPLASMIC FRACTION FROM *DICTYOSTELIUM*

Complete mixture	49.2*
- GDP-mannose + mannose	0
- GDP-mannose + mannose-6-phosphate	0
- $MgCl_2$ + $MnSO_4$	0
+ $MnSO_4$ (10^{-3}M)	0
Protein preheated	0

*pmol × mg protein^{-1} × min^{-1}.

Assay composition as given in Fig. 3.

The assay of the fractions demonstrates a different subcellular distribution of the various glycosyltransferases. Mannosyltransferase is present mainly in the endoplasmic reticulum fraction that sediments at 17,000×g and was separated from the mitochondria by zonal centrifugation. The same fraction contains

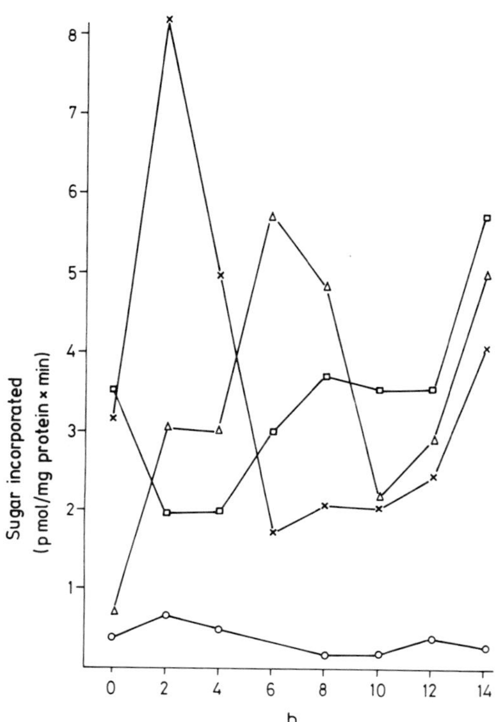

Fig. 3. Specific activities of the glycosyltransferases during the differentiation of the Dictyostelium cells. The abscissa indicates the time elapsed after transferring the amoebae into nutrient free buffer. -x- mannose-transfer; -o- fucose-transfer; -□- N-acetylglucosamine-transfer; -Δ- glucose-transfer. Composition of the samples: 0.012 M Tris/HCl pH 7.5 (0.06 M phosphate pH 6.6 f. glucose), 10^{-2} M divalent cations (Mg^{++} f. mannose and fucose, Mn^{++} f. N-acetylglucosamine, no extra addition f. glucose), 5-8 μM labeled nucleotide sugar (spec. activities approx. 200 mCi/mmol) in a total volume of 0.5 ml. Protein concentration 2-3 mg/sample. The aggregation of the cells begins between 8 and 10 h.

relatively high activity of fucosyltransferase, although fucosyltransferase activities in all fractions are much lower than the activities of the other glycosyltransferases. N-acetylglucosaminyltransferase, however, has a maximum activity in the fractions containing most of the ribosomal material, and only low activity in the mannosetransferase rich fraction. Also, most of the glucosyltransferase was found in the 40,000 and the 100,000×g sediments. No glycosyltransferase is present in the high speed supernatant.

Results obtained with a cyclic AMP phosphodiesterase linked to particles as a marker enzyme for cytoplasmic membrane (10) let us suppose that part of the mannosyltransferase-bearing membranes are derived from the cytoplasmic membrane. These data are, however, preliminary.

Figure 3 shows the specific activities of the glycosyltransferases during the development of the cells from the vegetative to the aggregating stage. The activities for mannose and fucosetransferase were assayed in the endoplasmatic reticulum fraction sedimenting with the mitochondria; the activities for glucose and N-acetylglucosamine transferase in the crude fraction sedimenting at 40,000×g.

Immediately after the induction of differentiation by transferring the cells into nutrient-free medium, a steep increase in the mannose as well as in the glucose transferase activity was observed. After having reached a maximum, the activities decrease until the aggregation period begins, and then rise again. N-acetylglucosamine transferase does not show the sharp increase in the first hours of differentiation but starts to increase in activity about 2 hrs before the aggregation. The developmental behavior of the fucose transferase is not sufficiently clear because the activity of this enzyme is very low during the entire period of development. It is evident that the transfer of glucose, mannose, and N-acetylglucosamine is controlled by the differentiation of the cells. However, the system has a limitation. The acceptors for our monosaccharides are endogenous, high molecular weight substances which are located in the same subcellular particles as the enzymes. Up to now, we have not been able to present suitable low molecular weight substances which could replace the endogenous acceptor in the transferase reaction, as described by Roth (12) for the cell surface linked galactose transferase in fibroblasts. Therefore the term "increased activity" means increased activity of the total transfer system. We are not yet able to distinguish between increased enzyme activity or increased acceptor concentration. A very important step would be to distinguish between transferases located in membranes of the endoplasmatic reticulum and in membranes of the cell surface. According to the hypothesis of Roseman (11) and according to the findings of Roth (9) and Bosmann (13), the glycosyltransferases of the cell surface could perform a direct role in cell contact formation. Experiments to clarify the presence of glycosyltransferases on purified cell membranes of *Dictyostelium* cells are in progress.

REFERENCES

1. BONNER, J. T. (1967) The Cellular Slime Molds, Princeton University Press.
2. SUSSMAN, M., AND LEE, F. (1955) *Proc. Nat. Acad. Sci.* *41*, 70.
3. KONIJN, T. M., VAN DE MEENE, J. G. C., CHANG, Y. Y., BARKLEY, D. S., AND BONNER, J. T. (1969) *J. Bacteriol.* *99*, 510.
4. GERISCH, G. (1970) Verhandlungsber. d. dtsch. zool. Gesellschaft, G. Fischer Verlag.
5. BEUG, H., GERISCH, G., KEMPFF, S., RIEDEL, V., AND CREMER, G. (1970) *Exp. Cell. Res.* *63*, 147.
6. GERISCH, G., MALCHOW, D., WILHELMS, H., AND LÜDERITZ, O. (1969) *European J. Biochem.* *9*, 229.
7. WATTS, D. J., AND ASHWORTH, J. M. (1970) *Biochem. J.* *119*, 171.
8. BAUER, R., RATH, M., AND RISSE, H. J. (1971) *European J. Biochem.* *21*, 179.
9. ROGGE, H., AND RISSE, H. J., unpublished results.

10. MALCHOW, D., NÄGELE, B., SCHWARZ, H., AND GERISCH, G. (1972) European J. Biochem. 28, 136.
11. ROSEMAN, S. (1970) in Chemistry and Physics of Lipids, North Holland Publ. Co., Amsterdam.
12. ROTH, S., AND WHITE, D. (1972) Proc. Nat. Acad. Sci. 69, 485.
13. BOSMANN, H. B. (1972) Biochem. Biophys. Acta 279, 456.
14. PENNINGTON, R. J. (1961) Biochem. J. 80, 649.

A POSSIBLE ROLE OF PERMEABILITY CONTROLS IN REGULATION OF CELL DIVISION

P. M. Bhargava

THE RATIONALE

Our interest in models of regulation of cell growth based on control of permeability derives from three basic considerations:

(*a*) All types of cells from higher organisms in which the phenomenon of malignancy is encountered are auxotrophic for certain nutrients (the "essential nutrients").

(*b*) The cells need to take up the essential nutrients from the environment, both for cell maintenance and for growth. In other words, these nutrients must be transported from the environment into the cell, across the cell membrane, in resting as well as in dividing cells.

(*c*) Available data on the rates of turnover, for example of macromolecules, in resting cells, and on the rates of growth of dividing cells of higher organisms, indicate that the rate of uptake of essential nutrients in growing cells must be many-fold higher than that in resting cells.

THE QUESTIONS ASKED

Two specific questions may now be asked:

(*a*) What is the order of difference between the rates of uptake of essential nutrients obtained in resting and in dividing cells of the higher organisms, and what is the nature of the change

---

Abbreviations: KRP buffer - Krebs Ringer original phosphate buffer; MEM - minimal essential medium of Eagle.

(qualitative and quantitative) in these rates during transition from the resting cell to the dividing cell stage and vice versa?

(b) If the change in the rates of uptake of essential nutrients occurring when a resting cell is triggered into cell division (and vice versa) is significant, could this change be the primary functional event caused by the "agent" triggering the division, which event leads the cell into the division cycle? In other words, can a viable model be constructed based on control of the above change in the rates of uptake, which would explain regulation of division of normal cells in higher organisms in a general way, and which would also, concurrently, provide a plausible explanation of the phenomenon of malignant transformation?

In this chapter, I should like to summarize some of the work which has been recently carried out by my colleagues and myself, partly at the Regional Research Laboratory, Hyderabad, and partly at the Institut du Radium, Orsay, which attempts to provide an answer to the first of the above two questions. Towards that end, I should like to very briefly discuss the rationale for the development of models attempting to explain regulation of cell division on the basis of control of permeability, and present the salient features of one such model.

CHANGES IN THE RATES OF UPTAKE OF ESSENTIAL AMINO ACIDS DURING TRANSITION FROM THE RESTING TO THE DIVIDING CELL STAGE AND VICE VERSA

We have attempted to gain information on these changes in five different model systems.

(a) Studies on synchronized cells. We have studied the uptake of several labelled essential amino acids during the first cell cycle of an asparagine-dependent strain of BHK cells (1), starved of asparagine for 24 hr and then initiated into synchronous cell division by transfer to an asparagine-containing medium (2). During asparagine starvation, the cells stayed viable, accumulated in early G_1 (or G_0) phase, and started into synchronous cell division on provision of asparagine. The first cell cycle following asparagine starvation could, therefore, be taken to represent a transition of the cell population from the resting cell stage to the dividing cell stage. The uptake of labelled thymidine began between 4 and 7 hr (no detectable uptake being observed in the first four hours) after initiation of the cells into the division cycle by replenishment of asparagine; DNA synthesis, as measured by the incorporation of labelled thymidine, started at 6-8 hr, reached a peak at 12-14 hr, and then fell, reaching a minimum value at 17 hr, just after which the number of cells in the culture increased two-fold. The first cell cycle, therefore, took about 17 hr; during this cell cycle, the total cellular protein increased generally (though not always) linearly, usually after a short initial lag. Figure 1 describes the above characteristics of the system observed in a typical experiment.

We found that the rate of total uptake of the amino acids, size of their intracellular free pool, and the rate of their incorporation into protein were, at the end of the first cell cycle, on an average, eight-fold of that at the beginning of the cell cycle. The increase in these parameters during the cell cycle was not linear but proceeded in spurts followed by periods during which there was no further increase. Typical results obtained with one amino acid (valine) are presented in Figs. 2, 3, and 4. Details of this investigation are described elsewhere (2).

(b) Control by serum of the uptake of essential amino acids by chick embryo fibroblasts in tissue culture. We have observed that the uptake of essential amino acids by growing chick embryo cells in tissue culture is drastically reduced 30-45 min after serum deprivation (ref. 3; representative results are given in Fig. 5A and Table I); this deprivation did not affect the uptake in the first 30-45 min, indicating that the drastic reduction in uptake after 30-45 min following removal of serum was a consequence of an

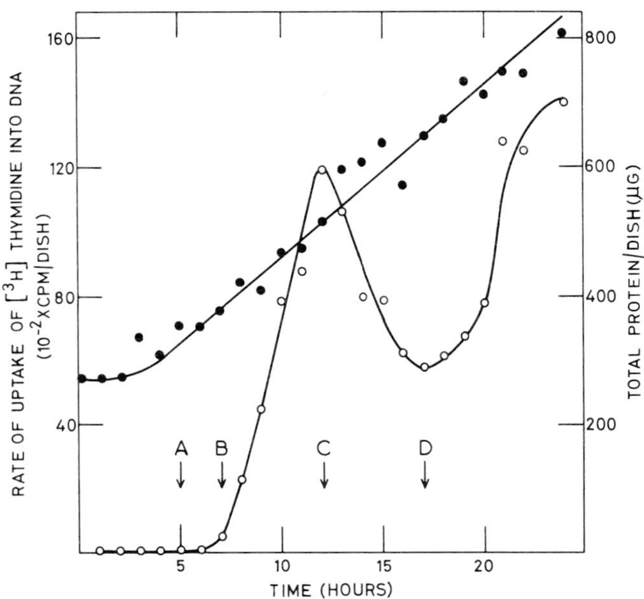

Fig. 1. Changes in the total protein content and in the rate of DNA synthesis (from [^{14}C]thymidine) during the first cell cycle of BHK cells grown synchronously. An asparagine-dependent strain of BHK cells was synchronized by maintenance in an asparagine-free tissue culture medium for 24 h at 37°; 0 h represents the time when synchronous growth of cells began after the "resting" period of 24 h, following resuspension of the cells in an asparagine-containing medium (the growth medium). A separate petri dish, into which 5 ml of the cell suspension (1.2×10^6 cells) was seeded, was used for each time point. At the specified time, a set of dishes was taken out, the cells were washed and incubated for 30 min at 37° in serum and asparagine-free growth medium containing [^{14}C]thymidine (300,000 cpm) and the ^{3}H-labelled amino acid. After incubation, the cells were washed and processed for estimation of total protein and of radioactivity in DNA. o, protein content; •, rate of DNA synthesis. For explanation of A, B, C and D, see the legend to Fig. 2. Control experiments showed that deprivation of serum for periods up to 30 min did not affect the uptake of amino acids.

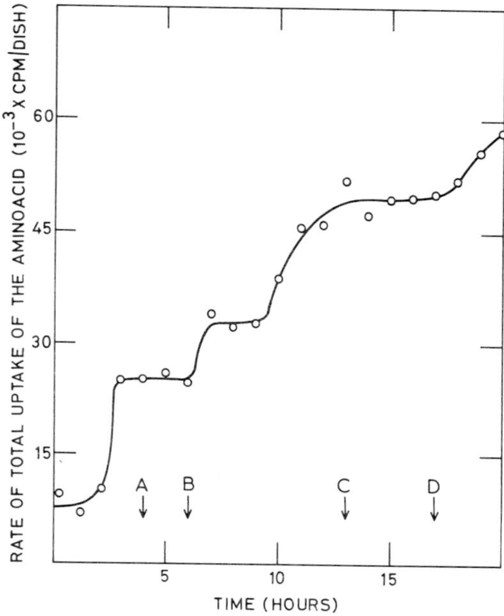

Fig. 2. Changes in the rate of the total uptake of valine during the first cell cycle of BHK cells. The cells, at the stated period, were incubated with the [^{3}H]valine (14.3×10^6 cpm) for 30 min, as described in the legend to Fig. 1. After incubation, the cells were washed and the distribution of the labelled amino acid between the acid-soluble and the acid-insoluble fractions determined as usual. The uptake values given are the total of the acid-soluble and the acid-insoluble fractions. A, B, C and D refer to the time points at which, respectively, the uptake of [^{14}C]thymidine began, the incorporation into DNA of the thymidine taken up began, the incorporation reached the peak, and this incorporation reached the lowest value after the peak.

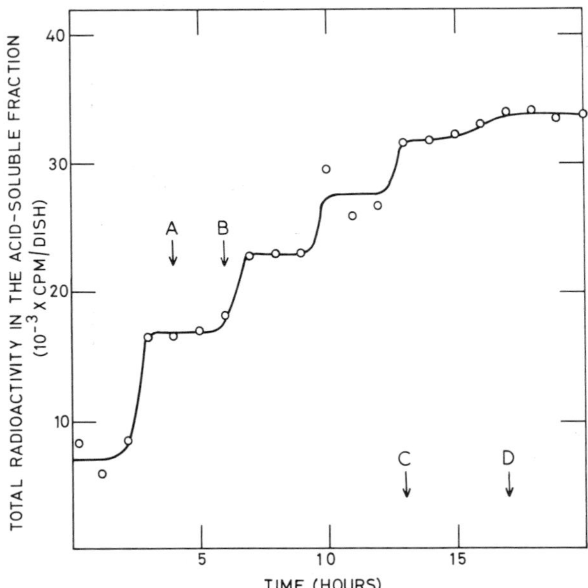

Fig. 3. Changes in the rate of accumulation of [^{3}H]-valine in the acid-soluble fraction during the first cell cycle of synchronized BHK cells. For details and explanations, see the legends to Figs. 1 and 2. Under the experimental conditions used, the externally added labelled amino acid was equilibrated with its intracellular pool, so that the data presented on the Y-axis could be taken to be a measure of this pool.

Fig. 4. Changes in the rate of incorporation of [^{3}H]-valine into the acid-insoluble fraction during the first cell cycle of synchronized BHK cells. For details and explanations, see the legends to Figs. 1 and 2.

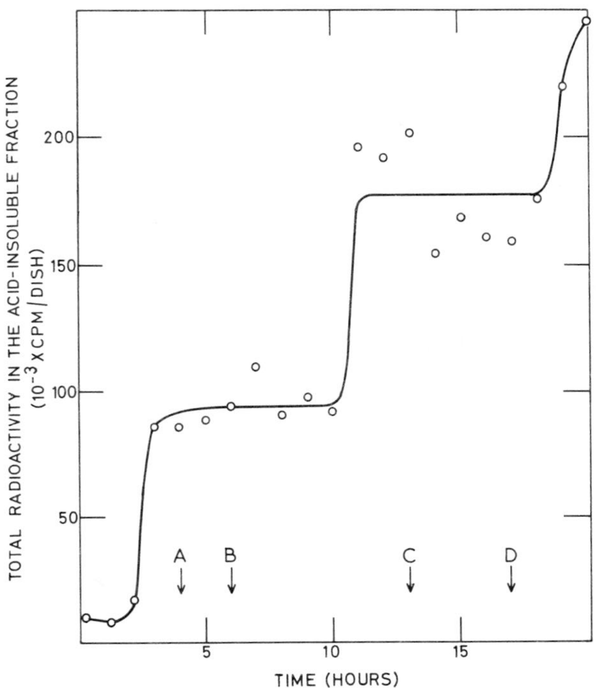

Fig. 4. (Legend on opposite page)

TABLE I

COMPARISON OF THE UPTAKE OF VALINE BY GROWING NONCONFLUENT CHICK EMBRYO CELLS AND BY THE NONGROWING HYPERCONFLUENT CELLS, IN THE PRESENCE AND IN THE ABSENCE OF SERUM[a]

^{3}H-amino acid	Nature of the culture	Medium of incubation	Time of incubation (min)	Uptake (cpm/mg protein) Acid-soluble	Acid-insoluble	Total (acid-soluble + acid-insoluble)
Valine	Growing (nonconfluent)	+ Serum	42	174700 (44.1)	221400 (55.9)	396100
			82	202300 (29.3)	487900 (70.7)	690200
			180	194500 (16.9)	954100 (83.1)	1149000
		- Serum	42	155800 (38.2)	252200 (61.8)	408000
			82	105200 (26.4)	293700 (73.6)	398900
			180	57540 (14.1)	351900 (85.9)	409400
	Nongrowing (hyperconfluent)	+ Serum	42	41350 (43.6)	53530 (56.4)	94880
			82	48450 (29.7)	114800 (70.3)	163300
		- Serum	42	39200 (42.8)	52330 (57.2)	91530
			82	32850 (31.4)	71650 (68.6)	104500
			180	24190 (23.9)	76890 (76.1)	101100

[a]Replicate cultures of normal chick embryo fibroblasts [95.0 ± 2.0 μg protein/dish in the case of growing (nonconfluent) cells, and 840.7 ± 24.3 μg protein/dish in the case of nongrowing hyperconfluent cells; this corresponded to about 10^6 and 10^7 cells, respectively] were washed with serum-free growth medium, and incubated in 2.5 ml of either the original (serum-containing) growth medium removed and stored before washing, or the serum-free growth medium, with [^{3}H]valine (50 μC), for the specified period; the monolayers were washed again and removed from the dish by trypsinization, and the uptake into the acid-soluble and acid-insoluble fractions estimated in the usual way. A separate petri dish was used for each time point and each cell density.

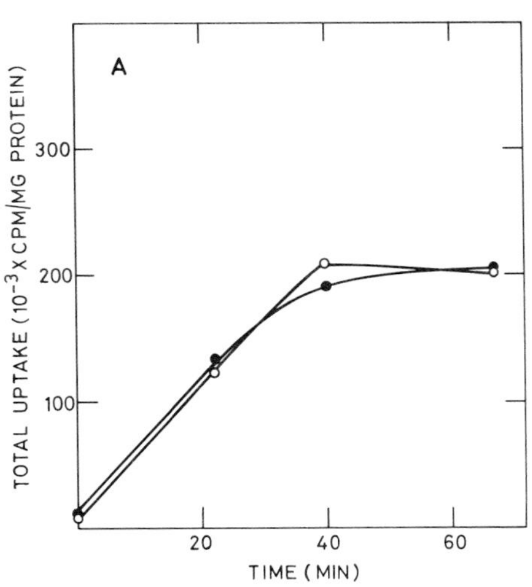

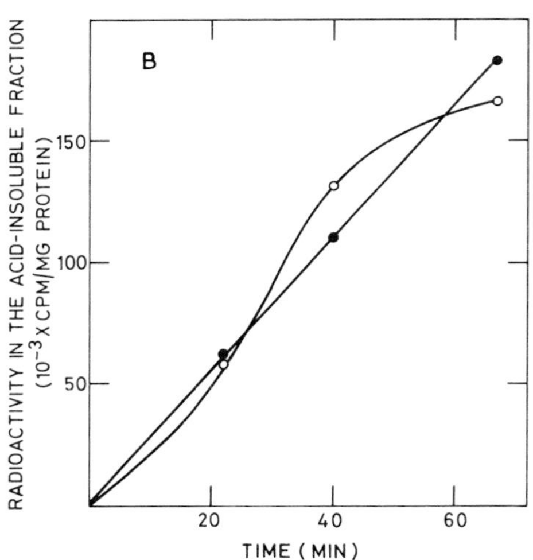

(Fig. 5A, 5B, above)

Fig. 5. The time-course of uptake of leucine by normal chick embryo fibroblasts, at low and high cell densities The cells, grown in the usual way in a serum-containing medium, were washed on the petri dish with serum-free (legend continues, following page)

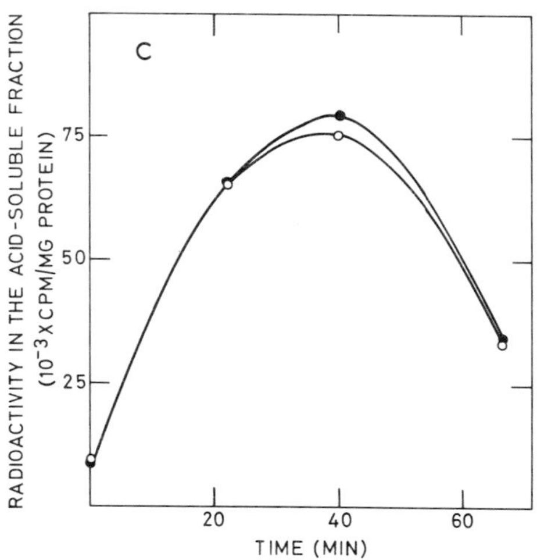

(Fig. 5C)

(Fig. 5 legend, continued:)
growth medium incubated for the specified time in the serum-free medium with 25 μCi of [^{3}H]leucine (the amount added was too small to alter significantly the concentration of the amino acid in the tissue culture medium), washed again, removed from the dish by trypsin, and processed for the estimation of the uptake (total uptake and that into the acid-soluble and acid-insoluble fractions taken separately) in the usual way. A separate petri dish was used for each time point and each cell density. The amount of protein per petri dish varied between 214 and 374 (mean 301.8 ± 12.7) μg for cells at low cell density (o), and 585 and 891 (mean 688 ± 20.5) μg for cells at high cell density (●). Low cell density: 3×10^6 cells; high cell density (just below the level of confluence required for complete cessation of cell division): $6\text{-}7 \times 10^6$ cells. A, total uptake (acid-soluble + acid-insoluble); B, radioactivity in the acid-insoluble fraction; C, radioactivity in the acid-soluble fraction.

event which took some time to occur after the commencement of serum-deprivation. This event did not seem to lead to any gross damage to intracellular metabolic machinery, as the incorporation of the amino acids taken up into protein continued at a significant rate after uptake stopped (Fig. 5B), with consequent (and equivalent) depletion of the acid-soluble pool (Fig. 5C) during this period. For growth of most cells in tissue culture, serum is known to be essential. These studies, therefore, strongly suggest that (i) serum controls, directly or indirectly, enhancement in the rate of uptake of essential amino acids that is obtained during growth of cells in tissue culture, over and above the rate of uptake in resting cells; and (ii) conditions under which the cells can no longer grow, e.g., the absence of serum, lead to a rapid reduction in the rate of uptake of these amino acids.

(c) Changes in the rate of uptake of essential amino acids during transition of growing nonconfluent cells in tissue culture to nongrowing hyperconfluent cells. Table I shows that when growing chick embryo cells reach a level of confluency when they can no longer divide, there is a several-fold reduction in the rate of transport of valine; the average reduction, for several amino acids, was about five-fold, both in the presence and in the absence of serum. This study (3) also, therefore, shows that transition from the dividing cell stage to the resting cell stage is accompanied by a drastic reduction in the rates of transport of essential amino acids.

(d) Comparison of the rates of uptake of essential amino acids obtained in the normal liver, a malignant solid hepatoma and the Zajdela ascitic hepatoma. We have found that the uptake of essential amino acids per mg acid-insoluble material in the Zajdela ascitic hepatoma cells, at 60-120 min, using extracellular amino acid concentrations normally required for growth is, on an average, 3-4 times more than in normal liver; the rates of uptake obtained in a dimethylaminoazobenzene-induced solid hepatoma were in between those observed for normal liver and the Zajdela cells (4). The results for one amino acid, arginine, are given in Table II. We also calculated the ratio (I/E) of intracellular concentration of the amino acid taken up to its extracellular concentration.

TABLE II
A COMPARISON OF THE IN VITRO UPTAKE OF ARGININE BY NORMAL LIVER SLICES, SLICES FROM A MALIGNANT SOLID HEPATOMA, AND ZAJDELA ASCITIC HEPATOMA CELLS[a]

Tissue or cell type	[^{3}H]Arginine taken up (nmoles/mg TCA ppt.)								
	Acid-soluble fraction			Acid-insoluble fraction			Total (acid-soluble + acid-insoluble)		
	60 min	120 min	180 min	60 min	120 min	180 min	60 min	120 min	180 min
Normal liver	1.34	1.88	2.73	0.02	0.13	0.13	1.36	2.01	2.86
Solid hepatoma	1.13	2.04	3.48	0.08	0.32	0.33	1.21	2.36	3.81
Zajdela cells	4.53	4.53	4.22	0.34	0.90	0.87	4.87	5.43	5.09

[a]Slices (210-270 mg. wet wt.) from normal liver or a malignant solid hepatoma, or the Zajdela cells (69 × 10^6), were incubated in 3 ml of a tissue culture medium with high specific activity [^{3}H]arginine (47.3 × 10^6 cpm) for the stated period; the final concentration of arginine in the incubation medium was 1.2 μmoles/ml. At the desired time, the slices or the cells were washed with Dulbecco's phosphate buffered saline until they were free of the radioactive incubation medium, treated with TCA, and the radioactivity in the acid-soluble and acid-insoluble fractions estimated in the usual way.

TABLE III
COMPARISON OF THE ABILITIES OF NORMAL LIVER PARENCHYMAL CELLS, PARENCHYMAL CELLS FROM A MALIGNANT SOLID HEPATOMA, AND ZAJDELA ASCITIC HEPATOMA CELLS, TO CONCENTRATE AMINO ACIDS[a]

Amino acid	Tissue or cell type	External concentration of the labeled amino acid (nmoles/ml incubation medium)	Internal concentration of the labeled amino acid (nmoles/ml cellular volume)			Internal concentration / External concentration (I/E)		
			60 min	120 min	180 min	60 min	120 min	180 min
Leucine	Normal liver	800	409	436	522	0.6	0.6	0.8
	Solid hepatoma	800	884	1675	1649	1.2	2.7	2.6
	Zajdela ascitic hepatoma	800	3366	7584	6728	4.8	13.3	11.7
Valine	Normal liver	800	375	478	785	0.5	0.7	1.3
	Solid hepatoma	800	1288	2124	2170	1.9	3.5	3.6
	Zajdela ascitic hepatoma	800	5776	8482	11260	8.7	15.1	21.5
Lysine	Normal liver	800	787	1130	1224	1.3	2.1	2.3
	Solid hepatoma	800	1403	3812	3924	2.1	9.0	10.4
	Zajdela ascitic hepatoma	800	8044	8522	8358	13.1	15.3	14.9
Arginine	Normal liver	1200	313	438	636	0.3	0.4	0.6
	Solid hepatoma	1200	660	1186	2024	0.6	1.1	2.0
	Zajdela ascitic hepatoma	1200	4746	4744	4418	4.3	4.4	4.1

[a]The values for internal concentration were calculated from the data for the acid-soluble fraction obtained as in Table II, the average volume of parenchymal cells in normal liver and in the solid hepatoma (taken to be 10800 μ^3 in either case) and of Zajdela cells (900 μ^3), the amount of cells/slices used, and the number of parenchymal cells per unit weight of normal liver and of the solid hepatoma (taken to be 81 × 10^6 and 32.4 × 10^6 cells/g wet wt., respectively).

at various time points, accounting for the lower cellularity in the case of the solid hepatoma. For the Zajdela cells this ratio was 7.2-21.0 times (average 13.8 times) more than for normal liver; the values for the solid hepatoma were in between those for normal liver and the Zajdela cells (Table III). The results obtained in this study (4) also suggested that the rate of total uptake of an amino acid, its I/E ratio, proportion of the amino acid taken up utilized for protein synthesis, the rate of incorporation of the amino acid into protein, and the ability of the cells to grow, may be correlated; all of these parameters were many-fold higher for the rapidly dividing malignant liver cells (the Zajdela cells) than for resting (normal) liver cells.

(e) Changes in the rates of uptake of essential amino acids following dispersion of liver to a single cell suspension. Dispersion of a tissue to a single cell suspension is commonly used to prepare cellular material for monolayer or suspension tissue culture; the dispersed cells grow until a high cell density is reached in the culture when they stop growing even though nutrients are not limiting. In such cases, the rate of division of cells in the organized tissue is often much slower than that obtained for the dispersed cells growing in tissue culture. It would, therefore, appear that dispersion of the tissue to a single cell suspension leads to creation of conditions favorable for growth. Dispersion of a non-rapidly dividing tissue to a cell suspension may, therefore, be taken to be analogous to a transition from the resting cell stage to the dividing cell stage. For example, although in adult mammalian liver the mitotic activity is very low, rapidly growing cultures of adult parenchymal cells have been reported to arise from primary liver cell suspensions (5). We have, therefore, compared the uptake of essential amino acids by liver cells in suspension with that obtained in undispersed adult liver tissue. We have found that the rate of uptake of these amino acids by liver cells in suspension (prepared by a modification of the method of

Jacob and Bhargava, ref. 5a) at low cell concentrations, is several fold higher than that obtained in liver slices (Figs. 6 and 7; ref. 6).

We have also observed that the rate of uptake in the cell suspension is dependent on cell concentration; the rate decreased with increasing cell concentration, until it reached the value obtained in intact liver tissue, beyond which there was no further decrease on increasing the concentration of cells (Figs. 6 and 7; Table IV). These observations suggest that creation of conditions which favor division of cells leads

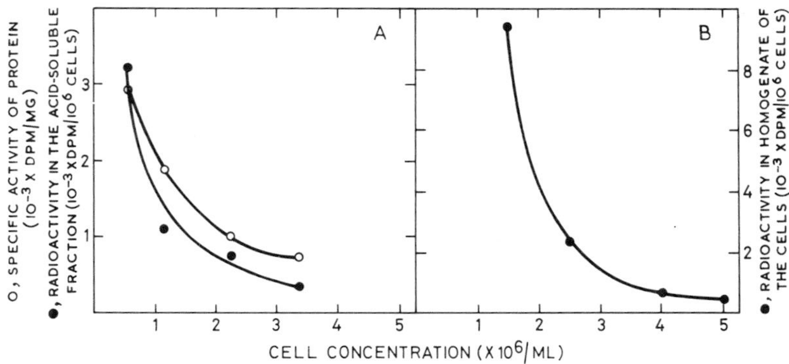

Fig. 6. Effect of cell concentration of the transport of [^{14}C]phenylalanine and [^{14}C]leucine in a liver parenchymal cell suspension. The cells were incubated (A) in 4.0 ml of Ca^{2+}-free KRP buffer for 3 h at 37° with [^{14}C]phenylalanine [0.303 μmoles (11.1×10^5 dpm)/ml]; or (B) in 5.0 ml of Ca^{2+}-free KRP buffer for 2 h at 37° with [^{14}C]leucine [0.382 μmoles (11.1×10^5 dpm)/ml]. The incubation mixture was separated into the cell and the medium fractions by centrifugation. In A, the accumulation of the labelled amino acid in the free amino acid pool of the cells and its incorporation into protein were separately estimated; in B, the total transport was estimated by measurement of radioactivity in a homogenate of the washed cells. Similar results were obtained when the incubation medium was a tissue culture medium (cf. Table IV).

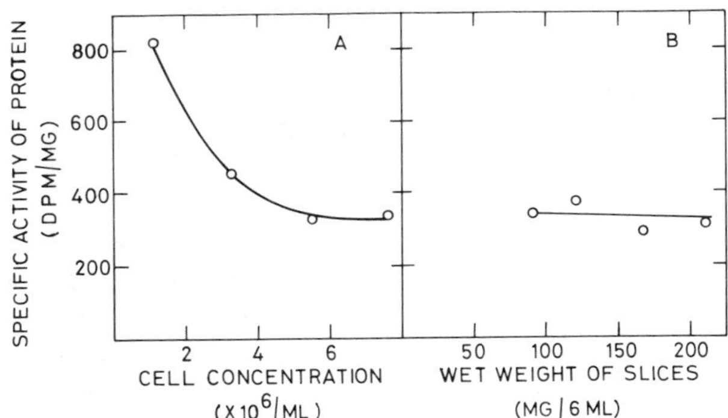

Fig. 7. Effect of cell/tissue concentration on the incorporation of [^{14}C]histidine into protein in (A) liver parenchymal cell suspensions, or (B) perfused liver slices. The cells or the tissue slices were incubated in 6.0 ml of Ca^{2+}-free buffer for 3 h at 37° with [^{14}C]histidine [5.6 nmoles (4.4×10^5 dpm)/ml] and the radioactivity incorporated into cell/tissue protein estimated as usual. Similar results were obtained when the incubation medium was a tissue culture medium.

to a several-fold increase in the rate of uptake of essential amino acids. On the other hand, when the same cells are maintained under conditions which do not favor growth, such as high cell density, the rate of uptake of the amino acids is reduced to a level obtained in resting cells (Fig. 7).

We have also investigated the mechanism of the above-mentioned cell concentration effect (6; P. M. Bhargava and K. S. N. Prasad, unpublished). We found that when cells at low concentration were incubated in a conditioned medium obtained from the high concentration cells, the rate of uptake in the low concentration cells was about the same as that obtained in the high concentration cells; the high concentration cells seemed to secrete or leak out into the medium of incubation a substance (or substances) which reduced the enhanced uptake in the low concentration cells to the level obtained in the high concentration

TABLE IV
EFFECT OF CONDITIONED MEDIA ON THE TOTAL UPTAKE OF LABELLED AMINO ACIDS BY LIVER CELL SUSPENSIONS AT "LOW" AND AT "HIGH" CELL CONCENTRATIONS[a]

Container(s)	Medium of incubation	Total uptake of the amino acid (dpm/mg protein)					
		Experiment 1		Experiment 2		Experiment 3	
		0.5×10^6 cells/ml	5.0×10^6 cells/ml	0.5×10^6 cells/ml	5.0×10^6 cells/ml	0.5×10^6 cells/ml	5.0×10^6 cells/ml
Separate	KRP buffer/MEM	28500	7610	37880	13470	26020	7910
Separate	Conditioned medium from 0.5×10^6 cells/ml	24130	6570	23510	13810	13060	7820
Separate	Conditioned medium from 5.0×10^6 cells/ml	15930	6680	9190	13240	7900	6760
Separate	Pronase-treated conditioned medium from 0.5×10^6 cells/ml	--	--	20830	--	--	--
Two-compartment chamber	KRP buffer	--	--	15790	13950	--	--

[a]The conditioned media were prepared by incubating suspensions of rat liver parenchymal cells at two different concentrations (0.5×10^6 and 5.0×10^6 cells/ml) in Ca^{2+}-free KRP buffer (Expts. 1 and 2) or MEM (Expt. 3), at 37° for 1.5 hr. The cells were removed by centrifugation; the supernatant was dialyzed at 4° for 8 hr against 5 changes of the KRP buffer in Expts. 1 and 2 only. In Expts. 2 and 3, portions of the conditioned medium were treated with Enzit-pronase. A fresh batch of cells in suspension was then prepared in KRP buffer (Expts. 1 and 2) or MEM (Expt. 3); an appropriate volume containing the required number of cells was dispensed in a series of centrifuge tubes and sedimented at 200 × *g*. The supernatant was decanted and the cells suspended in the required medium (Ca^{2+}-free KRP buffer, MEM, conditioned medium from 0.5×10^6 cells/ml or 5.0×10^6 cells/ml, or pronase-treated conditioned medium). The cell suspension was then transferred quantitatively to 25 or 50 ml Erlenmeyer flasks (separate but same-size containers were used for the two cell concentrations), or to a two-compartment cell (the two compartments were separated by a 1.0 micron Millipore filter; one compartment contained 0.5×10^6 cells/ml and the other 5.0×10^6 cells/ml). A mixture of [^3H]-labeled amino acids (phenylalanine, histidine, and lysine) in 0.2 ml of the KRP buffer (or water) was then added and the cells incubated for 1.5 h at 37°. The total incubation volume was, for separate containers, 10 ml in Expts. 1 and 3, and 5 ml in Expt. 2, and for the two-compartment cell, 15 ml in each compartment. The final concentration of the labelled amino acids in Expts. 1 and 2 was: phenylalanine, 1.15 pmoles (7.4×10^6 dpm)/ml; histidine, 1.01 pmoles (7.4×10^6 dpm)/ml; and lysine, 2.22 pmoles (7.4×10^6 dpm)/ml. In Expt. 3, the final concentrations were phenylalanine, 0.2 μmoles (7.4×10^6 dpm)/ml; histidine, 0.2 μmoles (7.4×10^6 dpm)/ml; and lysine, 0.4 μmoles (7.4×10^6 dpm)/ml. After incubation, the cells were washed and dissolved in 2-4 ml of 0.6 N NaOH for determination of the total uptake.

cells or in the organized tissue. This substance was not dialysable and was sensitive to heat and pronase treatment, suggesting that it was proteinoid (Table IV). In a two compartment cell in which one compartment contained cells at high concentration and the other contained cells at low concentration, the two compartments being separated by a Millipore filter which would allow such a substance but not cells to diffuse from one compartment to the other, the uptake in the compartment containing the low concentration cells was the same as that in the high concentration cells (Table IV). These observations indicate the presence in normal adult liver of a material which interferes with any enhancement in the rate of uptake of essential amino acids over and above that obtained in resting liver cells. Liver cells can apparently synthesize/secrete this material *in vitro*, and a critical concentration of it in the extracellular environment seems to be necessary for it to be active, as only the conditioned medium from the high concentration cells showed the transport-inhibitory activity. We have also shown the presence of this activity in the "extracellular material fraction" (7) obtained following dispersion of adult rat liver to a single cell suspension. This further supports the view that the low-level uptake obtained in resting cells of adult liver is a consequence of the presence of this material on the cell surface; dissociation of cells by the mechanical method we used could release this material from the cell surface and lead to its recovery in the extracellular material fraction.

Conclusions. The above observations suggest the following:

(i) The rate of uptake of essential amino acids in dividing mammalian cells is several-fold higher than that in resting mammalian cells.

(ii) Transition of mammalian cells from the resting cell stage to the dividing cell stage leads to a several-fold increase in the rate of uptake of essential amino acids.

(iii) Conditions which favor growth also favor high rates of uptake of amino acids, and those which do not favor growth allow the amino acids to enter the cells only at relatively low rates equal to those obtained in resting cells.

(iv) The reduced rate of uptake in resting cells appears to be a consequence of a negative control of permeability exercised by a substance/substances (*I*), probably proteinoid in nature, which acts on the cell surface. *I* can prevent the rise in the rate of uptake over and above that obtained in resting cells, but does not affect the low-level uptake obviously necessary for maintenance of the resting cells; this, in turn, suggests that there may be two different systems of uptake of essential amino acids, one which is operative in resting cells (and may also be operative in dividing cells) and the other which is operative only in dividing cells, and is susceptible to a negative control by I.

MODELS ATTEMPTING TO EXPLAIN REGULATION OF CELL DIVISION ON THE BASIS OF CONTROL OF PERMEABILITY

Although the above-mentioned investigations and related studies of other workers do not establish that an increase in the rates of uptake of essential nutrients is the first and primary functional event which occurs when a resting cell is triggered into division, they do suggest this possibility. This view is supported by the observation that following triggering of cells into the division cycle in systems other than those studied by us as well (for example, the triggering of lymphocytes by plant agglutinins, of liver cells by partial hepatectomy, and of a variety of cells by various hormones), the earliest observable change appears to be a substantial increase in the rate of uptake of essential nutrients (8-19).

One advantage in assuming that regulation of cell division occurs through control of uptake of essential nutrients and that enhancement in the rate of this uptake can trigger a programmed set of events which could take the cell through the division cycle, is that it allows one to construct relatively simple models which would not only explain the general characteristics (from the point of view of regulation of growth) of the various states in which cells of higher organisms are found (such as embryonic, resting, dividing, and malignant), but also provide a mechanistic basis for transition from one of these states to the other, such as embryonic to adult, normal adult resting to normal adult dividing, normal adult dividing to normal adult resting, normal to malignant, and malignant to normal. Further, since all cell systems where such types of growth regulatory controls are involved are auxotrophic for a variety of carbon-containing nutrients, and since the rate of uptake required to be achieved before the cells may go through the division cycle appears to be about an order of magnitude higher than that required for cell maintenance, models of growth regulation based on control of uptake of essential nutrients would, perhaps, represent the _simplest_ set of such models. Models of this type could, therefore, provide a specially useful starting point for further investigation, as long as they are definitive enough to allow clear-cut, experimentally testable predictions to be made.

We have been engaged in constructing such models and I should now like to present the salient features of one of them.

The model. The model, schematically presented in Fig. 8, postulates four chemical (_I_, Anti-_I_, SF1, and SF2) and two structural (Sites A and Sites B) entities. Sites A and B are transport sites on the membrane for essential nutrients; there may be a different set of Sites A and B for different nutrients. Sites A are open in resting cells and need a serum factor, SF1, for operation at V_{max}. Sites B are closed in resting cells but open in dividing cells; this control is achieved through a substance _I_ - probably a conjugated, bivalent protein with a high rate of turnover, a high degree of histogenetic specificity (so that _I_ for one cell type in the organism would not act on other cell types in the same organism), but a relatively low affinity for Sites B (see later) - which, when bound to Sites B, closes these sites. Removal of _I_ from Sites B would, therefore, result in an increased uptake of nutrients which could lead to a programmed operation of the events which take the cell through the division cycle and which culminate in cell division. One can postulate several possible logics for this programme [for example, the size and profile of the amino acid pool, which change in a definitive manner during the cell cycle (2), may exercise a control on the nature of the proteins synthesized].

A second serum factor, SF2 [which would explain the need for serum in tissue culture of normal cells,

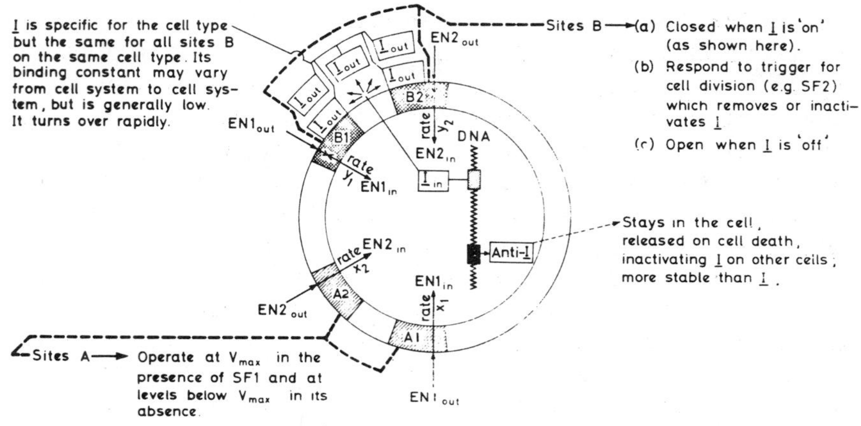

MALIGNANT TRANSFORMATION = INHERITABLE INTRACELLULAR EVENT WHICH INTERFERES WITH THE PRODUCTION OR ACTIVITY OF I

Fig. 8. A model for the regulation of cell division through regulation of permeability to essential nutrients. EN1 and EN2, essential nutrients; Sites A1 and A2, membrane sites for the transport of various ENs in resting cells (the model does not preclude their being open in dividing cells); Sites B1 and B2, membrane sites for the transport of ENs in dividing cells; rate x, the maximal rate of transport of an EN through a Site A; rate y, the maximal rate of transport of an EN through a Site B, probably ranging between 5x and 10x (see text); SF1, serum factor necessary for transport of an EN through Site A at the maximal rate; I, an inhibitor of transport through Sites B, which comes out of the cell and acts from outside. Anti-I, an intracellular factor, antagonistic to I and normally incapable of coming out of the cell; SF2, a factor present in serum, antagonistic to I.

and the control of permeability described above by serum in such cell cultures (3)], and other external triggers for cell division such as certain hormones, are postulated to destroy or inactivate I or prevent its binding to Sites B, thus opening up Sites B and enhancing the uptake of essential nutrients required by the model to dispatch a cell through the division cycle. A similar effect may be obtained by the release of another factor, Anti-*I*, from the cells, e.g., as a result of shock or tissue damage as in partial hepatectomy; Anti-*I*, like *I* but unlike SF2 (and SF1) which it would resemble functionally, is specific for the cell type from which it is derived.

In the model, growth ceases as a consequence of re-establishment of *I* function on the membrane. For example, cessation of growth in high density, confluent cultures could be a consequence of synthesis of enough *I* by the large number of cells now present to nullify the effect of SF2.

Malignant transformation is defined as an interference at the genetic level with the production or activity of *I*, so that the capacity for transition from the dividing state to the resting state is lost. Such an interference could occur in several ways. In the case of chemical carcinogens, this may be due to a mutation caused by them in the gene for *I* or for one of the components of the site on the membrane to which *I* is bound. In the case of viral carcinogens, the oncogenic virus may code for a product which, among other possibilities, may interfere with the binding of *I* to Sites B. Chemical carcinogens have been shown to bind DNA (20-22) and act as mutagens (23,24), and oncogenic viruses appear to code for products which migrate to the cell surface (25-27). In malignant cells, therefore, the uptake would be constitutively high (4) and the cells will, consequently, always be on the move *towards division* through the cell cycle, sometimes slowly but always surely, as appears to be the case. The malignant cells would also show surface changes due to exposure or modification of Sites B and removal of *I* from the cell surface. Such changes, implicit in the model, could explain the known differences - immunological (*inter alia*, 25, 28, 29) or other, such as certain special features of interaction with Con A (*inter alia*, 30) - in the surface properties of resting normal and malignant cells. Malignant cells, with Sites B open constitutively, would also be expected not to need SF2 for growth in tissue culture. The lower requirement of serum for growth of malignant cells in culture is well known; the small amount of serum required by them may be needed to provide SF1 to keep Sites A open for the uptake of certain essential trace nutrients, the rate of uptake of which may not be altered following triggering of cells into the division cycle and which, therefore, may not need Sites B. Reversal of malignancy (that is, transition of malignant cells to normal cells), e.g., by somatic cell hybridization of malignant cells with normal cells (31-44), would imply re-establishment of I function in the hybrid. I^+ (normal) phenotype would be dominant in the hybrid if malignancy was a result of deletion of I function, as may be the case, as mentioned above, in chemical carcinogenesis, but not in viral carcinogenesis. This view is supported by the reported ability of normal cells to inhibit the growth of malignant cells in tissue culture (45).

The suggestion that I may be a conjugated protein is based on our observations on liver cells briefly described above, and the reported ability of several types of enzymes, such as proteases and hyluronidase, to initiate non-growing cells (e.g., in a confluent culture) into cell division (46-52) and to increase cellular permeability (53,54). The assumption that I is bivalent allows one to propose that I has a low affinity for Sites B when only one of these valencies is satisfied (as is likely in cell suspensions or in low density cell culture where, therefore, one would need relatively high concentrations of I in the medium to block all Sites B, and where washing the cells would remove I from Sites B relatively easily), but a high affinity for Sites B when both the valencies are satisfied [as would happen in adult tissues and in high density, monolayer cell cultures, from both of which, therefore, I will not be removed by washing but could be removed by separation (dispersion) of the cells].

The model, it will be seen, provides a basis for the transition of normal resting cells to normal dividing cells and of normal cells to malignant cells, and *vice versa*, and a plausible explanation of the nature of growth control obtained in these cells. In embryonic tissues, among other possibilities suggested by the model, the gene for *I* may be turned off, or the concentration of SF2 in the cellular environment may be high (it is known that growth promoting activity of foetal sera is higher than that of adult sera). The transition from the embryonic to the adult state may represent a progressive activation of the *I* gene or inactivation of the gene for SF2.

Concluding remarks. We are at present attempting to purify, characterize and/or identify the six entities postulated in the above model (*I*, Anti-*I*, SF1, SF2, Sites A and Sites B). The material with *I*-type activity obtained from liver that we have described above may be a good candidate for *I*. Such materials may be responsible for the chalone-type (55-59) growth and transport inhibitory activity of normal tissue extracts; purified preparations of *I* should be able to inhibit the uptake of essential nutrients in homologous tumor cells, and thereby inhibit their growth. Tissue-specific antigens may indeed represent another possible group of contenders for *I*.

The growth stimulatory factor specific for liver cells, shown to be present in the plasma of partially hepatectomized animals (60), would seem to be a possible candidate for Anti-*I*; another would be the factor secreted by certain serum-independent mutant cell lines, which can replace serum for growth of the parent, serum-dependent cells (61).

The model suggests precise assays for SF1 and SF2 which should assist in their isolation if they exist; they could be identical with some of the growth and transport stimulatory factors already described in serum (*inter alia*, 62-64).

Isolation and characterization of *I*, Anti-*I*, SF1 and SF2 should make it possible to locate Sites A and Sites B.

REFERENCES

1. MONTAGNIER, L., GRUEST, J., AND BOCCARA, M. (1971) Colloques Internationaux C.N.R.S. la l-asparaginase No. 197, 159.
2. BHARGAVA, P. M., ALLIN, P. A., AND MONTAGNIER, L. J. Membrane Biol., communicated.
3. BHARGAVA, P. M., AND VIGIER, P. J. Membrane Biol., communicated.
4. BHARGAVA, P. M., SZAFARZ, D., BORNECQUE, C., AND ZAJDELA, F. J. Membrane Biol., communicated.
5. IYPE, P. T. (1971) J. Cellul. Physiol. 78, 281.

5a. JACOB, S. T., AND BHARGAVA, P. M. (1962) Exp. Cell Res. 27, 453.

6. BHARGAVA, P. M., SIDDIQUI, M. A., KUMAR, G. K., AND PRASAD, K. S. N. J. Membrane Biol., communicated.
7. IYPE, P. T., BHARGAVA, P. M., AND TASKER (1965) Exp. Cell Res. 40, 233.
8. WHITNEY, R. B., AND SUTHERLAND, R. M. (1973) Biochim. Biophys. Acta 298, 790.
9. KAY, J. E. (1972) Exp. Cell Res. 71, 245.
10. AVERDUNK, R. (1972) Hoppe-Seyler Zeit. Physiol. Chem. 353, 79.
11. VANDENBERG, K. J., AND BETEL, I. (1973) FEBS Letters 29, 149.
12. FERRIS, G. M., AND CLARK, J. B. (1972). Biochim. Biophys. Acta 273, 73.
13. ORD, M. G., AND STOCKEN, L. A. (1972) Biochem. J. 129, 175.
14. SMITH, D. M., AND SMITH, A. E. S. (1971) Biol. Reprod. 4, 66.
15. RIGGS, T. R., AND PAN, M. W. (1972) Biochem. J. 128, 19.
16. HJALMARSON, A., AND AHREN, K. (1965) Life Sci. 4, 863.
17. ZIBOH, V. A., WRIGHT, R., AND HSIA, S. L. (1971) Arch. Biochem. Biophys. 146, 93.
18. GUIDOTTI, G. G., BORGHETTI, A. F., LUNEBURG, B., AND GAZZOLA, G. C. (1971) Biochem. J. 122, 409.

19. GOLDFINE, I. D., GARDNER, J. D., AND NEVILLE, D. M. (1972) J. Biol. Chem. 247, 6919.
20. MILLER, E. C., AND MILLER, J. A. (1966) Pharmacol. Rev. 18, 805.
21. FARBER, E. (1968) Cancer Res. 28, 1859.
22. BHARGAVA, P. M. (1970) in Control Processes in Multicellular Organisms (Wolstenholme, G. E. W., and Knight, J., eds.), p. 158, Churchill, London.
23. GROVER, P. L., COOKSON, M. J., AND SIMS, P. (1971) Nature New Biology 234, 186.
24. AMES, B. N., GURNEY, E. G., MILLER, J. A., AND BARTSCH, H. (1972) Proc. Natl. Acad. Sci. U.S.A. 69, 3128.
25. BURGER, M. M. (1971) in Current Topics in Cell Regulation (Horecker, B. L., and Stadtman, E. R., eds.), Vol. 3, p. 135, Academic Press, N.Y.
26. LENGEROVA, A. (1972) Adv. Cancer Res. 16, 235.
27. ECKHART, W. (1972) Ann. Rev. Biochem. 41, 503.
28. AOKI, T., STEPHENSON, J. R., AND AARONSON, S. A. (1973) Proc. Natl. Acad. Sci. U.S.A. 70, 742.
29. MUKHERJI, B., AND HIRSHAUT, Y. (1973) Science 181, 440.
30. NOONAN, K. D., AND BURGER, M. M. (1973) J. Cell Biol. 59, 134.
31. WEISS, M. C., TODARO, G. J., AND GREEN, H. (1968) J. Cellul. Physiol. 71, 105.
32. HARRIS, H., MILLER, O. J., KLEIN, G., WORST, P., AND TACHIBANA, T. (1969) Nature 223, 363.
33. EPHRUSSI, B., DAVIDSON, R. L., AND WEISS, M. C. (1969) Nature 224, 1314.
34. BARSKI. G. (1970) Nature 227, 67.
35. KLEIN, G., BREGULA, U., WIENER, F., AND HARRIS, H. (1971) J. Cell Sci. 8, 659.
36. BREGULA, U., KLEIN, G., AND HARRIS, H. (1971) J. Cell Sci. 8, 673.
37. WIENER, F., KLEIN, G., AND HARRIS, H. (1971) J. Cell Sci. 8, 681.
38. HARRIS, H. (1971) J. Natl. Cancer Inst. 48, 851.
39. MACPHERSON, I. A. (1971) Proc. Roy. Soc. B. 177, 41.
40. HARRIS, H. (1971) Proc. Roy. Soc. B. 179, 1.

41. VAN DER NOORDAA, J., VAN HAAGEN, A., WALBOOMERS, M. M., AND VAN SOMEREN, H. (1972) J. Virol. 10, 67.
42. SELL, E. K., AND KROOTH, R. S. (1972) J. Cellul. Physiol. 80, 453.
43. WIBLIN, C. N., AND MACPHERSON, I. (1973) Intnl. J. Cancer 12, 148.
44. LEVISOHN, S. R., AND THOMPSON, E. B. (1973) J. Cellul. Physiol. 81, 225.
45. STOKER, M. G. P. (1967) J. Cell Sci. 2, 293.
46. SINCLAIR, R., REID, R. A., AND MITCHELL, P. (1963) Nature 197, 982.
47. BURGER, M. M. (1970) Nature 227, 170.
48. VASILIEV, J. M., GELFAND, I. M., GUELSTEIN, V. I., AND FETISOVA, E. K. (1970), J. Cellul. Physiol. 75, 305.
49. CECCARINI, C., AND EAGLE, H. (1971) Nature New Biology 233, 271.
50. VAHERI, A., RUOSLAHTI, E., AND NORDLING, S. (1972) Nature New Biology 238, 211.
51. YAMAMOTO, K., OMATA, S., OHNISHI, T., AND TERAYAMA, H. (1973) Cancer Res. 33, 567.
52. NOONAN, K. B., AND BURGER, M. M. (1973) Exp. Cell Res. 80, 405.
53. MALLUCCI, L., WELLS, V., AND YOUNG, M. R. (1972) Nature New Biology 239, 53.
54. KUBOTA, Y., UEKI, H., AND SHOJI, S. (1972) J. Biochem. 72, 235.
55. BULLOUGH, W. S., AND LAURENCE, E. B. (1968) Nature 220, 134.
56. BULLOUGH, W. S., AND DEOL, J. U. R. (1971) Europ. J. Cancer 7, 425.
57. FRANKFURT, O. S. (1971) Exp. Cell Res. 64, 140.
58. MAUGH, T. H. (1972) Science 176, 1407.
59. HENNINGS, H., AND HOUCK, J. C. (1973) FEBS Letters 32, 1.
60. MORLEY, G. D., AND KINGDON, H. S. (1973) Biochim. Biophys. Acta 308, 260.
61. SHODELL, M., AND ISSELBACHER, K. (1973) Nature New Biology 243, 83.

62. HOFFMAN, R., RISTOW, W. J., VRIER, J., AND FRANK, W. (1973) Exp. Cell Res. 80, 275.
63. JULLIEN, M., GLAT, C., AND HAREL, L. (1972) Exp. Cell Res. 73, 530.
64. STAUSS, P. R., AND BERLIN, R. D. (1973) J. Exp. Med. 137, 359.

THE ROLE OF MEMBRANES IN BACTERIAL DIFFERENTIATION

N. Vasantha and Kunthala Jayaraman

INTRODUCTION

There are well defined stages of sporulation documented by the use of cacogenic mutants and physiological blocks. However, the triggers that bring out the morphogenetic change-over remain largely unelucidated. One of the cellular byproducts of the sporulating *Bacilli* are the polypeptide antibiotics, whose physiological role in the producing organism is unknown. Several experimental evidences have revealed the pleiotropic regulation of polypeptide production and sporulation, thus implicating the antibiotics in the process of sporogenesis. The inner membrane fraction of *B. polymyxa* harbors one of the key enzymes in the biosynthetic pathway of the antibiotic, Polymyxin (2,4-Diaminobutyric acid activating enzyme). The density-gradient profile of this fraction undergoes variations during the growth cycle of *B. polymyxa*. The correlation of membrane profiles with the physiological manifestation of antibiotic expression prior to the onset of sporulation will be discussed.

Sporulation in Bacillus species represents a switchover of an actively dividing, fragile vegetative cell to a non-dividing and heat resistant spore. The process of normal cell division is stopped, thus setting the stage for differentiation. The series of events accompanying this differentiation process are well documented and intensively reviewed (1-4). Thus the commitment of cells to sporulate obviously includes membrane-associated phenomena and attempts are made in this study to correlate the variation of the nature of membranes at different stages of development in *Bacillus polymyxa*.

The initial molecular trigger for the onset of this differentiation process is unknown and the possible candidates are the early spore specific products that appear at stage t_0 of sporulation. These include formation of a new proteinase and production of polypeptide antibiotics (3,1). It has been demonstrated by our studies that the polypeptide antibiotics may play a causative role in the process of differentiation in *Bacillus polymyxa* (5). It has further been established that the antibiotic synthesizing complex is localized on the heavy membrane complex of the cells. One of the marker enzymes that have been useful in the study of membrane is the 2-4 Diaminobutyrate activating enzyme which has been closely associated with polymyxin production (6). The membrane-associated biosynthetic machinery of the antibiotic is monitored during the differentiation process.

MATERIALS AND METHODS

Bacterial culture conditions. *Bacillus polymyxa* 2459 was a generous gift from Pfizer Co. USA. The highly sporulating culture grown in Tryptose broth was stored as heat killed spore stocks. For each experiment, cultures were started from the spore stocks. They were grown in synthetic medium (7) at 30°C with shaking.

Isolation of crude membrane fraction. Cells at a desired growth stage were harvested by centrifuging them at 15,000 g for 10 min in Janetzki K-24 refrigerated centrifuge and washed with 5 mM potassium phosphate buffer pH 7.5. They were suspended (20 mg wet wt/ml) in 10 mM Tris-HCl buffer pH 7.5 containing 10 mM KCl, 2 mM EDTA, 2 mM 2-mercaptoethanol (TKM-EDTA) and 20% sucrose and incubated with lysozyme (1 mg/ml) for 20 minutes at 30°C with gentle shaking. The spheroplasts thus obtained were osmotically lysed by diluting tenfold with the TKM-EDTA buffer and were centrifuged at 39,000 g for 30 min in Sorvall RC2-B refrigerated centrifuge. The pellet containing unlysed whole cells and membrane fraction was washed and suspended in TKM buffer containing 2 mM Mg^{++}.

Incorporation of precursor of polymyxin in the membrane fraction. Cells of *B. polymyxa* 2459 grown in synthetic medium were harvested and suspended in fresh synthetic medium (5 ml) containing chloramphenicol (200 μg/ml) and incubated for 30 min at 37°C. C^{14}-Threonine (5 μc) was added and the cells were incubated for 60 min at 37°C. This channelized the incorporation of the precursor exclusively into the polymyxin fraction (7). The cells were centrifuged and washed with 5 mM phosphate buffer, pH 7.5, twice and the crude membrane was prepared as mentioned.

Purification of membrane fractions. A 5 ml discontinuous sucrose gradient was prepared with 60% to 20% sucrose in 5 mM potassium phosphate buffer pH 7.5 (60% - 2.0 ml, 50% - 1.5 ml, 40% - 0.5 ml, 20% - 0.2 ml) and the sample (less than 10% in volume of the total gradient) was layered and spun at 19,000 g for 30 min. The tubes were punctured and fractions were collected. Their absorbancy at 260 mμ was read with a Beckman DU2 spectrophotometer and the radioactivity was determined in a gas flow proportional counter. L-2,4-Diaminobutyrate activating enzyme activity in these fractions was monitored by the ATP-PP^{32} exchange procedure (6).

DNAse treatment. The crude membrane preparation was treated with pancreatic DNAse (50 μg/ml) for 10 min at 0°C. The treated fractions were then subjected to sucrose density gradient for purification purposes.

RESULTS

Precursor incorporation studies into the membrane complex. When the cells of *B. polymyxa*, grown to their mid log phase, are treated with chloramphenicol and pulsed with C^{14}-Thr, the label is predominantly chanelized to the polypeptide antibiotic synthesis. When the membrane fraction was purified, it was found that the DAB-activating enzyme activity coincided with the C^{14} Threonine incorporation in the membrane complex (Fig. 1). In contrast the leucine activating enzyme activity was separated from this and was associated with slower sedimenting fraction.

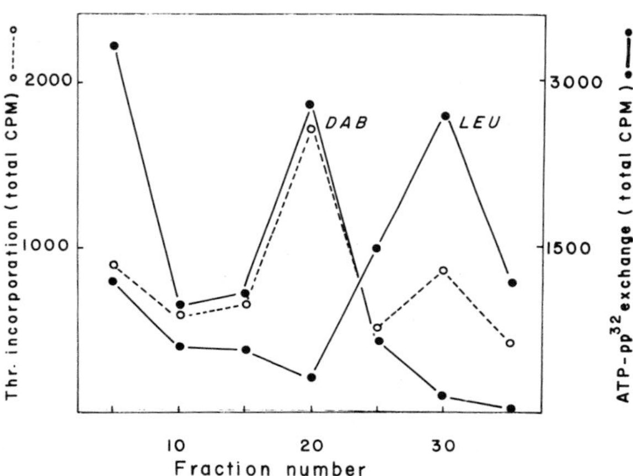

Fig. 1. Separation of the membrane complex involved in polymyxin synthesis. The purification of the heavy membrane complex was followed as outlined in Materials and Methods. The membrane complex was layered on 20%-60% sucrose gradient and spun at 30,000 g for 30 minutes. The fractions were monitored for DAB-dependent and leucine dependent ATP-PP^{32} exchange activity (o——o). The incorporation of C^{14} Threonine was also monitored (o---o).

<u>Membrane profile in vegetative cells</u>. The purified membrane complex from the early log cells, sedimented as a single, slow-sedimenting band (Fig. 2a). However, this membrane fraction upon DNAse treatment split into a heavy and a light fraction

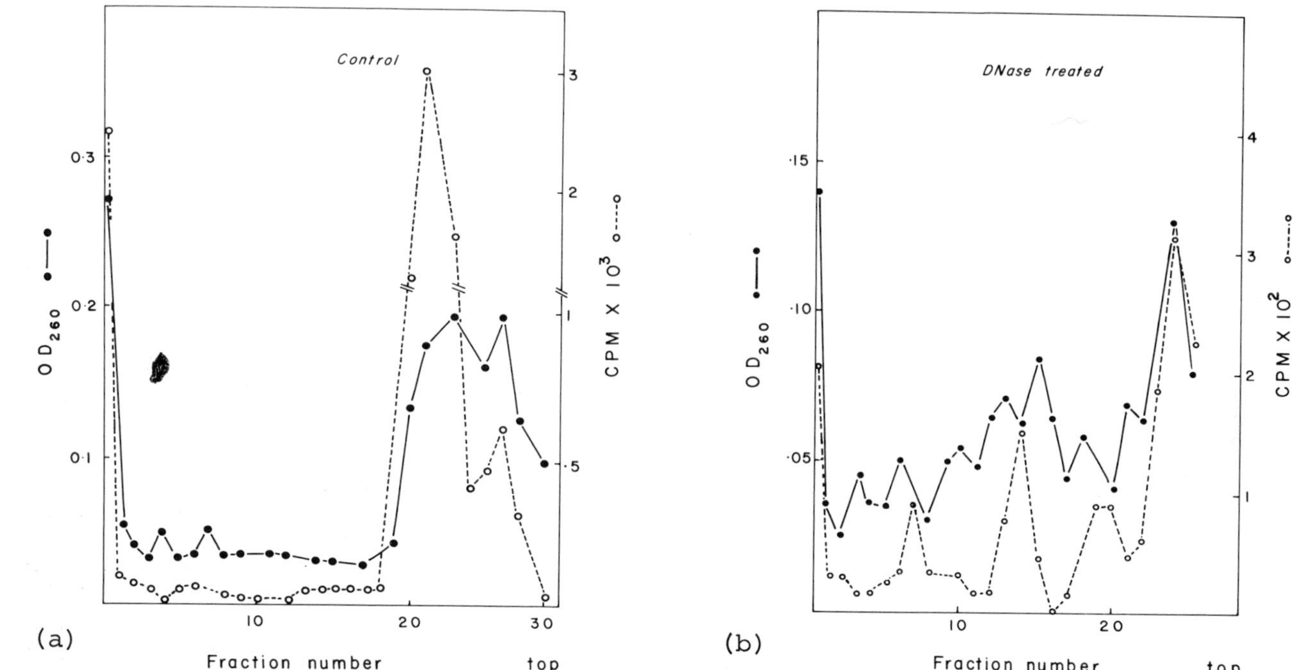

Fig. 2. Sedimentation pattern of the membrane fractions at the early log phase of the cells of <u>B. polymyxa</u>. (a) The C^{14} Threonine incorporation (o---o) and the O.D.$_{260}$ (o——o) profiles were monitored for the membrane complex layered on 20%-60% sucrose density gradient spun at 19,000 g for 30 minutes. (b) The same membrane preparation was treated with DNAse 50 ug/ml before layering it on the gradient.

(Fig. 2b). The C^{14}-Thr incorporation and OD profile were undistinguishable.

The DAB-dependent ATP-PP^{32} exchange activity which was present at a very low level in the slow sedimenting normal membrane fractions showed a remarkable enhancement of the enzyme activity upon treatment with DNAse. Further, the activity profile coincided with the faster sedimenting fraction (Fig. 3).

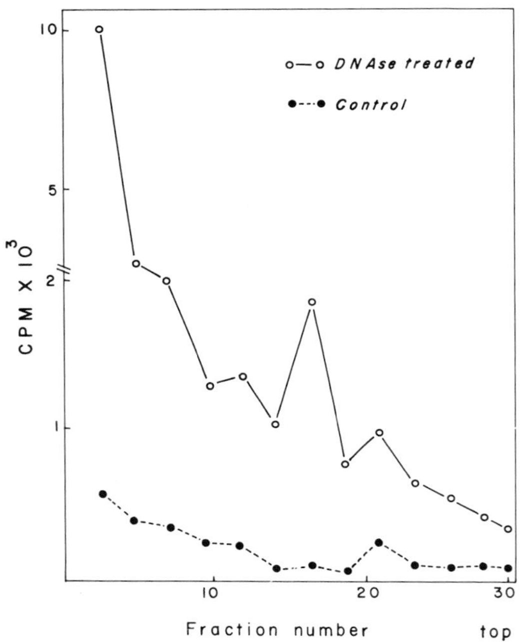

Fig. 3. DAB-dependent ATP-PP^{32} exchange activity in the membrane preparations of early log cells of B. polymyxa. This enzyme activity was monitored in sucrose density gradient fractions as outlined in Materials and Methods.

Membrane profile in cells committed to the process of sporulation. The membrane fraction isolated from the cells at their late log phase revealed a diverse pattern in their membrane components. There appeared two distinct fractions in these preparations capable of Threonine incorporation in the absence of protein synthesis (Fig. 4a). This profile remained unaltered upon DNAse treatment except for the enhancement of the incorporation in the heavier fraction, with the concommitant lowering of the incorporation in the lighter fraction (Fig. 4b). This indicates that the lighter fraction is transformed to the faster sedimenting fraction upon DNAse treatment. The DAB activating enzyme activity was present at a high level and its activity profile coincided with these two membrane fractions. This enzyme activity showed enhancement upon DNAse treatment, although not to the same extent as with the early log cell membrane fraction (Fig. 5).

DISCUSSION

The results of these studies show that the membrane fractions undergo a structural change from the vegetative stage to their committed situation. What is surprising is the result that the cells are capable of activating the Diaminobutyrate even at their early stages of growth cycle although this activity seems to be enmasked in the membrane preparations. This enmasking presumably is due to the combination of the enzyme-membrane complex with DNA as the DNAse treatment seems to uncover the potential enzyme activity.

The membrane profile shifts from a slowly sedimenting single band pattern to a multiple species pattern at the late log phase and a similar dual-banding pattern is achieved by the DNAse treatment of the early membrane complex. It is known that under careful preparative conditions the DAB activating complex is present as a poly-disperse molecule ranging in its molecular weight from 100,000 to 300,000 (6). It is likely that upon DNAse treatment the DAB-activating complex becomes dislodged from its membrane binding and aggregates into a heavy molecular weight complex. A similar situation has been

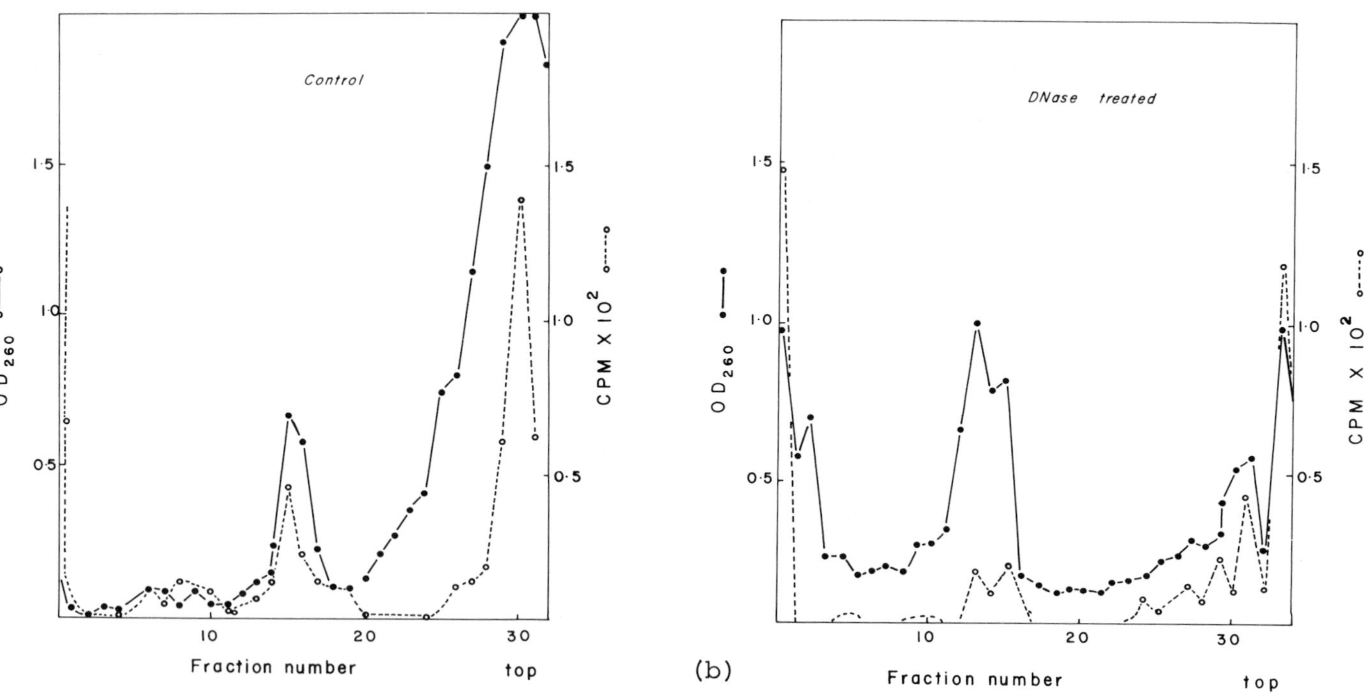

Fig. 4. Sedimentation profile of the membrane fraction of the late log phase cells of B. polymyxa. (a) The activity profiles were monitored as outlined in Fig. 2a. (b) The same membrane preparation was treated with DNAse 50 ug/ml before layering it on the gradient.

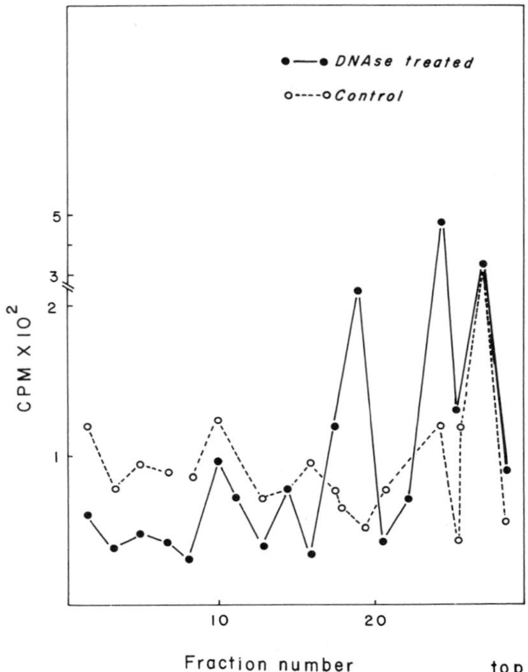

Fig. 5. DAB-dependent ATP-PP^{32} exchange activity in the membrane preparations of the late log cells of B. polymyxa. This enzyme activity was monitored in sucrose density gradient fractions as outlined in Materials and Methods.

observed with the treatment of membranes with Streptomycin (6,9). In view of these results it is tempting to speculate that the DAB activating complex is a membrane-associated protein and perhaps we are dealing with a DNA-membrane complex protein which could play an enormously interesting role in the process of cellular differentiation.

Our previous studies have revealed that the cells at their early log phase have the capacity to recognize their own antibiotic when added externally. Upon addition of the antibiotic the vegetative growth ceases and the cells are eventually committed to sporulation. This sensitivity to the antibiotic is, however, lost at a later growth stage when its own production of the antibiotic has commenced. It is possible that the membrane complex containing DAB activating enzyme represents the "receptor sites" for the antibiotic trigger, for the commitment to sporulation.

REFERENCES

1. MANDELSTAM, J. (1969) Symp. Soc. Gen. Microbiol. 19, 377.
2. FREESE, E. (1972) in Current Topics in Developmental Biology (Moscona, A. A., and Monroy, A., eds.) Vol. 7, p. 85, Academic Press, N.Y.
3. SADOFF, H. L (1972) in Spores V (Halvorson, H. O., Hanson, R., and Campbell, L. L., eds.), p. 157, Publication office Am. Society for Microbiol.
4. SZULMAJSTER, J. (1973) Symp. Soc. Gen. Microbiol. 23, 45.
5. JAYARAMAN, K., AND KANNAN, R. (1972) Biochem. Biophys. Res. Commun. 48, 5, 1235.
6. JAYARAMAN, K., MONREAL, J., AND PAULUS, H. (1969) Biochim. Biophys. Acta 185, 447.
7. PAULUS, H., AND GRAY, R. (1964) J. Biol. Chem. 239, 865.
8. CONOVER, T. E., PRARIE, R. L., AND RACKER, E. (1963) J. Biol. Chem. 238, 2831.

SUBJECT INDEX

A 4
B 5
C 6
D 7
E 8
F 9
G 0
H 1
I 2
J 3